T0325203

EFFECTIVE INDUSTRIAL MEMBRANE PROCESSES: BENEFITS AND OPPORTUNITIES

This volume is the proceedings of the 2nd International Conference on Effective Industrial Membrane Processes: Benefits and Opportunities, organised by BHR Group Ltd and co-sponsored by the Institution of Chemical Engineers and the European Society of Membrane Science and Technology, held at the Scandic Crown Hotel, Edinburgh, UK, 18-21 March 1991.

ACKNOWLEDGEMENTS

The valuable assistance of the Technical Advisory Committee and panel of referees is gratefully acknowledged.

Technical Advisory Committee

Dr M. Turner (Chairman)	University College London
Dr M.A. Cook	SmithKline Beecham Pharmaceuticals
Mr W. Lavers	Consultant
Dr M.S. Le	ICI Chemicals and Polymers
Prof. P. Meares (ex-officio)	University of Exeter
Dr R. Paterson	University of Glasgow
Dr G. Pearce	Kalsep
Mr D. Pepper	PCI Membrane Systems Ltd
Mr P.G. Redman	Institution of Chemical Engineers
Mr S. Ruszkowski	BHR Group Ltd

Overseas Corresponding Members

Prof. E. Drioli	Napoli University, Italy
Prof. A.G. Fane	University of New South Wales, Australia
Mrs J. Farrant	Membrane Technology and Research Inc, USA
Mr R.L. Riley	Separation Systems International, USA
Mr F.F. Stengaard	DDS Filtration, Denmark
Prof H. Strathmann	The European Membrane Society, Germany

EFFECTIVE INDUSTRIAL MEMBRANE PROCESSES: BENEFITS AND OPPORTUNITIES

Edited by

M K Turner

The Advanced Centre for Biochemical Engineering
University College London, UK

Published by
ELSEVIER APPLIED SCIENCE
LONDON and NEW YORK

Organised and sponsored by BHR Group Ltd

Co-sponsored by the Institution of
Chemical Engineering and the European
Society of Membrane Science and Technology

ELSEVIER SCIENCE PUBLISHERS LTD
Crown House, Linton Road, Barking, Essex IG11 8JU, England

Sole Distributor in the USA and Canada
ELSEVIER SCIENCE PUBLISHING CO., INC.
655 Avenue of the Americas, New York, NY 10010, USA

WITH 55 TABLES AND 154 ILLUSTRATIONS

British Library Cataloguing in Publication Data

Effective industrial membrane processes : benefits
and opportunities.
I. Turner, M.K.
660.2842

ISBN 1-85166-723-7

Library of Congress CIP data applied for

PREFACE

The aim of the Technical Advisory Committee, in planning the content of this meeting, was to illustrate the range of separation processes in which the use of membranes was practical and effective at an industrial scale. As Professor Strathmann reveals, the market for process equipment built around membranes is now worth about $5x10^9$ annually, and it seemed important to review this technology, and to point the direction of future technical advances.

All but the most critical reader should find some items of interest. The Committee would admit to not fulfilling all of thier aims, although those delegates who attended the meeting in Edinburgh judged it a success. In the event it provided representative examples of processes from the food and beverage industry, from water treatment, and from the chemical industry, of which the removal of alcohol from fermented beverages, shipboard desalination and solvent recovery are three.

The major uses of charged membranes and sterile processes are not covered, nor is the largest market, $1.2x10^9$ annually, for artificial kidney dialysis. However, it is interesting to see artificial kidney now finding an alternative use as a reactor for the production of monoclonal antibodies. We are also reminded by Professor Michel of the importance and efficiency of natural membranes in the kidney under conditions where fouling is crucial to their performance and enhances their selectivity.

In contrast, fouling and poor selectivity are two problems which limit the application of synthetic membranes. Several papers consider the methods of pretreating liquid streams to reduce fouling, or of cleaning the membranes to reverse it. The electron microscope recently revealed that the fouling layer from these liquids penetrates deeper into the membrane than was once assumed. Selectivity in these liquid processes remains crude and unpredictable, raising the comment from Dr Carpenter that confidence in the viability of membrane processes only arises out of extensive testing at a pilot scale. Perhaps this explains why so much new discussion concerns the use of membranes for the rather crude separations needed to retain the waste released from traditional processes.

Gas separations are an exception. The selectivity of the membranes can now produce gases of sufficient purity at competitive costs, although the membrane step must still integrate effectively with the rest of the process.

The conference was interesting. For that, I would like to thank the Advisory Committee and its Overseas Members whose meetings I had the pleasure of chairing. I would also like to thank Tracey Peters and Carl Welch of the BHR Group who administered the Conference and organised its publication. I hope this book makes a useful addition to the literature.

PROFESSOR M.K. TURNER
The Advanced Centre for Biochemical Engineering
University College London

CONTENTS

SESSION E: APPLICATIONS IN THE FOOD AND BEVERAGES INDUSTRY

SESSION F: APPLICATIONS IN THE FOOD AND BEVERAGES INDUSTRY

SESSION G: BIOTECHNOLOGY APPLICATIONS

SESSION H: APPLICATIONS IN THE CHEMICAL INDUSTRY

SESSION J: GAS SEPARATION IN THE PETROCHEMICAL INDUSTRY I

SESSION K: GAS SEPARATION IN THE PETROCHEMICAL INDUSTRY II

ECONOMIC ASSESSMENT OF MEMBRANE PROCESSES

H. Strathmann
Universität Stuttgart, Institut für Chemische Verfahrenstechnik
Böblinger Straße 72, 7000 Stuttgart 1, FRG

1. Introduction

In recent years membranes and membrane processes have become industrial products of substantial technical and commercial importance. The worldwide sales of synthetic membranes in 1990 were in excess of US $ 2.0 x 10^9. Taking into consideration that in most industrial applications membranes account for about 40% of the total investment costs for a complete membrane plant, the total annual sales of the membrane based industry is close to US $ 5 x 10^9 [1]. Membranes and membrane processes have found a very broad range of applicaions. They are used today to produce potable water from seawater, to treat industrial effluents, to recover hydrogen from off-gases or to fractionate, concentrate, and purify molecular solutions in the chemical and pharmaceutical industry. Membranes are also key elements in artificial kidneys and controlled drug delivery systems. The growing significance of membranes and membrane processes as efficient tools for laboratory and industrial scale mass separations is based on the several properties, characteristic of all membrane separation processes, which make them superior to many conventional mass separation methods. The mass separation by means of membranes is a physical procedure carried out at ambient temperature, thus the constituents to be separated are not exposed to thermal stress or chemical alteration. This is of particular importance for biochemical or microbiological application where often mixtures of sensitive biological materials have to be separated. Furthermore membrane processes are energy efficient and rather simple to operate in a continuous mode. Up- or downscaling is easy and process costs depend only marginally on the plant size.

In spite of impressive sales and a growth rate of the industry of about 12 to 15% per year the use of membranes in industrial scale separation processes is not without technical and economic problems. Technical problems are related to insufficient membrane selectivities, poor transmembrane fluxes, general process operating problems and lack of application know-how. Economic problems originate from the multitude of different membrane products and processes with very different price structures in a wide range of applications which are distributed on a very heterogeneous market consisting of a multitude of often very small market segments for

Effective Industrial Membrane Processes — Benefits and Opportunities, pp.1–19

individual products. This has led to relatively large production volumes for some products, such as hemodialysers, disposable items used only once for a few hours and sold in relatively large and uniform market segments. Other membranes, such as certain ceramic structures, used in special applications in the food, chemical or pharmaceutical industry are expected to last for several years in operation and can only be sold in relatively small quantities to small market segments. Consequently production volumes for these items are low and prices high. The rapid development of new membrane products and processes opening up new applications makes it difficult to predict the growth rate of the market with reasonable accuracy. The demand for efficient separation processes in the chemical, food and drug industry as well as in biotechnology and for solving challenging environmental problems, however, has not only led to a rather optimistic view but also to a considerable amount of speculation concerning the future development of the membrane based industry.

In this paper the different membrane products and processes are evaluated in terms of their state of development and their technical and economic relevance. Their potentials and limitations are pointed out. The present market for membranes and related products is analysed in terms of areas of application, technical requirements, economic limitations and regional distribution. The structure of today's membrane industry is examined in terms of its product lines, operating strategies and regional distribution. Main parameters affecting the present and future utilization of membranes and membrane processes, such as membrane performance and costs, process reliability, longterm experience etc. are described. Different strategies used by various companies for handling their membrane related business are illustrated and evaluated.

The critical needs in terms of basic and applied research, capital investments, and personal education for a further growth of the membrane industry are discussed. Various research topics are identified as being crucial for the future development of membranes and membrane processes. They are analysed for their technical relevance, commercial impact, prospects for successful realization and the financial effort this will require.

2. Fundamentals of Membranes and Membrane Processes

To understand better the significance, both technical and economic, that membrane processes have in solving mass separation problems, some basic aspects concerning membranes and their function shall be briefly reviewed at this point.

2.1 Definition of a Membrane, its Function and Structure

In a most general sense, a membrane is an interphase separating two homogenous phases and affecting the transport of different chemical components in a very specific way. A multitude of different structures is summarized under the term "membrane". Often a membrane can be described easier by the way it functions than by its structure. Three different modes of transport can be distinguished in synthetic membranes, as indicated in Figure 1. Finally a component may also be transported against its chemical potential gradient through the membrane if coupled to a carrier and an energy delivering reaction. However, this mode of mass transport, referred to as active transport, has no up-to-date technical or commercial relevance.

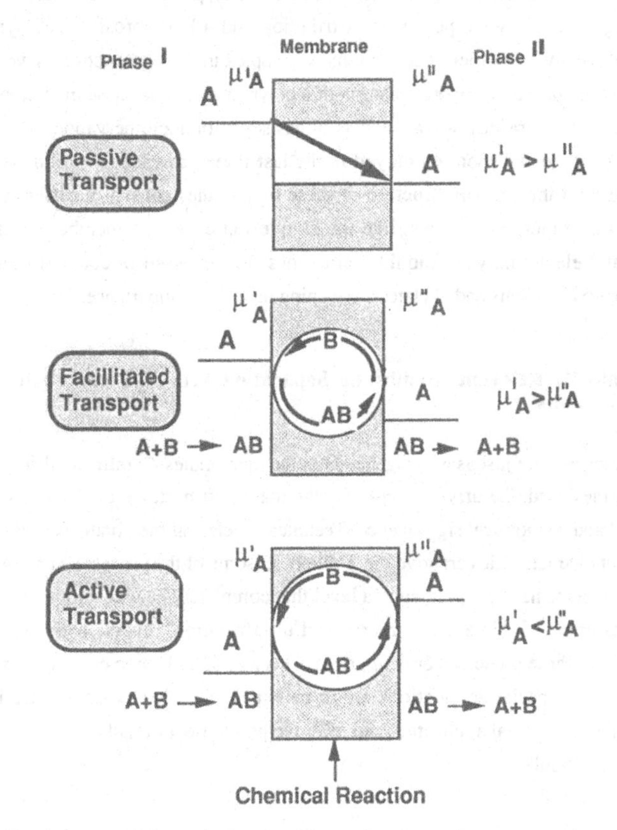

Fig. 1: Schematic drawing illustrating the various modes of mass transportation in synthetic membranes

2.2 Properties and Applications of Technically Relevant Membranes

Process design and chemical engineering aspects are important for the overall performance of a membrane separation process. The key element, however, is the membrane. Various structures are used as membranes. Some are very simple, such as microporous structures, others are more complex containing functional groups and selective carriers. A summary of technically relevant membranes, their structure and area of application is given in Table I.

The most simple form of a synthetic membrane are porous plates or foils. They are produced by pressing and sintering a polymeric, ceramic or a metal powder. These membranes have relatively large pores, a wide pore size distribution and a low porosity. The symmetric or asymmetric phase inversion membranes are more complex in their production as well as in their structure. They are produced by precipitation of a polymer solution. These membranes are used today in micro- and ultrafiltration as well as in gas separation and pervaporation. Composite membranes are more and more employed in the last three processes mentioned above. The selective layer and the support structure of these membranes consist of different materials. Membranes with functional groups, like the simple ion exchange membranes, are used in electrolysis and electrodialysis. Liquid membranes that are used in coupled transport with selective complexing agents and chelates are gaining increasing importance.

2.3 Technically Relevant Membrane Separation Processes and their Applications

Membrane processes are just as heterogeneous as the membranes. Significant differences occur in the membranes used, the driving forces for the mass transport, the applications and also in their technical and economical significance. Technically relevant membrane processes and their most important characteristics are given in Table II. In some of the processes the membranes as well as the processes have reached such a level that completely new developments can not be expected. Examples of this are the micro- and ultrafiltration, reverse osmosis, dialysis or electrodialysis. In these processes only improvements and optimization of existing systems and their adaptation to special applications are to be expected. Other processes are in the very beginning of their industrial application and offer the possibility of totally new developments of membranes and modules.

TABLE I Properties and applications of technically relevant synthetic membranes

Membranes	Basic Materials	Manufacturing Procedures	Structures	Applications
Ceramic membranes	Clay, silicate, aluminiumoxide, graphite, metal powder	Pressing and Sintering of fine powders	Pores from 0.1 to 10 micron diameter	Filtering of suspensions, gas separations, separation of isotopes
Stretched membranes	Polytetrafluoro-ethylene, polyethylene, polypropylene	Stretching of partially crystalline foil perpendicular to the orientation of crystallyts	Pores of 0.1 to 1 micron diameter	Filtration of aggressive media, cleaning of air, sterile filtration, medical technology
Etched polymer films	Polycarbonate	Radiation of a foil and subsequent acid etching	Pores of 0.5-10 micron diameter	Analytical and medical chemistry, sterile filtration
Homogeneous membranes	Silicone rubber, hydrophobic liquids	Extruding of homogeneous foils, formation of liquid films	Homogeneous phase, support possible	Gas separations, carrier-mediated transport
Symmetrical microporous membranes	Cellulose derivatives, polyamide, polypropylene	Phase inversion reaction	Pores of 50 to 5000 nonometers diameter	Sterile filtration, dialysis, membrane distillation
Integral asymmetric membranes	Cellulose derivatives, polyamide, polysulfone, etc.	Phase inversion reaction	Homogeneous polymer or pores of 1 to 10 nanometers diameter	Ultrafiltration, hyperfiltration, gas separations, pervaporation
Composite asymmetric membranes	Cellulose derivatives, polyamide, polysulfone, polydimethyl-siloxane	Application of a film to a microporous membrane	Homogeneous polymer or pores from 1 to 5 nanometers diameter	Ultrafiltration, hyperfiltration, gas separations, pervaporation
Ion exchange membranes	Polyethylene, polysulfone polyvinyl-chloride etc.	Foils from ion exchange resins or sulfonation of homogeneous polymers	Matrix with positive or negative charges	Electrodialysis, electrolysis

TABLE II Technically relevant membrane processes

Membrane separation process	Driving force for mass transport	Type of membrane employed	Separation mechanism of the membrane	Application
Micro-filtration	Hydrostatic pressure difference 50-100 kPa	Symmetrical porous membrane with a pore radius of 0.1 to 20 μm	Sieving effect	Separation of suspended materials
Ultra-filtration	Hydrostatic pressure difference 100-1000 kPa	Symmetrical porous membrane with a pore radius of 1 to 20 nm	Sieving effect	Concentration, fractionation and cleaning of macromolecular solutions
Reverse osmosis	Hydrostatic pressure difference 1000-10000 kPa	Asymmetric membrane from different homogeneous polymers	Solubility and diffusion in the homogeneous polymer matrix	Concentration of components with low molecular weight
Dialysis	Concentration difference	Symmetrical porous membrane	Diffusion in a convection-free layer	Separation of components with low molecular weight from macromolecular solutions
Electro-dialysis	Difference in electrical potential	Ion exchange membrane	Different charges of the components in solution	Desalting and deacidifying of solutions containing neutral components
Gas separation	Hydrostatic pressure difference 1000-15000 kPa	Asymmetrical membrane from a homogeneous polymer	Solution and diffusion in the homogeneous polymer matrix	Separation of gases and vapors
Pervaporation	Partial pressure difference 0 to 100 kPa	Asymmetrical solubility membrane from a homogeneous polymer	Solution and diffusion in the homogeneous polymer matrix	Separation of solvents and azeotropic mixtures

TABLE III Membrane modules, their properties and applications

Type of module	Membrane area per volume (m^2/m^3)	Price	Control of concentration polarization	Application
Pleated Filter Cartridge	800-1000	low	very poor	Dead end filtration
Tube	20-30	very high	very good	Cross-flow filtration of solutions with high solids content
Plate-and-frame	400-600	high	fair	Filtration, pervaporation, gas separation and reverse osmosis
Spiral-wound	800-1000	low	poor	Ultrafiltration, reverse osmosis, pervaporation gas separation
Capillary tube	600-1200	low	good	Ultrafiltration, pervaporation liquid membranes
Hollow fiber		very low	very bad	Reverse osmosis, gas separation

2.4 Membrane Modules and their Design

For a practical application membranes have to be installed into a suitable device generally referred to as membrane module. The membrane module design is closely related to the membrane process and its technical application. It is performed according to economical considerations, where the costs per installed membrane area are an important factor, and to other chemical engineering criteria, such as the flow of the feed solution at the membrane surface and the flow path of the permeate. The goal is to obtain an optimum flow to and from the membrane while minimizing pressure loss, concentration polarization and membrane fouling. For the different applications, different types of modules have been developed which generally fulfill the given demands. The main types of membrane modules used today on an industrial scale are listed in Table III.

3. Technically Relevant Membrane Applications and their Economical Significance

Membranes are used today in numerous applications, from medicine to wastewater treatment. The applications show great differences in their technical and economical significance.

3.1 Membrane Processes in Water Treatment

Water treatment is one of the most important areas of application of membranes. This includes the production of potable water from sea- and brackish water as well as the production of process water and the cleaning of industrial wastewater streams. Micro- and ultrafiltration as well as reverse osmosis and electrodialysis are the processes employed mainly in water treatment. Table IV shows the segmentation of the total market for membranes and membrane modules, respectively, in relation to the different processes.

TABLE IV Sales of the membrane industry (million US $ per year) in water treatment, divided by processes and applications

Processes / Applications	Micro-filtration	Ultra-filtration	Reverse osmosis	Electro-dialysis	Total
Desalting of seawater	-	-	20	-	20
Desalting of brackish water	-	-	35	35	70
Pretreatment of boiler and feed water	5	-	25	10	35
Ultrapure water	220	30	15	-	265
Sterile and lowpyrogen water	80	10	10	-	100
Industrial wastewater	5	20	15	15	55
Total	310	60	120	60	490

This table shows, that total sales of the membrane industry in the area of water treatment is 500 million US $ per year. Of all membrane processes, microfiltration plays the most important economical role because of its application in the production of ultrapure water in the semiconductor industry and of sterile and pyrogen-depleated water for the chemical and pharmaceutical industry. The main application of reverse osmosis is in the desalination of sea- and brackish water, but more recently reverse osmosis is also finding increasing use in the production of ultrapure water. The level that has been reached in reverse osmosis membrane performance today, makes it quite unlikely that membranes with revolutionary improved properties will be available in the near future. This, however, does not mean that further development is not necessary. The weak point of all available reverse osmosis membranes today is their poor thermal and chemical stability. All commercial membranes are either sensitive against hydrolysis or not stable in the presence of free chlorine or other strong oxidizing agents. This makes a rather extensive pretreatment of the raw water necessary. In ultrafiltration membrane fouling is a problem that has not been solved to date.

3.2 Membrane Processes in the Food Industry

Another interesting area of application for membranes is the food industry, as indicated in Table V, which shows the total annual sales of membranes and membrane modules in the food industry.

TABLE V Sales of the membrane industry (million US $ per year) in the food industry, divided by processes and applications

Processes / Applications	Micro-filtration	Ultra-filtration	Reverse osmosis	Electro-dialysis	Total
Dairy industry	-	35	-	10	45
Beverage industry	95	5	5	5	110
Meat processing	-	2	5	-	7
Starch industry	-	2	5	-	7
Total	95	44	15	15	169

Micro- and ultrafiltration, reverse osmosis and electrodialysis are mainly used in this application. Microfiltration dominates in the beverage industry for sterile filtration and clarification, while ultrafiltration and electrodialysis are used mainly in the dairy industry. The application of new membrane processes like pervaporation is just beginning. These processes will soon be used on an industrial scale. The overall membrane market in this area amounts to about 170 million US $ per year.

3.3 Membrane Processes in Medical Devices

Membranes have been used in medical applications for more than 25 years. The artificial kidney is the by far most interesting membrane application besides membrane-controlled therapeutical systems for the controlled release of drugs. Plasmapheresis also is now gaining importance, as shown in Table VI, which lists the annual sales of membranes in medical applications.

TABLE VI Sales of the membrane industry (million US $ per year) in medical devices, divided by processes and applications

Processes / Applications	Micro-filtration	Ultra-filtration	Dialysis	Total
Hemodialysis	-	-	900	900
Hemofiltration	-	120	-	120
Plasmapheresis	20	10	-	30
Total	20	130	900	1050

3.4 Membrane Processes in the Chemical and Petrochemical Industry

The use of membrane processes in the chemical and petrochemical industry is rapidly increasing. However, the sales in this area are comparatively small today, in spite of a large potential market. This is shown in Table VII, a summary of the annual sales of membranes and membrane modules in the chemical and petrochemical industry.

An interesting application is the separation of gases, such as the recovery of hydrogen from the ammonia synthesis and the separation of carbon dioxide from methane in enhanced oil recovery. The separation of oxygen and nitrogen by membranes is of interest for the production of oxygen-enriched air or inert gas. But membranes have to be improved in terms of selectivity and permeability before oxygen-nitrogen separation on a large scale becomes economically feasible.

Pervaporation and perstraction are relatively new membrane processes with numerous applications in the chemical and petrochemical industry.

TABLE VII Sales of the membrane industry (million US $ per year) in the chemical industry divided by processes and applications

Processes / Applications	Micro-filtration	Ultra-filtration	Reverse osmosis	Gas separation	Per-vaporation	Electro-dialysis	Electro-lysis	Total
Process water pretreatment	25	-	5	-	-	-	-	30
Fractionation of molecular mixtures	10	15	5	-	-	-	-	30
H_2-recycling	-	-	-	20	-	-	-	20
N_2-production	-	-	-	15	-	-	-	15
CO_2/CH_4 separation	-	-	-	5	-	-	-	5
O_2-enrichment	-	-	-	5	-	-	-	5
Separation of azeotropic mixtures	-	-	-	-	5	-	-	5
Desalting of process solutions, salt production	-	-	-	-	-	20	-	20
Chlorine alkaline electrolysis		-	-	-	-	-	70	70
Total	35	15	10	45	5	20	70	200

3.5 Membrane Processes in Biotechnology

Biotechnology is an industry where membranes offer great potential advantages. They are especially suitable for the separation of sensitive biological substances because the separation by membranes is a physical procedure which can be carried out at room temperature. All membrane processes including electrodialysis and pervaporation are of interest in biotechnology. Presently, however, sales of membranes for biotechnology applications are still rather low and in the order of 10 to 20 million US $. A reason for the relatively slow introduction of membrane processes in biotechnology is that the modern industrial biotechnology is still very young, and membranes and membrane processes have not yet been sufficiently adapted to the specific requirements of biotechnology. The list of industrial applications of membranes which is given here is far from complete. There are many other applications, especially in the laboratory, which contribute substantially to the total sales of the membrane industry.

4. The Membrane Market and its Expected Development

The world-wide sales of all membranes and membrane modules in 1990 were in excess of US-$ 2000 Million. There are, however, large differences as far as the different products and their regional distribution is concerned.

4.1 Process and Product Related Market Distribution

The distribution of the world membrane market according to processes is given in Figure 2. Here annual sales of membranes in the various processes and the expected growth of the market during the next five years are shown. The largest market for membranes today is that of dialysers used in artificial kidneys. Sales in other areas, such as ultrafiltration, electrodialysis and, in particular, gas separation and pervaporation, are presently of minor importance. The expected increase in sales in these processes is, however, significantly higher than in dialysis or microfiltration. While the annual increase in sales of all membrane products is expected to be in the order of 12 - 15 %, the sales increase will be significantly lower than that of the more recently developed processes, such as gas separation or pervaporation.

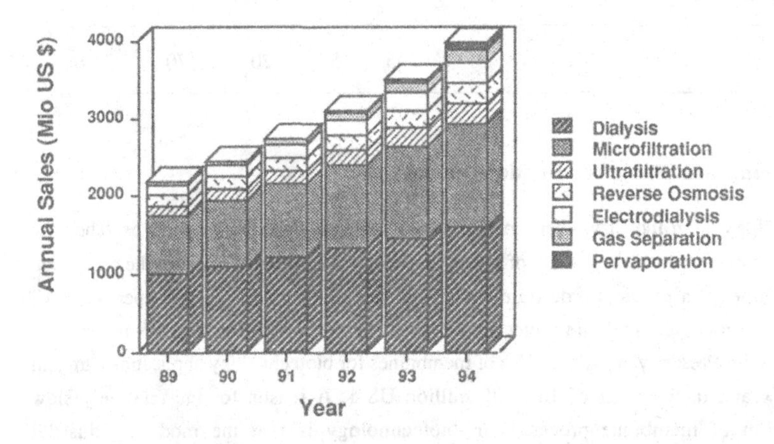

Fig. 2: Present and expected future annual sales of membranes according to the different processes

4.2 Regional Distribution of the Membrane Market

The membrane market is rather unevenly distributed, with about 75% of the market being located in the industrialized countries of Europe, USA and Japan, as indicated in Figure 3 which shows the regional distribution of the membrane market. 2)

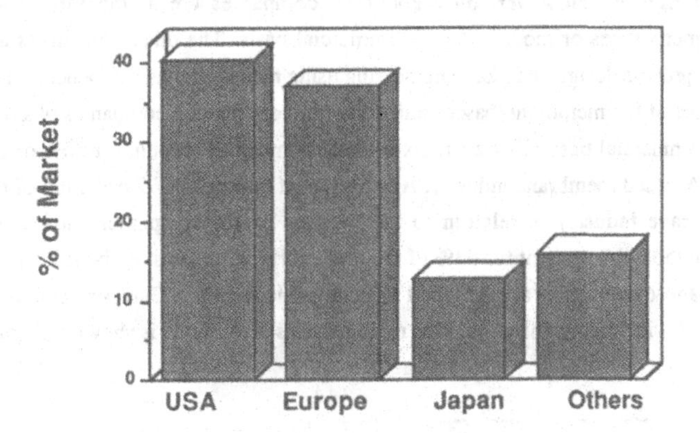

Fig. 3: Regional distribution of the membrane market

This is, however, a very global analysis. The distribution of the market can be totally different for the different membrane processes or membrane applications. For example, the market share of Germany in the area of the artificial kidney is 13%, and that for electrodialysis only 1% of the entire world market.

It is also obvious that the existence of a regional market is of great significance for the build-up of the related industry. Therefore, there is no manufacturer of ion-exchange membranes in Germany, but almost 80% of all hemodialysis membranes are manufactured here.

On the other hand electrodialysis is used mainly for brackish water desalination and in Japan for the production of bath salt. Both applications are of little significance in Europe, consequently there are no ion-exchange membrane manufacturers in Europe.

4.3 The Regional Distribution of the Membrane Industry

A look at the regional distribution of the membrane industry provides some interesting information. Almost all companies operate internationally and market their products world-wide. World-wide more than 100 companies are involved in one way or another in the membrane technology [3]. However, only about 60 companies are at the same time manufacturers of membranes or modules on a commercial basis. The other companies are mainly involved in process design and plant engineering using membranes as components. The USA has a large part of the membrane-based industry with more than 35 companies offering membranes on a commercial basis. However, six of these companies account for 80% of the sales of all the USA-based membrane industry. An analysis of the regional distribution of the sales in the membrane industry in relation to the location of the companies shows that companies based in the USA account for 63% of the total world-wide sales of about 2 billion US-$ per year. Japanese companies account for 17%, companies based in European countries for 20% of the world-wide membrane sales. This regional sales distribution is shown in Figure 4.

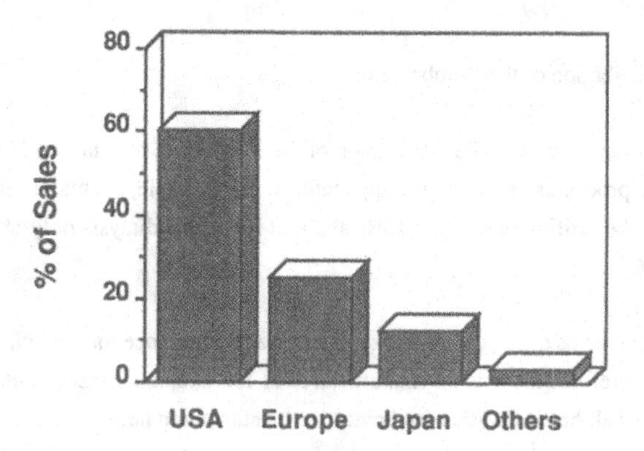

Fig. 4: Regional distribution of the membrane industry

4.4 Price Structure of Membrane Products and Processes

The prices of membrane products are closely related to their sales volume. For instance, the sales price for an ultrafiltration type membrane in an artificial kidney including the module is in the order of US-$ 20.- per 1 m^2 membrane area. The same membrane in an industrial ultrafiltration process will probably be in the order of US-$ 200.- per 1 m^2. A membrane used in pervaporation or gas separation, in principle not significantly more difficult to produce, will cost in excess of US-$ 500.-. The difference can be found in process application. To produce a 1 m^3 potable water from brackish water by reverse osmosis or electrodialysis will cost between US-$ 1 - 3.

To treat a m^3 of waste water or a process stream by reverse osmosis or ultrafiltration usually will be significantly higher and can easily reach US–$ 10 - 20 , depending on the feed solution constituent. Again here the process costs are highly effected by the market size.

5. The Membrane Industry and its Structure

The application of membranes is extremely wide-spread involving a multitude of different structures and processes. Similarly heterogeneous is the membrane industry. Being successful in the market requires, in addition to an appropriate membrane as the key element, a multitude of peripheral components, which consist of special hardware, such as pumps or monitoring equipment, and software, such as basic engineering, process design and especially application know-how.

The complexity of the membrane based industry is indicated in the schematic diagram of Figure 5, which shows on the horizontal level the various technologies required to penetrate the market. These technologies are: membrane research and development, membrane production, basic engineering, process design and marketing. In vertical direction the various membrane processes are indicated with fractional sales, which are highest in dialysis and microfiltration and lowest in pervaporation. Since every membrane process requires the various levels of technology, it is difficult for one company to cover all membrane processes on all levels of technologies and applications.

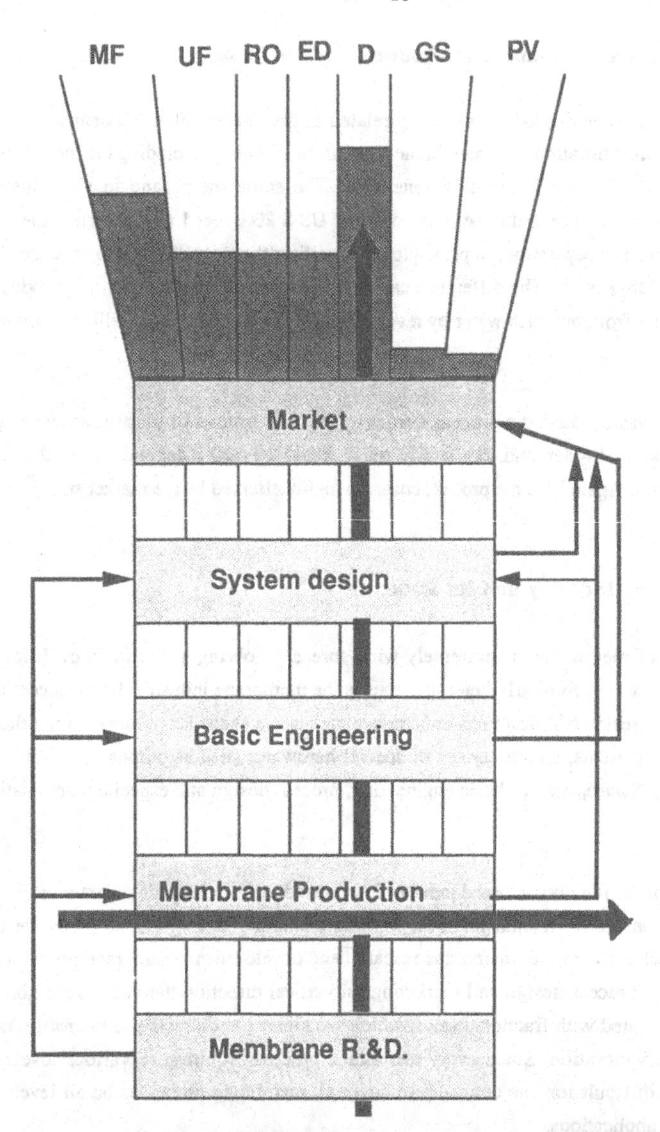

Fig. 5: **Schematic diagram illustrating the structure of the membrane industry MF = Microfiltration, UF = Ultrafiltration, RO = Reverse Osmosis, D = Dialysis, GS = Gas Separation, PV = Pervaporation**

Companies have often spezialized in certain areas. Some only produce membranes for various processes or applications and sell their products through various equipment manufacturers. This is indicated by the horizontal arrow in Figure 5. Other companies concentrate on one process, e.g. producing dialysers for the artificial kidney. This is indicated by the vertical arrow in Figure 5. These companies usually are relatively large, doing their own research, basic engineering, process design and marketing. Both business approaches seem feasible as very successful companies have shown.

6. Membrane-Related Research and Development

The R&D work in membrane science and technology is carried out at different levels and with different goals. A complete evaluation of the R&D activities is extremely difficult. All data concerning this topic are therefore questionable and can only be considered as an analysis of current trends. It is often difficult and arbitrary to decide which activities are research and development, "product development", or quality control.

The research and development activities carried out in the industry often differ significantly from such activities carried out at universities in terms of their goals. The research in industry is in general dictated by economical considerations and concentrated on the development and optimization of products and processes. The universities are more involved in basic research which is at least to some extent independent of economical considerations. Since the various membrane processes show great differences in their level of technical development, research work is concentrated on different topics in the different processes.

Table VIII shows some of the subjects on which R&D efforts are concentrated in membrane processes available today. Processes which are state-of-the-art and used industrially include microfiltration, ultrafiltration, reverse osmosis, dialysis, electrodialysis, and with some restrictions also pervaporation and gas separation.

These processes have the greatest economical relevance today. The R&D work in this area is concentrated on product improvement as well as process optimization and different aspects of application. Gas separation and pervaporation are of minor economical importance today, and the membrane sales for these processes are hampered by lack of process experiences. There is a number of membranes and membrane processes which exist today only on laboratory scale or as concepts, such as membranes with active transport properties, membranes in energy generation or energy conversion, etc. It seems too early to speculate about the future economical significance of these membranes.

TABLE VIII State of development of membrane processes with industrial use and present R&D efforts to improve membranes and processes in new applications

Process	Availability	Problems	Membranes	Goals of R&D-Work	Applications
Microfiltration	State-of-the-Art	Membrane life	Asymmetric, inorganic	Cross-flow filtration	Biotechnology
Ultrafiltration	State-of-the-Art	Membrane fouling	Surface modifications	Membrane cleaning	Biotechnology, chemical industry
Reverse Osmosis	State-of-the-Art	Chlorine stability, biofouling	Chlorine resistance	Water pretreatment	Biotechnology, wastewater
Dialysis	State-of-the-Art	Biocompatibility	Improved permeability	Process optimization	Biotechnology, food technology
Electrodialysis	State-of-the-Art	Alkaline stability	Ionselective and bipolar membranes	Multilple cell systems	Chemical industry, wastewater
Gas separation	Pilot plants	Membrane selectivity and flux	Membranes for specific gases	Process optimization	Air/solvent separation
Pervaporation	Laboratory loops	Membrane selectivity, process development	Membranes for specific materials	Process development and optimization	Wastewater treatment, biotechnology

Literature:

1) R.W. Baker, et al., "Membrane Separation Systems - A Research and Development Needs Assessment", National Technical Information Service, U.S. Department of Commerce, NTIS-PR 360, 1990.

2) Proceedings of Aachener Membran Kolloquium, Aachen, March 16-18, 1987, GVC VDI Gesellschaft für Verfahrenstechnik, Düsseldorf.

3) H. Strathmann, "Economic Evaluation of the Membrane Technology", in "Future Industrial Prospects of Membrane Processes", Eds.: L. Cecille, J.-C. Toussaint, Elsevier Science Publishers LTD, 1989.

SESSION A:

MASS TRANSFER AND CLEANING I

A CASE STUDY OF REVERSE OSMOSIS APPLIED TO THE CONCENTRATION OF YEAST EFFLUENT

Alan Merry (PCI Membrane Systems Ltd.)

ABSTRACT

In 1986 PCI Membrane Systems installed a reverse osmosis plant to concentrate an effluent arising from the manufacture of bakers yeast. The plant was installed in Europe, and contained 810 m^2 of membrane area. The membrane was a thin film composite membrane designated AFC99.

Following commissioning, and changes to the feed to the plant fouling of the membranes was experienced. A change to cellulosic membranes alleviated the fouling problems but restricted the water removal capacity of the plant.

The appearance of nanofiltration membranes gave the opportunity to both increase the capacity of the plant, and to overcome the fouling problems.

NOMENCLATURE

J	Flux	$l.m^{-2}.h^{-1}$.
K	Permeability coefficient.	$l.m^{-2}.h^{-1}.bar^{-1}$.
P	Applied pressure.	bar.
π_w	Osmotic pressure at membrane wall.	bar.
π_p	Osmotic pressure in permeate.	bar.

INTRODUCTION

The use of yeast by man in the making of bread reaches back into distant history. As we have become industrialised, so has the production of yeast both in pressed form, and dried form. As with most industrial processes an effluent stream is produced which must be treated before it can be disposed of. In 1982 PCI Membrane Systems became involved in concentrating this effluent as one of the steps in treatment. Over the following four years four plants were installed in Europe.

PROCESS

Yeast is produced in an aerobic fermentation on a molasses based medium. The yeast is harvested from the spent substrate by filtration, or centrifugation. The yeast is then washed. If the yeast is to be dried salt is added to the wash water to osmotically dehydrate the yeast. The spent substrate, and the spent wash water are normally evaporated to produce an animal feed. Details of the process vary from site to site, but these are proprietary information.

The overall process is shown in figure 1.

Effective Industrial Membrane Processes — Benefits and Opportunities, pp.23–32

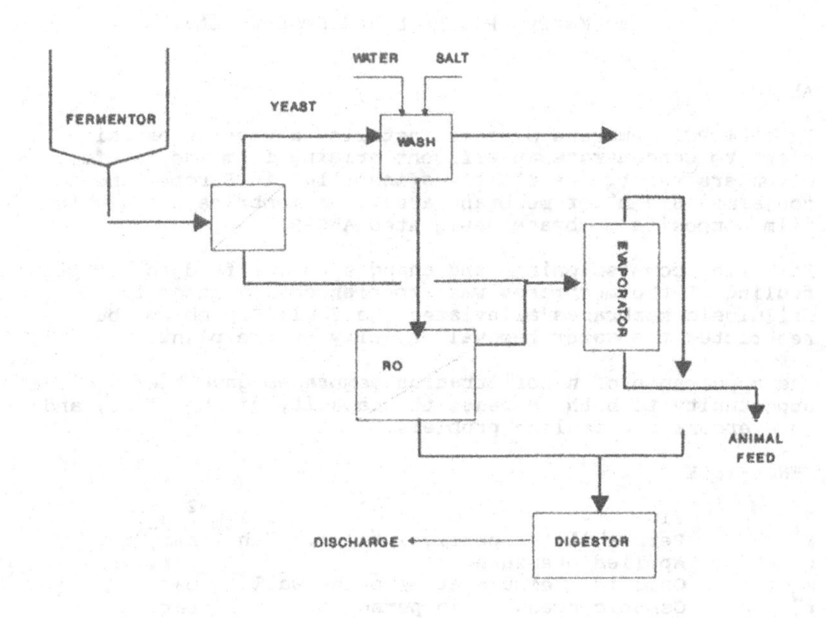

Figure 1 General process scheme for yeast effluent

THE PROBLEM

In the case discussed in this paper, the company was faced with the problem of disposing of an increased effluent flow due to an increase in their yeast production. The options were to increase the capacity of their evaporators, or to install some form of preconcentrator upstream of them. Reverse osmosis offered an economic solution as a preconcentrator, and it was this route that the company took.

TRIALS

In order to size the plant the company carried out two sets of trials at the factory. The first used a membrane area of 0.3 m^2, and the second used 5.2 m^2. A total of 97 runs were completed, although the duration of each run was only a few hours.

The data generated by the trials were the normal flux, concentration, and cod passage. In addition the effect of pressure and temperature were investigated, as was the relationship between osmotic pressure and conductivity.

The results indicated average fluxes of approximately 20.$l.m^{-2}.h^{-1}$. with an osmotic pressure of the feed of five to six bar at 10 mS conductivity.

THE PLANT

The plant was installed, and commissioned in 1986. It consisted of a six stage plant in a feed and bleed configuration. Each stage was fitted with 135 m^2 of AFC99 membrane. This membrane is a thin film polyamide composite giving better than 99% rejection of sodium chloride.

A variable capacity pump was fitted to the plant to accommodate variations in the feed flow and osmotic pressure. The nominal duty was to treat 20 $m^3.h^{-1}$. of effluent at 1.2% total solids content, producing 17.5 $m^3.h^{-1}$. of permeate. The actual performance was expected to fall in a range dependant on the actual solids content, the salt content, and the flow rate. The maximum solids specified in the concentrate was 9.6%. Figure 2 shows the design operating envelope for the plant.

Part of the original specification for the plant, and the feed stream is shown in table 1.

Table 1 Abridged feed specification

Total solids	0.96-1.44%
Ash	<5.3 $g.l^{-1}$.
Conductivity	<15 mS
COD	<8100 $mg.l^{-1}$.

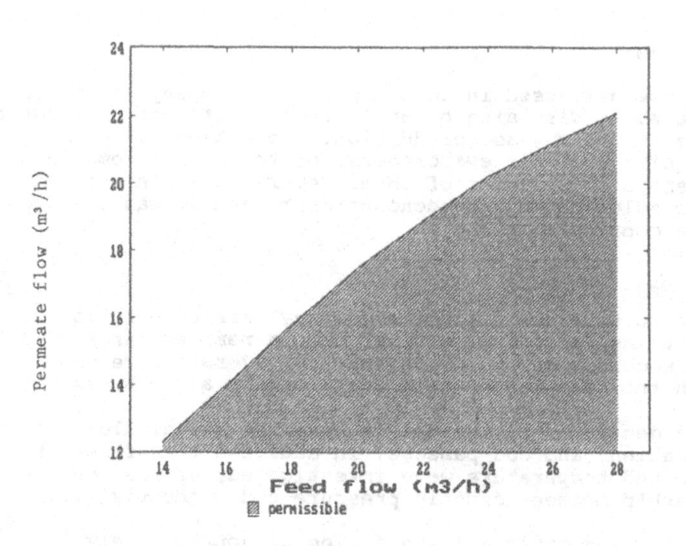

Figure 2 Operating envelope for original membranes

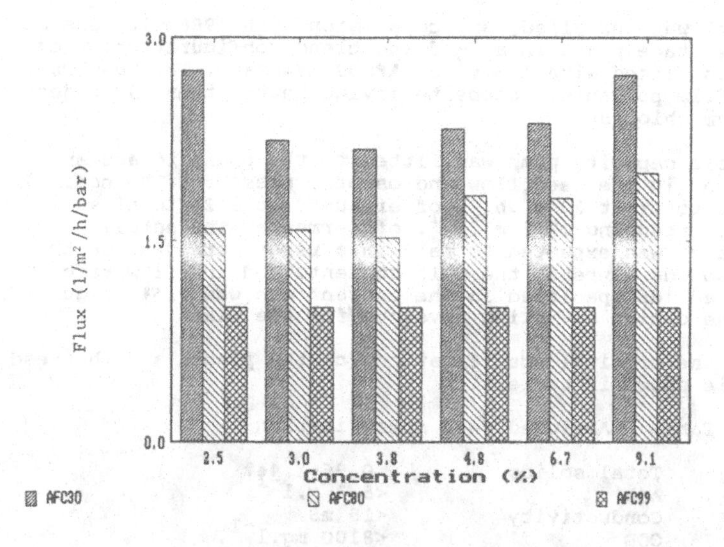

Figure 3 Comparative fluxes of test membranes AFC99=1

OPERATIONAL EXPERIENCE

Shortly after acceptance of the plant, the total solids in the feed to the plant rose to 1.8% to 2%, and the conductivities were found to be in the region of 18 mS. Consequently the water removal capability of the plant was reduced.

Over the first six months of operation the clean water flux fell to 25% to 50% of the original. Whilst some loss of clean water flux would be expected, the losses experienced indicated fouling. Despite this the plant was achieving the maximum permissible solids for over three months. Surveys across the plant indicated that the first stages of the plant were fouling more than the later stages. This "front end fouling" indicated that the fouling was due to adsorption of material in the feed, rather than precipitation of material from the feed,which would have manifest itself as fouling in the latter stages of the plant.

Membranes removed from the plant were coated in a brown deposit which could be removed by mechanical agitation. Analysis of the deposit showed a number of components including protein, calcium, and sodium to be present. The protein was thought to be the principal foulant. Up to that time the cleaning regime used was a nitric acid clean followed by a caustic detergent (Henkel P3 Ultrasil 11) at a concentration of 0.25%. Both cleans were of thirty minutes duration, and were carried out at about fifty degrees Celsius. It was obvious that this cleaning system was not effective under the conditions of operation at that time. A programme to investigate alternative cleaning was embarked upon. The initial steps were to extend the contact time with the caustic detergent, up to one hour. When this showed no effect higher concentrations were tried. A concentration of 2.5% showed some improvement in performance, but this concentration could only be used infrequently otherwise membrane life would be severely reduced, and the improvement was not sustained when less aggressive cleaning was employed again.

At this point enzymes were employed. The product Henkel P3 Ultrasil 50, which contains a blend of enzymes, surfactants, and complexing agents was tried. This product is formulated to clean proteins, and fats from membranes. The results were mixed, with on some occasions a significant improvement in plant performance being observed after the clean, but on other occasions no improvement was seen. The use of specific enzymes, such as amylase to attack starch, were also found to be not effective.

Alternative Membranes

With no obvious solution to the fouling problem the company decided to change the membranes, to a type made of cellulose acetate. This membrane limited the operating pressure to 37 bar, and the temperature to 37 degrees Celsius which intern limited the water removal capacity of the plant. These membranes were cleaned with an enzymic detergent, and then sanitised with hydrogen peroxide.

By the middle of 1988 the company decided that an increase in the water removal capacity of the RO plant was needed. They again approached PCI Membrane Systems to carry out trials on alternative membranes. Table 2 shows the characteristics of membranes that were tested.

Table 2 Characteristics of membranes tested in 1988 trials

Membrane	Water flux $l.m^{-2}.h^{-1}.bar^{-1}$	Retention % NaCl*
CPA15	>0.88	>89
CDA16	>0.88	>89
AFC80	>1.5	>80
AFC30	>1.75	>30 CaCl2

* Except AFC30

The CPA and CDA membranes are blended cellulose acetate membranes which are subject to similar operating constraints as the cellulose acetate membranes fitted in the plant. The AFC membranes are thin film composite polyamide. The chemistry of the AFC80 and AFC30 membranes is, however, sufficiently different to that of the AFC99 to confer different fouling characteristics to the membrane surfaces. In general these membranes exhibit a greater fouling resistance to proteins than AFC99.

The relative performance of the three types of AFC membrane on the effluent are shown in figure 3. The results are taken from a batch trial carried out on all three membrane types simultaneously. The operating pressure was 50 bar, and the temperature was 45 degrees Celsius. The potential difference is even greater at lower pressures as figure 4 indicates. Here the fluxes were measured at pressures of 13 bar. The membranes were also precompacted by running on water at 20 bar for 72 hours before the trials on effluent. The results clearly show the effect salt passage has in increasing the trans- membrane driving force as described by the equation:

$$J=K(P-(\pi_w-\pi_p)) \qquad (1)$$

Although the AFC30 shows a much higher passage of salt than the AFC99 the passage of COD does not differ proportionally in this instance. The average COD passage is shown in figure 5. Also the sulphate passage still remains reasonably low, which is important as the permeate is fed to a biological treatment plant where the presence of sulphate would cause hydrogen sulphide gas to be produced.

It was also thought that the lower chloride loading on the evaporators would reduce the corrosion.

Based on the evidence of the batch trials, the company fitted 2.6 m^2 each of AFC30, AFC80, and the cellulose acetate membranes into three stages of the plant for prolonged testing.

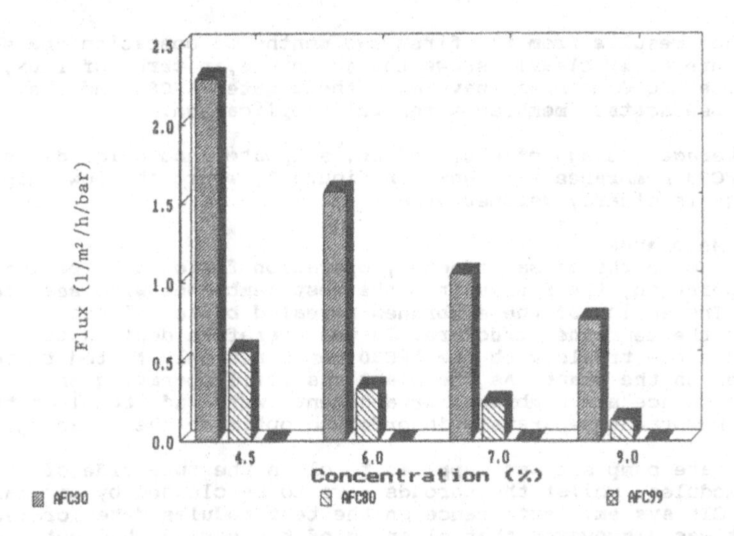

Figure 4   Comparison of AFC membranes on effluent

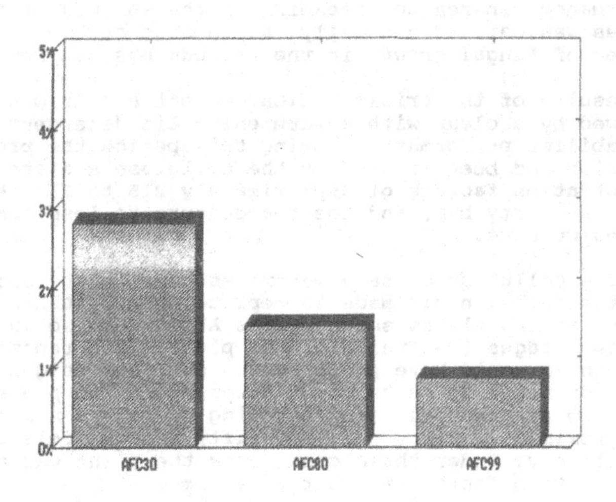

Figure 5   Passage of COD in small scale trials

Averaged results from the first two months of operation are shown in figure 6. It clearly shows the advantage,in terms of flux, that the AFC30 membrane have over the tighter AFC80 and the cellulose acetate membranes for this application.

The average passage of COD, sodium, sulphate, and chloride for the AFC30 membranes are shown in figure 7, where the low sulphate passage is clearly demonstrated.

AFC30 Membranes
After two months of satisfactory operation followed by period of non-operation, the fluxes from the test membranes were seen to fall. Inspection of the membranes revealed bacterial growth within the membrane structure. It was therefore decided to continue the trials with the AFC30 membranes only fitted to test modules on the plant. As the plant was still operating on cellulose acetate membranes arrangements were made to clean the AFC30 membranes separately in order to optimise the cleaning.

A separate pump set was supplied to clean the tube side of the test modules, whilst the shrouds were to be cleaned by the main plant CIP system. Performance on the test modules deteriorated, and it was discovered that cleaning of the shrouds had not been carried out. Inspection of the modules revealed a fungal growth in the shrouds. The membranes, and the modules were cleaned and sanitised by circulating a solution of 0.1% EDTA at 60 degrees Celsius though them for several hours, which restored their performance. Thereafter cleaning of the shrouds of the test modules was carried out daily, as is done on the main plant. The problem of fungal growth in the shrouds has not recurred

The results of the trials indicated that a nitric acid clean followed by a clean with a warm enzymatic detergent was suitable to stabilise performance. During this period the process condition had been limited by the cellulose acetate membranes to concentration factors of approximately 2.5 to 3, the pressure to less than forty bar, and the temperature to less than thirty-five degrees Celsius.

When the Cellulose acetate membranes came to the end of their life the decision was made to replace them with AFC30 membranes. However as the fluxes shown by the AFC30 were so much higher, only two stages (one third of the plant) were remembraned, although we would have preferred to see four stages in use to allow advantage to be taken of lower operating pressures. Operation recommenced with operating pressures and temperatures of approximately fifty bar, and fifty degrees Celsius respectively. Under these conditions the plant was achieving a concentration factor of about five times.

The plant had operated well for about a thousand hours when the clean water flux began to fall off. Various cleans were tried and a third stage, containing cellulose acetate membranes, was brought on line. Despite this the clean water flux continued to fall.

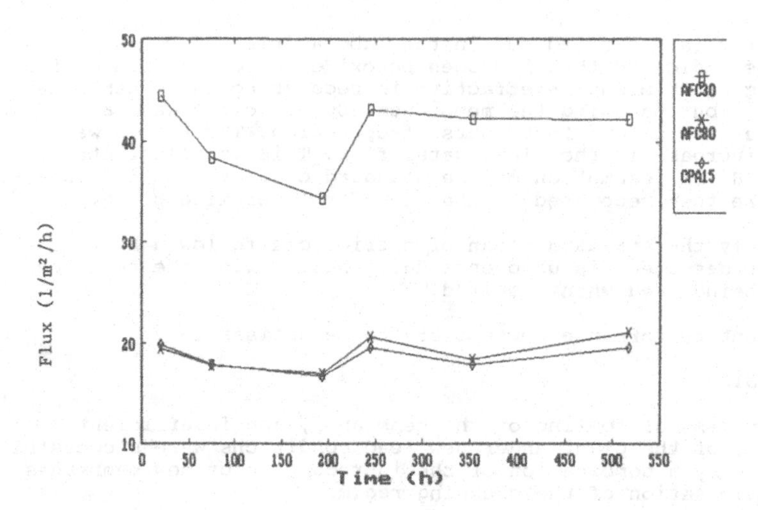

Figure 6 Performance of AFC membranes in large scale tests

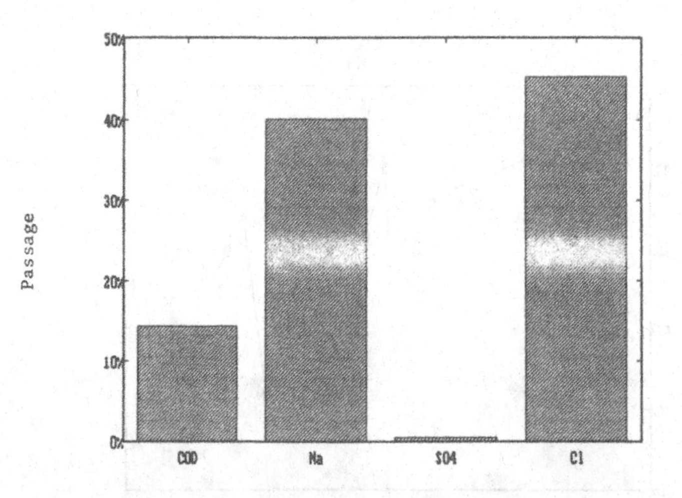

Figure 7 Solute passage for AFC30 membranes

However trials carried out in the laboratories of PCI Membrane Systems indicated that hydrogen peroxide in conjunction with a wetting agent might be effective in recovering the clean water flux without damaging the membranes. Such a clean was carried out on site just before 1500 hours of operation. The result was a sharp increase in the clean water flux. This is illustrated in figure 8. On resumption of the standard clean fluxes fell again, but were then recovered by the use of the peroxide clean.

Currently the standard clean of nitric acid followed by an enzymic detergent is used on a daily basis, with the peroxide clean being used when required.

The plant performance now appears to be satisfactory.

CONCLUSIONS

The problems of fouling of the membranes, and insufficient capacity of the plant under new feed conditions were successfully overcome by a combination of the introduction of new membranes and optimisation of the cleaning regime.

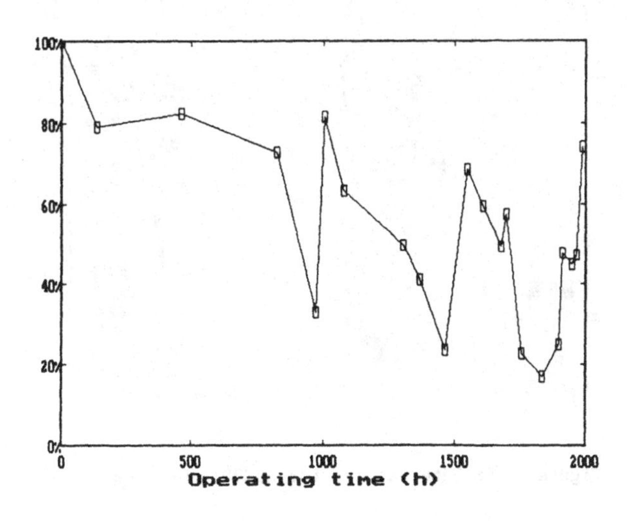

Figure 8 Clean water fluxes from plant

ENHANCED MASS TRANSFER WITH ULTRAFILTRATION AND MICROFILTRATION MEMBRANES

by D A Colman and C D Murton
BP Research (UK)

The efficiency of membrane separation processes can be substantially reduced by the occurrence of concentration polarisation which is a function of the mass transfer, membrane permeability and the permeate driving force. Operation in turbulent cross flow to achieve high mass transfer results in low net separation per unit length of membrane and, in filtration systems, the high axial pressure gradient can make it impossible to operate the whole of the membrane unpolarised. A technique by which high mass transfer is achieved through the pulsing of a low mean crossflow in a baffled channel has been developed and reported previously. This paper extends the results of the technique to the performance of ultrafiltration and microfiltration membranes with a fouling yeast suspension. Characteristic polarisation plots are reported with polarised fluxes equivalent to steady crossflows with Reynolds number ~6000 when the mean flow Reynolds number is <100. Potential operation of a module with a fully unpolarised membrane and significant separation is demonstrated and the estimated energy requirements shown to be at least comparable with steady flow systems.

NOMENCLATURE

d	-	channel height (mm, m)
L	-	channel length (m)
A	-	channel cross section area (m^2)
n	-	baffles per unit length (m^{-1})
f	-	pulse frequency (Hz)
x_o	-	pulse amplitude (mm, m)
x, $\dot{x}$ $\ddot{x}$	-	displacement and time derivatives (m)
q	-	mean flow (l/min, m^3/s)
q'	-	superimposed oscillation (l/min, m^3/s)
ΔP	-	transmembrane pressure (bar)
Cd	-	discharge coefficient around baffles
E_i	-	inertia energy (W)
E_p	-	pressure drop energy (W)
E_t	-	total energy (W)
ED	-	energy density (W/m^3)
ρ	-	fluid density (kg/m^3)
ν	-	fluid viscosity (m^2/s)
Th	-	Thomson Number = x_o/d
PuRe	-	Pulsatile Reynolds Number = $2\Pi fd^2/\nu$
Re	-	Mean Reynolds Number = $qd/A\nu$

Effective Industrial Membrane Processes — Benefits and Opportunities, pp.33–47

1. INTRODUCTION

1.1 Background

The efficiency of membrane separation processes can be substantially reduced by the occurrence of concentration polarisation which is a function of the mass transfer, membrane permeability and the permeate driving force. This polarisation is frequently minimised by operation with turbulent crossflow in order to achieve high mass transfer. However, this results in low net separation per unit length of membrane and consequently unfavourable module design implications. Additionally, in ultrafiltration and microfiltration systems, the high axial pressure gradient necessary for turbulent crossflow can make it impossible to operate the whole of the membrane unpolarised. This can be detrimental in fractionation duties (for example the clarification and sterilisation of beer, Reed et al 1989) where a polarised layer forming a secondary membrane can affect the transmission of valuable species into the permeate.

Turbulence promoters (Poyen et al 1987) and corrugated membrane surfaces (Racz et al 1986) have been used to increase the mass transfer at the membrane surface. Pulsatile flow in plain channels has also been investigated (Kennedy et al 1974). However, it is only recently that pulsatile flow combined with suitable channel geometries has been used to enhance mass transfer through the generation of vortex mixing.

1.2 Vortex Mixing

Except at very low Reynolds numbers (Re <10) flow past bluff bodies or surface irregularities will separate, generating an eddy or vortex on the downstream side. In steady confined flows these separation regions generally remain captive and have little effect upon the bulk stream flow. If however, the flow is decelerated, a separation vortex will grow, and upon flow reversal be ejected into the bulk flow. In pulsatile flow, in which bulk flow reversal occurs, repeated ejection of vortices from suitably designed baffles creates vortex mixing which has been shown to increase mass and heat transfer. Bellhouse et al using pulsatile flow across membranes with furrowed and dimpled surfaces to generate the vortices has demonstrated increases in membrane flux due to the enhanced mass transfer. Mackley (1987) on the other hand has placed orifice plate baffles in tubes and shown how vortex mixing can be generated with pulsatile flow. By selection of pulsatile flow parameters, this device has high heat transfer and plug flow characteristics. Howell (1989) using baffles similar to Mackley's design has demonstrated flow enhancement of ultrafiltration (UF) tubular membranes.

1.3 Work at BP

A technique by which high mass transfer is achieved through the pulsing of a low mean crossflow in a baffled rectangular channel has been developed by BP Research at Sunbury. It has been shown how mass transfer coefficients equivalent to steady flow Reynolds numbers in unbaffled channels of >10,000 can be achieved when the mean through flow Reynolds number is ~100 (Colman et al 1990). In addition, the effect upon an ultrafiltration membrane with a non fouling solute was also demonstrated.

This paper reports an extension of these results to the effect of the pulsed flow technique with a highly fouling yeast suspension on commercially available ultrafiltration and microfiltration flat sheet membranes.

2. THEORY

2.1 The System

In the geometric configuration for which results are reported, two flat sheet membranes are spaced d=6 mm apart forming a rectangular channel. Rectangular baffles 3 mm high span the centre of the channel transverse to the flow direction as shown in Figure 1. The baffles are supported from the sides of the channel and are spaced at 12 mm.

The flow through the channel comprises a mean flow q with a superimposed oscillation q' which is driven by a double piston pump. This oscillatory flow is sinusoidal and

$$q' = 2\Pi f x_o A \sin(2\Pi f t) \qquad 1)$$

where f is oscillation frequency, x_o is linear amplitude and A is channel cross sectional area. The resulting pulsatile flow can be characterised by three non dimensional parameters for amplitude, frequency and mean flow:

Thomson number $Th = x_o/d$
Pulsatile Reynolds number $PuRe = 2\Pi f^2 d / \nu$
Mean flow Reynolds number $Re = qd/A\nu$

2.2 Energy Requirement

The maximum energy required to pulse the flow can be estimated from the inertia energy E_i and the pressure drop energy E_p.

Over the first half of a cycle it is assumed that energy is only required to accelerate the flow and the inertia energy is dissipated on deceleration.

Whence

$$E_i = 2f \int_0^{\frac{1}{4f}} \rho A L \ddot{x} \dot{x} \, dt \qquad 2)$$

$$= 4\Pi^2 \rho A L x_o^2 f^3 \qquad 3)$$

The pressure loss through the baffles can be derived by assuming that each baffle acts like an orifice. Whence for each baffle:-

$$\text{Pressure drop} = \frac{q'^2 \ \rho \ (A^2 - A^2/4)}{C_d^2 \ A^4/2} \qquad 4)$$

where the discharge coefficient Cd = 0.61 is assumed. Whence evaluating equation 4, the energy due to pressure loss is

$$E_p = 2f \ 4.03 \ A\rho nL \int_0^{\frac{1}{2f}} \dot{x}^3 \, dt \qquad 5)$$

$$= 43 \ A\rho nL \ \Pi^2 \ x_o^3 \ f^3$$

where n is the number of baffles per unit length of channel (n = 83/m in the present system).

Thus the total energy $E_t = E_i + E_p$ 6)

and the energy density

$$ED = \Pi^2 \ \rho \ x_o^2 \ f^3 \ (4 + 43 \ x_o \ n) \qquad 7)$$

3. EXPERIMENTAL

3.1 Test Procedure

A membrane module was used which could house two 0.05 m^2 flat sheet membranes having a width of 100 mm and length of 500 mm. The module allowed for baffles to be inserted between the membranes or for the channel to be unbaffled, Figure 3.

Each test run was initiated by pumping distilled water through the system and raising the pressure to 2.5 barg (with the permeate offtake closed). With the cross flow set at 0.5 l/min, the pulsing pump was switched on and pulsations set at 8 Hz and 8 l/min amplitude. These flow conditions corresponded to a Thomson number Th = 0.74, Pulsatile Reynolds number PuRe = 1770 and Mean Flow Reynolds number Re = 82. By reducing permeate back pressure, the transmembrane pressure was increased in order to obtain the pure water flux. Freshly prepared yeast suspension was then added to the feed reservoir to obtain the required concentration and membrane performance data recorded regularly during the test (approximately every 3 minutes). The permeate flow was allowed up to one hour to stabilise at each condition before the transmembrane pressure was raised and the data retrieval process repeated. An example of the sequential operation to determine flux vs pressure plots is shown in Figure 5.

3.2 Materials Used

3.2.1 Feedstock

Suspensions of bakers yeast were used as feedstock for the filtration tests. This was selected in order to model, in the laboratory, high fouling feeds typical of the biotech and brewing industries. For the ultrafiltration tests, a 10% suspension was prepared in distilled water (5°C) and homogenised at 500 bar using an APV Gaulin Tempress. Two passes through this press yielded a suspension with >90% disrupted yeast cells. This was added to the distilled water in the test rig to obtain the required concentration (0.1% or 1% by weight).

The ultrafiltration tests used GR 40 PP polysulphone membranes from DDS. These had a specified molecular weight cut off of 100 k and pure water flux at 18°C of 4.5 $l/(m^2$ min bar). The microfiltration tests used Filtron Omega 0.16 μm cut off membranes, with a pure water flux at 18°C of 130 $l/(m^2$ min bar).

4. PULSATILE FLOW PHENOMENA

4.1 Vortex Mixing

The basic flow structure associated with pulsatile flow in the baffled channel geometry has been described previously (Colman et al 1990) and is shown in Figure 2 for PuRe = 250, Th = 0.45 and zero net throughflow. Four stages of development through half a cycle are shown and vectors to the side of each picture show the momentary relative magnitude and direction of the flow. On acceleration from zero flow, separation regions are formed on the downstream side of each baffle. These create a pair of vortices which grow in size and strength as the flowrate increases. As the flow decelerates, Figure 2.3, so these vortices continue to grow and secondary vortices can be seen to form close to the channel walls. At zero flow the vortices expand rapidly entraining fluid and filling the whole of the channel. As the flow reverses so the cycle is repeated in the opposite direction. This vortex creation, expansion and ejection causes fluid from the centre of the channel to be mixed with fluid from close to the walls. Thus the mass transfer to the walls is enhanced and embryonic concentration boundary layers forming at a membrane surface are destroyed.

The sequence shown in Figure 2 has a low PuRe and remains symmetric about the channel centre line. Increasing the frequency such that PuRe >500, symmetry is lost and the flow becomes chaotic, although the basic mechanism of vortex mixing remains.

4.2 Mass Transfer

The enhancement in mass transfer due to the pulsed flow has been discussed (Colman et al 1990) and is shown in Figure 4.

Mass transfer is plotted against mean flow Reynolds number for steady flow in a baffled and unbaffled channel and also for a pulsed flow superimposed on a steady flow in a baffled channel. To the left of the dashed line, which shows the point of zero reversal for the pulsation, the mass transfer coefficients of the pulsed flow are relatively constant. However, where there is no reversal (to the right) the effect of the superimposed oscillation diminishes and the mass transfer becomes mean flow dominated.

Thus the mass transfer to the walls of the channel can be made independent of the mean through flow provided PuRe.Th>Re.

In addition, mass transfer data from the unbaffled channel with steady flow Re = 100 - 10,000 is also plotted in Figure 4. This shows the local mass transfer measured 30 d from the leading edge of the electrode. The transition from laminar to turbulent flow is clearly seen at Re ≃2500. In the laminar region, agreement is close to the analytical Leveque type solution for mass transfer (Rousar et al 1971).

In turbulent flow, the slope of the plot corresponds to the Chilton Colburn relation for mass transfer (Gekas et al 1987). The over measurement is believed to be due to the turbulent front in the channel occurring only about 15 d upstream of the measuring point.

Comparing data in Figure 4, it can be seen that the mass transfer in the baffled channel with pulsing flow but low mean flow, Re ≃100, is comparable to that from a steady flow Re ≃6000 in the unbaffled channel.

It should be noted that all the measured mass transfer coefficients are for zero flow through the surface. In a membrane system such as ultrafiltration, the presence of flux will modify the actual mass transfer (Gekas et al 1987).

5. MEMBRANE PERFORMANCE

5.1 Ultrafiltration Tests

5.1.1 Characteristics of a Test

Data from a typical test run shown in Figure 5, from which equilibrium flux measurements are obtained at various transmembrane pressures (ΔP). This is typical of all test runs executed. At low ΔP the observed fluctuations in flux are due to the small fluctuations in ΔP in this unpolarised region of operation.

The equilibrium fluxes derived from these data are plotted against ΔP in Figure 6. The characteristic unpolarised/polarised plot is seen. In the unpolarised linear increase region, the membrane is flux limiting although the flux is less than the pure water flux. This is probably due to an increased membrane resistance caused by protein adsorption fouling within the pores.

As ΔP rises above 0.6 bar so the flux becomes polarised and mass transfer limited.

The change from unpolarised to polarised can be seen in the plot of flux against time, Figure 5. While the membrane is unpolarised at ΔP <0.6 bar flux increases with ΔP and there is no subsequent decline. However, at higher temperatures, there is an immediate increase in flux with ΔP, followed by a rapid decline to equilibrium, due to a polarisation layer build up. These increases in ΔP are each preceded by a short period of operation at 0.05 bar. This low pressure operation allows the polarised layer to be cleared from the membrane surface. It is noted from the near constant fluxes at this low pressure that there is apparently very little further fouling of the membrane occurring when it is polarised.

5.1.2 Pulsed Flow/Steady Flow Comparison

A comparison between steady flow in an unbaffled channel and pulsed flow in a baffled channel, each with the same mean crossflow of 0.5 l/min (Re = 82), is shown in Figure 6 for yeast concentrations of 0.1% and 1%.

Under steady flow operation, the membranes soon polarise as ΔP is increased at both concentrations due to the low mass transfer. This makes it very difficult to operate unpolarised. Membrane flux is thus effectively constant at each concentration and approximately a factor of 10 below the maximum (polarised) fluxes obtained with pulsed flow operation.

Normally steady flow would be operated with much higher crossflows and narrower channels to increase the mass transfer. Previous tests with Dextran have suggested that, for equivalent fluxes, a steady flow of around 40 l/min in the unbaffled channel would be necessary. This has not been confirmed since the maximum steady flow in the rig is 30 l/min. These results however, demonstrate how significant fluxes can be obtained with low mean crossflow and that the membrane can be operated fully unpolarised. This means that a significant separation can be effected in a single pass through a membrane module when the pulsed flow technique is used.

5.1.3 Continuous Unpolarised Operation

A test was carried out at 1% yeast during which ΔP was held at 0.2 bar for 5.5 hours, Figure 7. From the flux against pressure plots, it had been established that, under these conditions, the membrane would be unpolarised. After the initial 30 minutes, during which some fouling of the membrane occurred, an equilibrium flux of 0.21/(m^2 min) was established. During continued operation under these unpolarised conditions, no further reduction in flux was observed. A subsequent increase in ΔP to 1 bar showed a rapid decline to the polarised flux of ≈0.35 l (m^2/min). This is a significant result since it shows that once a non low fouling membrane has become initially fouled then no further flux reducing fouling occurs when it is operated unpolarised.

5.2 Microfiltration Tests

Data is also included from a test with a microfiltration membrane and non homogenised bakers years (ie: whole yeast cells) Figure 8.

The characteristic unpolarised/polarised curves are observed for each plotted concentration. At 0.01% yeast the unpolarised section follows the pure water flux closely as ΔP increases and before it becomes polarised. Unlike the UF tests, the same membrane was used for the higher concentration of 0.1% yeast as well. It is possible that the membrane had become fouled whilst operating polarised, causing the flux at unpolarised conditions to be below the unfouled pure water flux. However, in spite of this, unpolarised operation is clearly shown.

6. IMPLICATIONS FOR MODULE DESIGN

The above shows that pulsed flow in baffled channels can enhance the mass transfer in UF and MF membrane systems. Fluxes equivalent to those from high crossflow systems can be achieved when the net crossflow is very low and, when unpolarised, these fluxes are sustainable in high fouling media. For example in the test module with 0.1 m^2 of membrane and a crossflow of 0.5 l/min up to 25% of the feedflow went into the permeate in a single pass. Thus, since flux is independent of crossflow, with a few units in series, or lower crossflow, a significant concentration factor would be achievable; thereby reducing feedstock residence time and removing need for recirculation.

If unpolarised operation of UF or MF membranes is important in order to ensure good transmission of permeating species then these results show that the technique can be used to tune the membrane permeability, transmembrane pressure and mass transfer across the whole membrane surface to the feed concentration. This is not readily possible in high crossflow systems in which there is inevitably a significant axial pressure gradient on the feed side.

In the section on power requirements above an expression for the energy density of the baffled channel was derived, equation 7. Substituting the pulsation frequency, f = 8 Hz and amplitude x_0 = 4.42 x 10^{-3}m, the estimated energy density for this system is 2 kW/m^3. A flux comparable steady flow system in the unbaffled channel has a flow of 40 l/min (Re = 6500) and from Blasius friction factor an energy density of 2.1 kW/m^3. In this system however, the module length would be 40 m for similar separation and the axial pressure drop 0.76 bar, making unpolarised operation impossible over the whole membrane surface.

While the power required for the 8 Hz, 8 l/min pulsations used in this work is apparently comparable to that for the steady flow system it is noted that in the expression for energy density the dominating frequency and amplitude terms are to the third power. Slightly less intense pulsing may therefore reduce the energy density of a pulsed system significantly below that for similar steady flow as has been suggested by Howell 1989. This has not been investigated in this work.

The membrane flux enhancement technique described in this paper is the subject of a patent filing by the British Petroleum Company.

7. CONCLUSIONS

7.1 The effect of pulsed flows in baffled channels upon the performance of ultrafiltration and microfiltration membranes with a suspension of bakers yeast has been assessed.

7.2 By pulsing a low mean crossflow in a baffled channel (Re = 82), polarised fluxes a factor of 10 above the unpulsed, unbaffled case have been demonstrated that are equivalent to the flux from a steady high crossflow Re ~6000.

7.3 Flux against transmembrane pressure plots have been derived from both MF and UF membranes which show the characteristic unpolarised/polarised form.

7.4 By incorporating the technique into module design, it will be possible to control mass transfer and transmembrane pressure independently of the net crossflow. This will enable operation of membranes fully unpolarised and thus without a secondary membrane forming on the surface.

8. ACKNOWLEDGEMENTS

The authors would like to thank the British Petroleum Company for permission to publish this paper.

9. REFERENCES

1. Bellhouse B.J. et al, 'A High Efficiency Membrane Oxygenator and Pulsatile Pumping System and its Application to Animal Trials'. Trans. Amer. Soc. Artif. Organs 19, 1973.

2. Colman D.A. et al (1990) 'Enhanced Mass Transfer for Membrane Systems', I Chem E. Advances in Separation Processes.

3. Gekas V et al (1987) 'Mass Transfer in the Membrane Concentration Polarisation Layer under Turbulent Crossflow'. J. Mem. Sci 30.

4. Howell J.A. et al (1989) 'The Effect of Pulsatile Flow on UF Fluxes in a Baffled Tubular Membrane System', Chem Eng Res Des Vol 67.

5. Kennedy T.J. et al (1974) 'Improving Permeation Flux by Pulsed Reverse Osmosis'. Chem Eng Sci Vol 29.

6. Mackley M.R., (1987) 'Using Oscillatory Flow to Improve Performance'. The Chemical Engineer, February 433.

7. Poyen S. et al (1987) 'Improvement of Flux of Permeate in UF by Turbulence Promoters'. Inst Chem Eng Vol 27, 3.

8. Racz I.G. et al (1986) 'Mass Transfer, Liquid Flow and Membrane Properties in Flat and Corrugated Hyperfiltration Modules'. Desalination 60.

9. Reed R.J.R. et al (1989) 'Single Stage Downstream Processing of Beer Using Pulsed Crossflow Filtration'. Proceedings of the 22nd Congress, EBC, Zurich.

10. Rousar I et al (1971) 'Limiting Local Current Densities for Electrode located on the walls of a Rectangular Channel with Laminar Flow'. J. Electrochem Soc. 881 June.

Fig 1 CHANNEL GEOMETRY

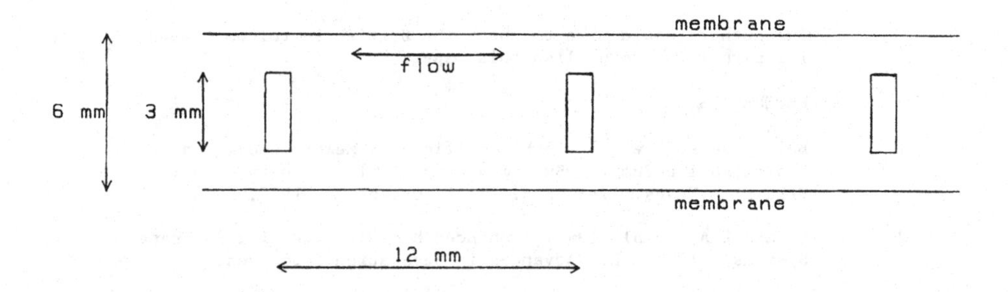

Figure 2.

12mm Baffled channel

Th = 0.9 PuRe = 250 Re = 0

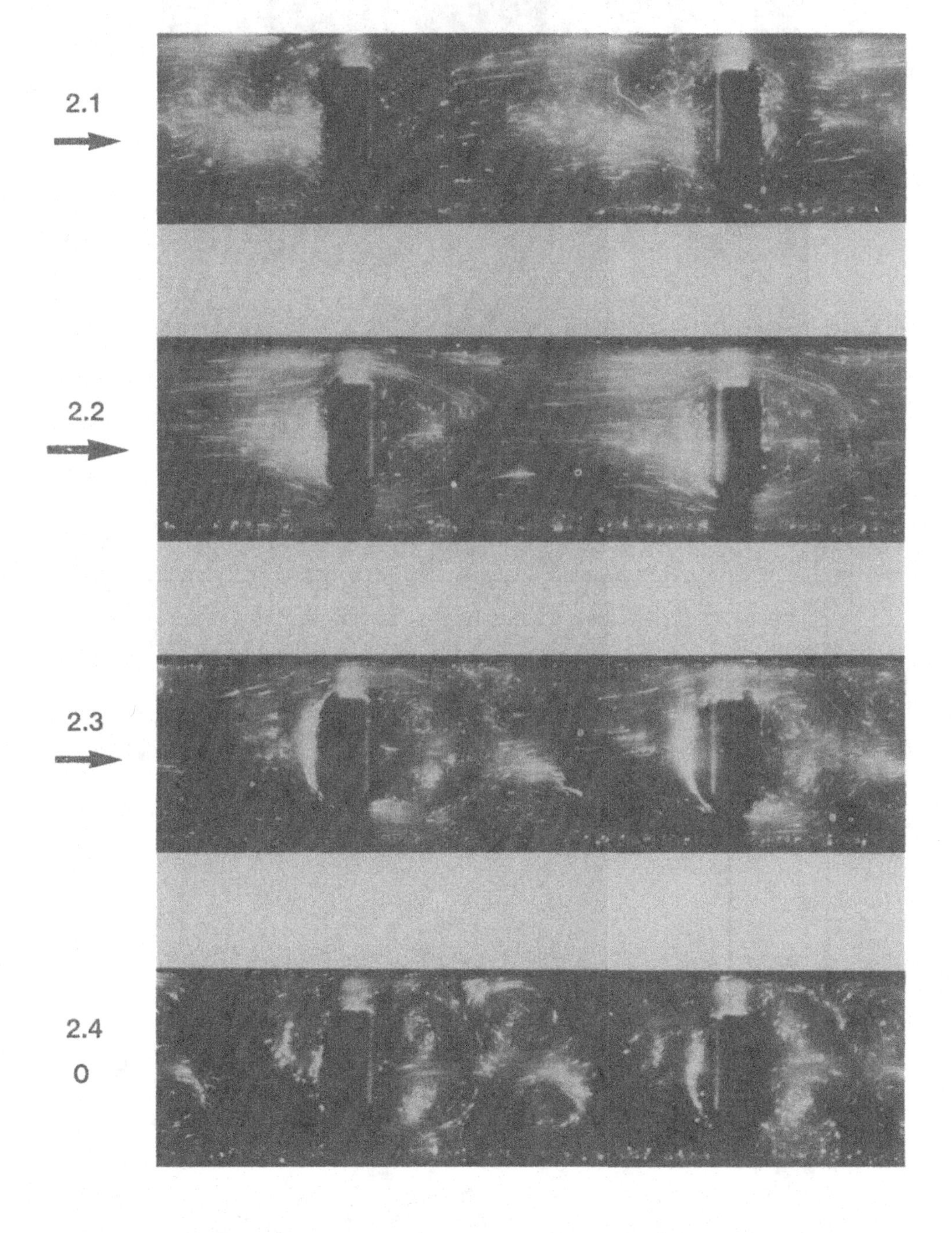

FIG 3 INSIDE OF ULTRAFILTRATION MODULE

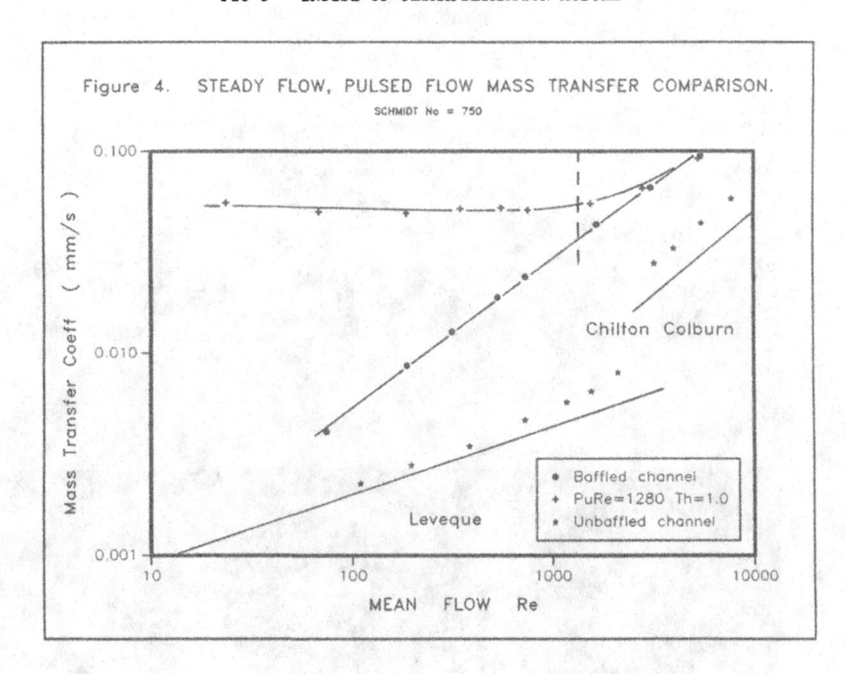

Figure 4. STEADY FLOW, PULSED FLOW MASS TRANSFER COMPARISON.

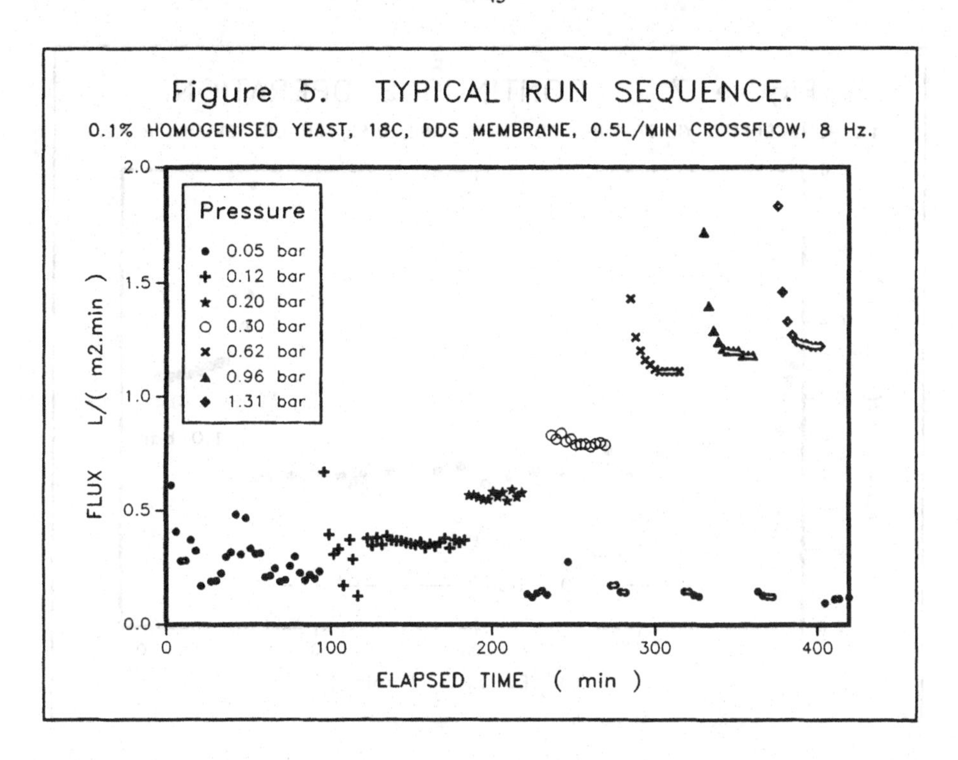

Figure 5. TYPICAL RUN SEQUENCE.
0.1% HOMOGENISED YEAST, 18C, DDS MEMBRANE, 0.5L/MIN CROSSFLOW, 8 Hz.

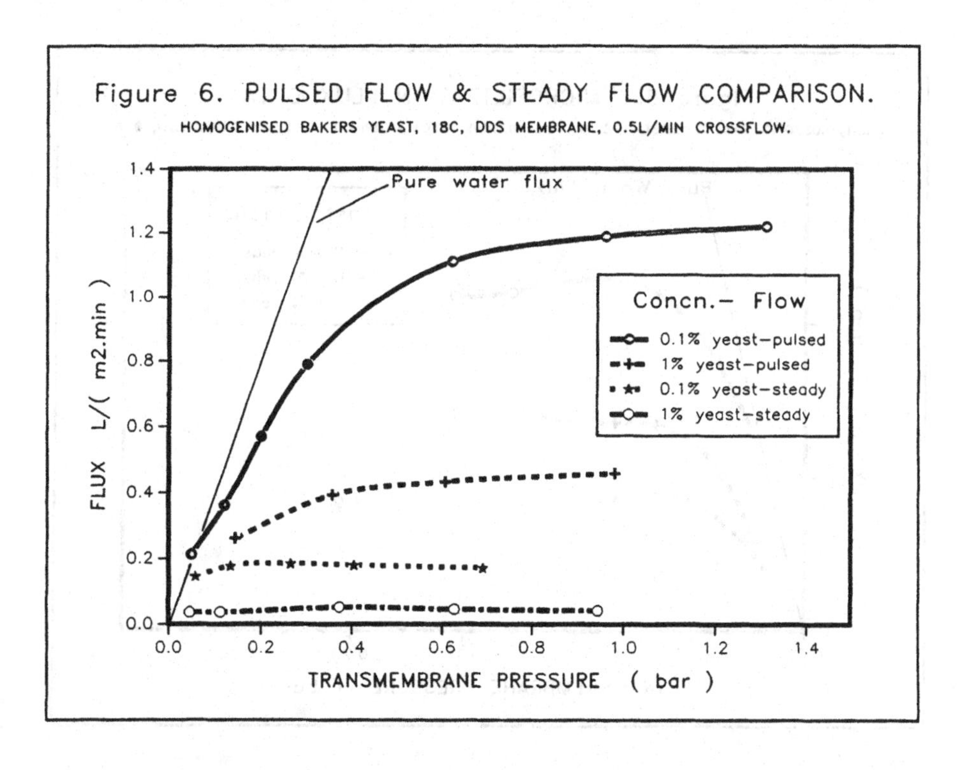

Figure 6. PULSED FLOW & STEADY FLOW COMPARISON.
HOMOGENISED BAKERS YEAST, 18C, DDS MEMBRANE, 0.5L/MIN CROSSFLOW.

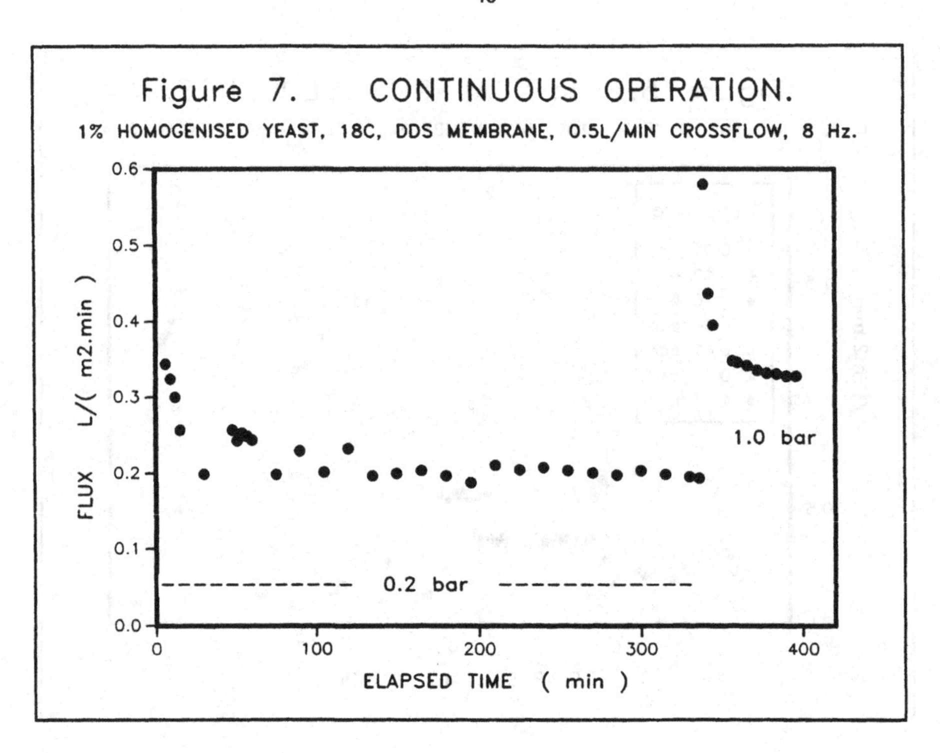
Figure 7. CONTINUOUS OPERATION.
1% HOMOGENISED YEAST, 18C, DDS MEMBRANE, 0.5L/MIN CROSSFLOW, 8 Hz.
FLUX L/(m2.min)
0.0
0.1
0.2
0.3
0.4
0.5
0.6
0.2 bar
1.0 bar
0
100
200
300
400
ELAPSED TIME (min)

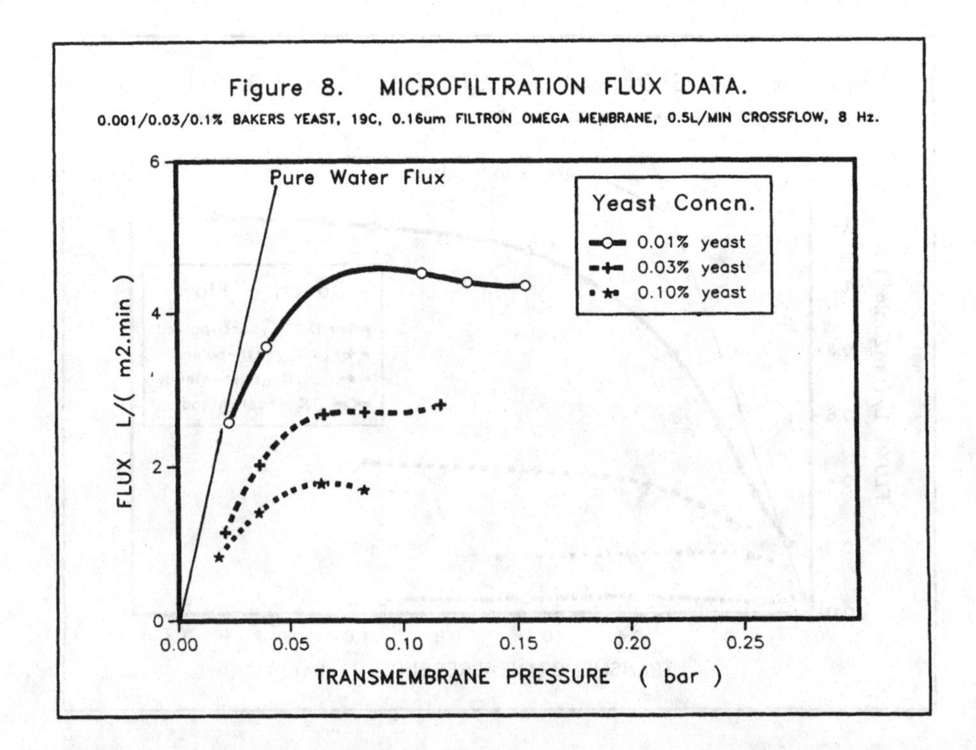
Figure 8. MICROFILTRATION FLUX DATA.
0.001/0.03/0.1% BAKERS YEAST, 19C, 0.16um FILTRON OMEGA MEMBRANE, 0.5L/MIN CROSSFLOW, 8 Hz.
Pure Water Flux
Yeast Concn.
0.01% yeast
0.03% yeast
0.10% yeast
FLUX L/(m2.min)
0
2
4
6
0.00
0.05
0.10
0.15
0.20
0.25
TRANSMEMBRANE PRESSURE (bar)

SESSION B:

MASS TRANSFER AND CLEANING II

HYDRODYNAMICS AND MEMBRANE FILTRATION

J.A. Howell and S.M. Finnigan

Membrane Applications Centre, Department of Chemical Engineering,
Bath University, Claverton Down, Bath, BA2 7AY.

Summary:

Many different approaches have been taken to combat fouling and concentration polarization. A great many of these deal with modifying the properties of the membrane surface or the hydrodynamics above it. This paper reviews the effects that hydrodynamic factors have on the performance of membrane filtration systems. Particular emphasis has been given to recent developments in mixing techniques.

Introduction:

With all cross-flow membranes, the presence of either a suspension or a solution causes, at constant transmembrane pressure, a decline in permeate flux. Firstly, there is a rapid decline in flux followed by a slower change over a period of several hours. Secondly, particularly in cases of ultrafiltration operated at sufficiently high pressures, a limiting flux condition is reached. The limiting flux is independent of transmembrane pressure and is much smaller than the pure water flux that would be achieved under similar hydrodynamic conditions. Aimar and Sanchez(1985) attributed the existence of the limiting flux to the achievement of a balance between the transmembrane pressure and the osmotic pressure differential between the soloution at the membrane surface and the permeate with the degree of concentration polarization controlled by the mass transfer coefficient in turn affected by variations of the physico-chemical properties and particularly the viscosity, in the boundary layer.

Concentration Polarisation:

One of the dominant features associated with fouling is the phenomenon of concentration polarization. The flux of permeate through the membrane causes a convective flow of both permeate and retained material towards the membrane surface. The retained material close to the surface is at a concentration greater than the bulk concentration and so a diffusive back-flow is generated. Steady-state conditions are rapidly obtained; Howell and Velicangil(1982) have computed a time of no more than a few seconds. Thus, the convective flow towards the membrane and the associated transmembrane flux is limited by the magnitude of the back-diffusion. In other words, one factor limiting permeate flux is the mass transfer of retained material back into the bulk flow. This transfer depends on the solution rheology and the hydrodynamics.

A concentration profile is established at the membrane surface due to the rejection of solute by the membrane, and normally called a concentration boundary layer or polarisation layer. The existence of this boundary layer is related not only to the limitation of permeate flux to far below the pure solvent flux through the membrane but also to membrane fouling and solute rejection (Le and Howell, 1985; Aimar et al., 1985; and Aimar and Sanchez, 1985). It is important to understand the detailed behavior of the concentration boundary layer in order that optimal design of ultrafiltration membrane process is possible.

At steady state the rate of solute arriving at the membrane is balanced by the sum of the solute leakage and the rate of back diffusion(1), i.e.:

$$J = k \ln \frac{[c_w - c_p]}{[c_b - c_p]} \tag{1}$$

Usually only an average flux can be measured, then from Eq.(1), an "apparent wall concentration" at the surface can be obtained. The flux is clearly a function of the mass transfer coefficient which is inflenced by the cross flow velocity amongst other factors. Hydrodynamics thus plays a big part in changing flux which is concentration limited. Obviously, this description of the concentration boundary layer is too simple. Polarisation is not evenly distributed over the membrane since the concentration boundary layer is developing from the inlet of the channel (Leung and Probstein, 1979; Clifton et

Effective Industrial Membrane Processes — Benefits and Opportunities, pp.49–60

al., 1984; Aimar et al., 1991). This means that the local wall concentration and local permeate flux vary along the length of a membrane, and the concentration boundary layer develops in thickness along the length of membrane.

Clifton et al(1984) built a hollow-fibre module which was divided into a number of compartments on the low-pressure side, and made the measurements of local permeation rates at different positions along the length of the fibre bundle. Additionally, they proposed a model assuming a completely osmotic limitation to flux, and used this integral model to calculate the growth of the polarisation layer in ultrafiltration with hollow-fibre membranes.

Aimar et al(1989) used an integral model similar to Clifton in an ultrafiltration module with a rectangular channel. The variations of flux with pressure were analysed by calculating the variation of wall concentrations along the membrane. The natural conclusion was that the average flux of a module with a short channel is larger than that with a long channel is useful for design and operation of ultrafiltration process when it is quantified.

A model for the flux which includes both osmotic limitation to permeate flux and long term fouling is as follows:

$$J = \frac{\Delta P - \Pi}{\mu(R_m + R_f)} \tag{2}$$

where R_m is initial membrane resistance,
R_f is the resistance of a fouling layer
and the osmotic pressure is given by

$$\Pi = a_1 c_w + a_2 c_w^2 + a_3 c_w^3$$

, where a_i is the ith osmotic virial coefficient
c_w is the membrane wall concentration.

The polarisation results in an increased osmotic pressure which decreases flux. The increased viscosity may also decrease flux by changing the back mass transfer coefficient. The wall concentration increases steadily with the development of the concentration boundary layer from the inlet of the channel to its outlet and as it increases so the local flux decreases. When the boundary layer is fully developed at steady state, the wall concentration should be at its maximum and equal to a constant and the flux will be significantly less than at the entrance. The shortest length at which the wall concentration is arbitrarily close to its maximum value may be termed the entrance length of the concentration boundary layer, designated L_e. It is possible as Clifton et al. (1984) have shown to calculate this entrance length which influences the design of the membrane system and affects scale-up calculations as shown by Aimar et al. (1989).

Once at the maximum value for the concentration which is governed by the osmotic resistance it is interesting to determine how the flux changes with bulk concentration. Equation (1) suggests that for a totally rejecting membrane the flux will be dependent on the logarithm of the bulk concentration. Experiments show that this is apparently true and that a limiting concentration at which the flux is zero can be found by plotting the flux against the log of the bulk concentration. However it has been shown (Pritchard et al, 1990) that the plot can deviate at high concentrations and intersect at a much higher concentration. This can be due to the change in the mass transfer coefficient as the viscosity rises. This only continues for a fixed period. As Field (1990) has shown for Newtonian fluids, and Pritchard et al (1990) for non-Newtonian fluids the mass transfer coefficient is increased if the ratio of the bulk viscosity to the wall viscosity increases. This does happen when the wall concentration is fixed and the bulk concentration increases during the period of the run. It is thus possible to obtain significantly higher concentrations towards the end of a concentration run.

Fouling:

Subsequently flux declines over time throughout the length of the membrane. There is a considerable range of opinion on the causes of the flux decline. The relevant properties of the membrane are in the first place, mechanical and chemical resistance and chemical compatibility rather than filtration properties. Murkes and Carlsson(1988) state that the fouling layer or layers which form on the membrane surface, sometimes referred to as a "secondary" or "dynamic" membrane, control the filtration process. In general, the flux declines and the rejection increases as the secondary membrane is formed. In addition to this reduction in throughput capacity and compromised separation capability, fouling results in increased power consumption, time consuming and expensive washing and cleaning operations and reduced membrane service life. This fouling layer may be accumulated as a time or time and concentration dependent process. As with concentration polarisation it is apparently reduced by improved hydrodynamics.

Improved Hydrodynamics:

Turbulent Flow

The hydrodynamic approach to improving the flux is either to reduce the concentration polarisation by increasing the mass transfer away from the membrane or to reduce fouling based on increasing the wall shear rate and/or scouring the membrane surface. This is achieved most easily by simply increasing the cross-flow rate either directly, so that the flow changes from laminar to turbulent, or indirectly, by modification of the channel geometry. Prior to the 1960's, the principle geometric configurations of available membranes were flat sheet or tubular structures. The number of configurations has expanded since then to include spiral wound, hollow fibre, thin channels and plate and frame modules. For a discussion of these modules and their relative advantages and disadvantages, the reader is referred to Michaels(1968), Harper(1980), Bell(1985), Le and Howell(1985) and Belfort(1988). In all cases, these units are constructed so that the feed solution flows tangentially over the membrane surface. Blatt et al(1970) stated that because of the extremely low diffusion coefficients of macromolecules and colloids in solution, the minimization of polarization(and realization of high ultrafiltration rates) is far more critical than in the case of microsolutes. For a given volumetric flowrate, the mass transfer coefficient can be maximized by maximizing the shear rate: modules operating in turbulent flow achieve this by using large recirculation rates which involve high pressure drops and may cause damage to liquids that are sensitive to temperature and mechanical treatment while for laminar flow modules, this can

---

Membrane Modification

1) Charged membranes: Hanemaaijer(1987), Reed and Dudley(1987), Gregor and Gregor(1978)
2) Membrane precoating: surfactants(Randerson, 1983, Fane et al, 1985); high MW polymers(Michaels et al, 1983); PEG(Le and Howell, 1983); Polyacrylonitrile or carbon coatings(Bauser et al, 1982)
3) Protective covers: Belfort and Marx(1979)
4) Langmuir-Blodgett layers: Speaker(1985)
5) Immobilized enzymes: Howell and Velicangil(1977, 1981); Wang et al(1980)
6) Electric Fields: Wakeman(1986), Bowen and Sabuni(1987)
7) Membrane electrets: Wallace and Gable(1974)
8) Piezoelectric membranes: Benzinger et al(1980)

Table I: Common techniques used for membrane modification to improve filtration performance.

be achieved with little or no recirculation and without appreciable pressure drop.
Michaels(1968) stated that laminar flow systems cannot accept process streams which contain significant amounts of coarse suspended matter. Such streams must either be prefiltered, settled or centrifuged to remove large particles prior to ultrafiltration or processed in wide conduit turbulent flow systems.

Turbulence/Convection Promoters:

One interesting approach is the use of a fluidized bed. Van der Waal et al(1977) describe how the irregular flow of liquid between the particles themselves and not the movement of the particles themselves is responsible for the improved mass transfer. An erosive action which removes the gel or fouling layer may also occur due to the impulse of the particles. However, the final selection of fluidized bed particles must be a compromise as larger and heavier particles cause a greater enhancement of the mass transfer but also increase the susceptibility of damage to the membrane. De Boer et al(1980) stressed that membrane damage may be reduced by appropriate startup procedure and the nature of the fouling layer itself which may provide a protective effect.

Lowe and Durkee(1971) took a slightly different approach for reverse osmosis of orange juice concentrate which involved pulsing of the concentrate along with plastic spheres from one end of the feed channel to the other. This provided a threefold improvement in flux and significant flux deterioration was not observed.
More attention has been devoted to fixed or static turbulence promoters. Static rods(Peri and Dunkley, 1971), wire spirals(Thomas and Watson, 1968), metal grills(Poyen et al, 1987) are examples of some of the many different types of turbulence promoters available. These alter the flow field in two ways: obstructing the flow increases the average flow velocity over that in an otherwise empty tube and the shear rate in the neighborhood of the membrane wall is increased. At sufficiently high Re numbers, secondary flows and turbulent eddies may be established which enhance mixing at the membrane surface and therefore reduce concentration polarization and/or fouling.

Some general conclusions concerning the use of turbulence promoters can be drawn from the literature studied:

a) the maximum increase in the rate of forced convection and the degree of flux enhancement is dependent upon Re. This dependency on Re is system and/or feed specific(Thomas and Watson(1968), Copas and Middleman(1974), Hiddink et al(1980).

b) optimum spacing between promoters and optimum distance from the transfer surface depends on the particular flow configuration(Thomas and Watson, 1968).

c) most of the convection promoters studied occupy a sizeable volume fraction(typically 20-50%). This will increase the frictional pressure drop by factors as large as several hundred resulting in reduced volumetric flowrates. However, turbulence promoters generally produce the same flux as conventional units at a much lower velocity. Operation at this optimum velocity means for the same flux the frictional pressure drop may be of similar or even smaller magnitude for turbulence promoters compared with normal systems.

d) ideally such devices should introduce no stagnant regions, cause no damage to the membrane, act continuously and be economically justifiable in terms of energy, installation and maintenance costs.

Rotating Membranes:

A rotating module design represents another approach of minimizing the concentration polarization problem. The major advantage of a rotating unit is that the permeate flux becomes independent of the circulation flow, as the shear rate at the membrane surface is controlled by the rotational velocity. This means higher viscosity or concentrated feeds can be treated in single pass flow, reducing circulation pumping costs. Hallström and López-Leiva(1978) describe their own rotary module consisting of an external fixed pressure shell and an internal rotary perforated stainless steel tube which acts as the support for a semipermeable membrane. Between the external shell and the inner rotating tube a narrow slit is formed-this 0.7 mm wide annular space forms the holdup volume for the feed/concentrate. This width is typical of such devices and means scale-up to industrial filtration may be difficult. For the ultrafiltration of skim milk they found that significant flux improvements occurred as rotational speed increased. No limiting flux behaviour was observed within the experimental range of velocity gradients(up to 8000 ls^{-1}). With this device, membranes may be located on both the outer stationary cylinder as well as the inner rotating drum.

Rebsamen(1981) describes how the velocity gradient in the annular gap is determined by two important flow directions; namely, the field of vortices caused by the rotation of the inner cylinder and the flow forced by the axial pressure difference along the annular gap. The two overlying velocity fields are governed by stability criteria. An immediate transition to turbulence does not occur above a critical shearing intensity-instead the flow in the annular gap builds a regular three dimensional vortex field classified as laminar field, thus achieving radial mixing of the suspension. These Taylor vortices can move along the annulus in ideal plug flow which prevents reverse flow or bypassing in the annular gap.

Long term tests have shown that as static residual layers are missing, the output capacity only decreases slightly.

Murkes and Carlsson(1988) state that this filter can be used for clarification, ultrafiltration and microfiltration purposes and also in thickening operations. Kroner and Nissenen(1988) studied direct concentration(single pass thickening, constant flux) and compared it with recycle batch concentration (constant pressure) under similar conditions for Baker's yeast. The average flux after 90 min in the recycling mode was slightly higher at 160 $lm^{-2}h^{-1}$ compared with 150 $lm^{-2}h^{-1}$ for direct thickening. The flux rate for direct thickening remained stable over a 23 hr period with a slight increase in pressure of 0.5 bar. In their opinion, much longer operational times should be possible by using the internal back pressure generated by the centrifugal force from rotation to clean the membrane periodically.

Backflushing:

Breslau et al(1975) discusses two techniques that may be used with hollow fibres when the feed stream runs through the lumen of the fibres. Such an approach requires pressurization of the shell and is only possible if the membrane is an integral part of the support structure.

Backflushing serves to clean the membrane surface by forcing permeate or other fluid back through the fibre which loosens and lifts off the cake accumulated on the inside of the fibre. A reservoir is required to accumulate the filtrate and the backwash fluid should contain no suspended matter which might foul the outer spongelike structure of the fibre.

Recycling, the second approach, is simply accomplished by closing off the permeate ports. As the feed stream continues to circulate throughout the fibres, the permeate that is continually produced results in a pressure buildup in the cartridge shell until an equilibrium pressure is reached which is very close to the average of the inlet and outlet pressures of the feed stream. This means that inside the first half of the fibre the pressure is greater than in the shell while the converse is true for the second half. The reverse backflushing action is coupled with a high shear rate of fluid across the inside of the fibre wall, effectively removing material loosely adhering to the membrane surface. Both normal ultrafiltration and backwashing occur but at greatly reduced rates due to the reduced transmembrane pressure. Both backflushing and recycling are often more effective if carried out in conjunction with cleaning agents.

A novel approach to backwashing is discussed by Fane and Fell(1987). The Memtec system is only feasible for hollow fibre membranes with a relatively low bubble point. The feed suspension is pumped across the outside of the fibre and the filtrate passes out through the lumen. The flux declines as colloids deposit on and within the membrane. This effect is reversed by pulsing the lumen with gas (air, nitrogen, etc) and backwashing with a gas/permeate mixture. The gas pulse expands the fibre and opens the pores allowing fouling material to be flushed out. This is more effective in controlling fouling than a liquid backwash for particulate systems.

Periodic flow reversal is another operational ploy for reducing fouling. Goel and McCutchan(1976) used this method in a tubular reverse osmosis system with Colorado river water as feed material. The average fluxes attained were 10-15% greater and the flux decline decreased betweeen cleaning runs. Maximum recovery based on $CaSO_4$ solubility was increased to the theoretical maximum of 87%. These improvements were attributed to added turbulence due to flow reversal, movement of the high salt concentration region from one end of the flow path to the other at short intervals and the gypsum crystals being denied time for growth.

Pulsed Flow:

In an excellent review article, Edwards and Wilkinson(1971) discuss how heat and mass transfer rates can be enhanced in laminar pipe flows by the imposition of a fluctuating pressure gradient. The studies discussed here concentrate on those that have investigated the use of pulsed flow for improving fluxes in <u>membrane filtration</u>. Kennedy et al(1974) summarized the main findings of Edwards and Wilkinson's work relevant to membrane filtration processes:
Pulsed flow in pipes will:
a) enhance mass and heat transfer;
b) modify the laminar/turbulent transition;
c) heighten the migration of solid particles away from the wall;
d) shift the maximum velocity under laminar flow conditions to the wall region.

Kennedy et al(1974) observed flux increases of up to 70% for pulsing frequencies up to 1 Hz in reverse osmosis of a 10 wt% sucrose solution. They attributed the gain in flux to a reduction in concentration polarization due to convection. Most of the experiments of this paper were in the turbulent or laminar-turbulent transition regimes.

Milisic and Bersillon(1986) investigated the use of pulsed flow as an anti-fouling technique in cross-flow filtration of a 0.1-1.0 gl^{-1} bentonite solution in a rectangular channel. Pulsations were produced by an air-driven valve located upstream from the filtration cell, fully automated for this purpose. Unlike backflushing, neither filtrate nor much energy is required for this system to be operated. The flux was increased by as much as 5 times compared with the flux in a standard run and there is an optimum range of values for the frequency and pulse duration; higher fluxes being favoured by higher frequency and shorter pulse duration. Pulsed flow does not appear to solve the important problem of membrane clogging by colloids or macromolecular material likely to occur in natural water.

Bauser et al(1982) state that pulsed flow may be used to improve membrane performance under experimental conditions where a non-linear relationship between flux and wall shear rate exists. They applied a periodic sequence of pumping pulses keeping the mean flow constant by simultaneous adjustment of the frequency and amplitude. Results for the microfiltration of whey under conditions of constant transmembrane pressure showed 25% improvements in flux after one hour and 38% after 2-3 hours. Similar results have been obtained for blood serum filtration, the maximum gain in this case being about 30%.

Subsequently, Bauser et al(1986) took a different approach, applying a pulsatile negative pressure to the filtrate side of the test module. Gains of about 50% were achieved with feasible pressure amplitudes and frequencies for the ultrafiltration of whey. Long term tests over several days detected no membrane damage due to pulsed flow.

Pulsed flow may be induced by other means such as vibration of a porous plate above the membrane surface(Charm and Lai, 1971), pump vibration(Nakao, 1979) or ultrasound(Semmelink (1973), Fairbanks (1973), Lozier and Sierka (1985)). High frequency vibrations, both electrically and acoustically produced, (Hermann, 1982) have also been investigated for the mitigation of fouling with little if any success. However Wakeman (1986) has used alternating current electrical fields with good success to remove a fouling layer in microfitlration. Bowen and Sabuni (1987) using a different approach also successfully employed electrical fields to enhance microfiltration.

Dimpled/Furrowed Membranes:

Bellhouse et al(1973) rejected pulsed flow by itself as a means of improved mass transfer. Most of their work(1973-1987) has concentrated on the development of membrane lungs for oxygen and carbon dioxide transfer between air and blood. These membranes consist of a large number of small, partly spherical dimples concave to the fluid channel(Dorrington et al, 1986). Alternatively they may be furrowed(Bellhouse et al, 1973). When pulsed flow is used in this system, significant improvements in gas permeation rates were observed. Sobey(1980) investigated the mechanism of mixing in the membrane oxygenator device of Bellhouse et al(1973). The dimensions chosen for the membrane oxygenator were shown to be near optimum in terms of mixing performance. It appears that in steady flow vortices form in the furrows, but remain trapped there, and little or no fluid exchange occurs between the vortices and the mainstream. For vortex mixing to be effective, the flow must be pulsatile and reversing. On flow reversal, these vortices are ejected from the furrows and immediately replaced by a set of counter-rotating vortices. It is this combination of vortex motion in the hollows and vortex ejection which was thought to eliminate fluid boundary layers and augment mass transfer.

In a practical mass transfer device there will be a mean flow superimposed on the oscillatory flow. Sobey(1980) also investigated the influence of the ratio of net forward to maximum flow, NFR, on the flow patterns. When this ratio is small the basic mixing mechanism remains unaltered. Alternatively if this ratio is large, then the flow becomes unidirectional and no vortex ejection occurs. When both flow components are of the same order of magnitude, (NFR=0.4-0.6), the flow patterns become complicated and it is impossible to decide *a priori* whether high or low convective mixing would be obtained. These results were verified by Stephanoff et al(1980) using flow visualization.

A drawback of this technique was demonstrated by Abel et al(1981) in their haemodialyzer work. Pulsed flow enhanced diffusion in both the radial and axial directions. The latter is undesirable, as it reduces the mean concentration difference between blood and dialysate and raises the apparent resistance to transfer. This may be overcome by increasing mean velocities and increasing the dialyser flow path length-the utilization of long and narrow blood and dialysate channels is an effective means of achieving both changes together.

Wyatt et al(1987)successfully applied this technique to the harvesting of microorganisms using E.coli and a 0.2 micron polysulfone membrane. Volumetric efficiencies of the order of 300-400 $lm^{-2}h^{-1}$ were achieved with no pressure applied to the system. The application of low pressures to the retentate line also increased fluxes. An increase in pressure from 0-56 mm Hg increased the percentage of permeate obtained from 49-94% using repeated single pass filtration. Optimal permeate flux was achieved with a dimpled membrane with pulsed flow. Fluxes increased with increasing frequency over the range 2-5 Hz. With <u>both</u> flat and dimpled membranes, water fluxes after each experiment were the <u>same</u>, equalling about 25% of the initial clean water flux for no pulsing; with pulsing, the corresponding values were 51 and 75% respectively. They described a number of other potential applications: removal of cell debris; the harvesting of shear sensitive cells; prefiltration of water and media; ultrafiltration(eg. to separate enzymes); and the clarification of solvents.

Racz et al(1986)stated that the low flow rates and pulsating flow means the mechanism of vortex mixing is not directly applicable to reverse osmosis. They used self-made membranes with periodically spaced semi-cylindrical corrugations in a rectangular channel. Turbulence promotion was thought to be achieved by the fluid eddies following each other quickly, leaving insufficient length of channel through which the fluid can flow undisturbed thus reducing the buildup of a boundary layer at the membrane/salt solution interface. The membranes were self-made. They were cast on a polyester nonwoven support, had fluxes of about 30 $lm^{-2}h^{-1}$ and a retention of about 75% using feed containing 5 gl^{-1} Na_2SO_4 at $P_{tm}=40$ bar.

These values are a different order of magnitude from those used by Bellhouse et al(1973-1987). Corrugations were shown to achieve the same mass transfer at much lower velocities than flat membranes. Pressure drop and pumping power were also reduced under these conditions. Further results were given for mass transfer and hydrodynamics by van der Waal and Racz(1989). Van der Waal et al(1989) applied this technique to polysulfone ultrafiltration membranes and obtained similar results. Corrugating these membranes improved the flux without changing the rejection behaviour. It appears that a critical distance between corrugations exists below which the presence of corrugations results in decreased performance compared to a flat membrane. They concluded from these studies that the corrugations increase mass transfer in a more effective way than an increase in the flowrate. Racz also felt that this technique would be applicable to microfiltration and adaptable to tubular systems.

Application of Pulsatile flow with Baffles

Following the work of Mackley(1987) with vortex mixing in baffled tubular reactors, Finnigan and Howell(1989A) applied this technique to a tubular membrane system by fitting geometrical inserts of doughnut or disc shape to create a periodically grooved channel. Membrane performance for these systems alone and in combination with pulsed flow was investigated and the results were compared with a conventional system operating under the same conditions of cross-flow velocity and transmembrane pressure. Ultrafiltration was carried out for a 10-25 gl^{-1} solution of a purified 95% whey protein Bipro using FP100 ultrafiltration membranes. Preliminary studies(1989A, 1989B, 1989C, 1989D) were carried out under laminar flow conditions and transmembrane pressures up to 2.4 bar using a "snapshot" technique, fully described in Finnigan and Howell(in press, (A)). Subsequent work(Finnigan and Howell, in press, (B)) investigated a number of different baffled systems and the operational range was extended to include turbulent flow conditions and transmembrane pressures up to 5.5 bar.

A significant improvement in flux was observed with the baffled systems under both steady and pulsed flow conditions. The relative improvement reached a maximum in the Re range 750-2200 and 350-1550 at C_b=10 and 25 gl^{-1} respectively. At a higher Reynolds number of 6450, fluxes were greater than or equal in magnitude to fluxes corresponding to fully turbulent flow conditions(Re=16000-50000) in a conventional system. In pulsed flow, comparative fluxes could be obtained at relatively low net cross-flow velocities when the pulsed flow Reynolds number, Re_p=6450, where Re_p is calculated from the maximum velocity in pulsed flow. At P_{tm}=4 bar, fluxes varied from 60-70 $lm^{-2}h^{-1}$ for a conventional system at Re=16000-50000 to 75-95 $lm^{-2}h^{-1}$ for the different disc baffled systems. The "decoupling" of flux from net cross-flow velocity offers the opportunity for use of this system in a single pass, continuous mode of operation for thickening purposes or to avoid the pumping costs associated with recirculation.

Flow visualization was used to study the flow patterns in the conventional and baffled systems under pulsed and steady flow conditions. In steady flow, baffles increased local mass transfer rates by promoting turbulence and interrupting development of the boundary layer. Vortex mixing occurred with pulsed flow in the baffled systems enhancing mass transfer and preventing the development of velocity and concentration boundary layers.

The frequency and amplitude(Finnigan and Howell, in press, (C)) needed to be above certain minimum values for an optimum improvement in flux to be observed. At the same Re_p value, it was more effective to improve fluxes using short strokes rather than long strokes, as the frequency was higher in the former situation. In general, a greater improvement in mass transfer, mixing and flux was observed with "short, fast" strokes rather than "long, slow" strokes. Further improvements in flux were obtained by increasing Re_p(higher frequencies and/or amplitudes (lower St)) until the onset of pressure dependent behaviour.

Disc shaped baffles with a centre to centre baffle spacing of 1.6 times the tube diameter(DI1.6) were found to give the best all-round performance of the baffled systems investigated. Baffles were shown to be dissipating energy more effectively than a conventional system (Finnigan and Howell, in press, (D)). An optimum flux/power range was identified for the baffled systems corresponding to Re=700-1450. Within this range, the power consumption was a maximum of 1 Wm^{-2}. Fluxes of similar magnitude could be obtained in a conventional system but only at a much greater power consumption of approximately 20 and 45 Wm^{-2} at P_{tm}=2 and 4 bar respectively.

Preliminary experiments(Finnigan and Howell, in press (E)) have also been carried out looking at the effect of feed concentration on ultrafiltration fluxes for two different feed solutions using the DI1.6 baffles in steady and pulsed flow and comparing the filtration performance with an empty tube system operating at high cross-flow velocities.

Colman and Mitchell(1990) investigated vortex mixing generated by pulsed flow to enhance membrane performance. A system was designed in which 3 mm high baffles were spaced away from the surface of a flat sheet membrane in a 6 mm high rectangular channel. An interbaffle spacing of 12 mm was found to be optimal for mass transfer. The flow structure associated with pulsed flow in this baffled system shows the same sequence of vortex creation, expansion and ejection each cycle, as described by Sobey(1980) and Mackley(1987). As frequency is increased the flow becomes chaotic but this basic vortex mixing mechanism remains unchanged. The RTD was shown to exhibit plug flow characteristics with low axial dispersion. A mass transfer coefficient, measured at zero flux, equivalent to steady flow at Re>10000 in an empty tube was achieved by using pulsed flow in baffled channels when the net cross-flow rate is Re=100-200. Thus, the mass transfer coefficient can be made independent of the net cross-flow velocity and is relatively constant, provided flow reversal occurs. When there is no flow reversal, the effect of the superimposed oscillations diminishes and the mass transfer becomes mean flow dominated.

A flux, equivalent to that from turbulent cross-flow was achieved using this technique with pervaporation membranes. This technique was also applied to the ultrafiltration of a 1 wt% solution of Dextran T500(MW=500000) using DDS GR40 PP membranes(MWCO=100000). Dextran is a low fouling solute and no permanent fouling of the membrane was observed. Fluxes were enhanced by a factor of three when pulsed flow at a pulsed Reynolds number, Re_p, of 800, is superimposed on a low net cross-flow(Re=200) in a baffled channel relative to the flux in an unbaffled channel with steady flow(Re=200). The pulsed flow flux is equivalent to the flux for steady flow in the baffled channel at Re=1000 and greater than the flux in the unbaffled channel at the maximum cross-flow velocity attainable in the test rig(Re=3000). Limiting flux behaviour is demonstrated by each system and is reached at approximately 0.5 and 1.0 bar for the unbaffled and baffled systems respectively. By incorporating this technique into membrane module design, it will be possible to control mass transfer to the membrane surface independently of the net cross-flow and permeate driving force. No assessment of the power requirement was made.

Conclusions:

The vortex mixing technique in baffled systems shows considerable potential for application to membrane filtration systems:
a) good radial mixing is achieved with the radial and axial velocity components being of similar magnitude;
b) near plug flow characteristics can be obtained with low axial dispersion thus maintaining axial concentration gradients along the length of the module;
c) the mixing effect, mass transfer and flux can be decoupled from the net cross-flow rate;
d) energy consumption within this system is expected to be small;
e) fluxes in pulsed flow in the baffled system are similar in magnitude to steady flow turbulent fluxes in an unbaffled tube;

Developments in this field are awaited with interest.

References:

Abel, K., Jeffree, M.A., Bellhouse, B.J., Bellhouse, E.L., Haworth, W.S., (1981), "A Practical Secondary-Flow Hemodialyzer", **Trans. Am. Soc. Artif. Intern. Organs**, 27, 639-643

Aimar, P., Lafaille, J.P., and Sanchez, V., (1985), "Influence of adsorption on protein ultrafiltration using organic/inorganic membranes", in **Fouling and Cleaning in Food Processing**, Second International Conference on Fouling and Cleaning in Food Processing, ed. D. Lund, E. Plett, C. Sandu, Wisconsin, 466-475

Aimar, P., and Sanchez, V., (1985) "Membrane fouling and limiting phenomena in ultrafiltration", *ibid*., 486-495

Aimar, P., Clifton, M.J., and Howell, J.A., "Concentration polarization build-up in hollow fibres: a method of measurement and its modelling in ultrafiltration", **J. Memb. Sci.**, (in press)

Aimar, P., Howell, J.A., and Turner, N.M., (1989), "Effects of the boundary layer development on the flux limitation in ultrafiltration, **Chem. Eng. Res. Des.**, 67, 255-261

Aref, H., (1984), "Stirring by Chaotic Advection", **J. Fluid Mech.**, 143, 1-21

Bauser, H., Chmiel, H., Stroh, N., Walitza, E., (1982), "Interfacial Effects with Microfiltration Membranes", **J. Memb. Sci.**, 11, 321-332

Bauser, H., Chmiel, H., Stroh, N., Walitza, E., (1986), "Control of Concentration Polarization and Fouling in Medical, Food and Biotechnical Applications", **J. Memb. Sci.**, 27, 195-202

Belfort, G. and Marx, B., (1979), "Artificial Particulate Fouling of Hyperfiltration Membranes", **Desal.**, 28, 13-30

Belfort, G., (1988), "Membrane Modules: Comparison of different Configurations Using Fluid Mechanics", **J. Memb. Sci.**, 35, 245-270

Bell, G., (1985), "Membrane Separation Processes", presented at **Downstream Separation Processes in Biochemical Engineering**, IChemE, Glasgow, Nov 21

Bellhouse, B.J., Bellhouse, F.H., Curl, C.M., MacMillan, T.I., Gunning, A.J., Spratt, E.H., MacMurray, S.B., Nelems, J.M., (1973), "A High Efficiency Membrane Oxygenator and Pulsatile Pumping System, and its Application to Animal Trials", **Trans. Am. Soc. Artif. Intern. Organs**, 19, 72-79

Benzinger, W.D., Toal, M.G., Sprout, O.S., Hinde, G.M., (1980), "Development of Non-Fouling Piezoelectric Ultrafiltration Membranes", **Final Report to the Office of Water Research and Technology**, Aug, 35pg

Blatt, W.F., Dravid, A., Michaels, A.S., Nelsen, L., (1970), "Solute Polarization and Cake Formation in Membrane Ultrafiltration: Causes, Consequences and Control Techniques", in **Membrane Science and Technology**, ed J.E. Flinn, Plenum Press, New York, 47-97

Bowen, W.R. and Sabuni, H., (1987), "Electrically Enhanced Membrane Filtration", in **Proceedings from the Workshop on Concentration Polarization and Membrane Fouling**, Twente University, Enschede, May

Breslau, B.R., Agranat, E.A., Testa, A.J., Messinger, S., Cross, R.A., (1975), "Hollow Fibre Ultrafiltration", **Chem. Eng. Progress**, 71(12), 74-80

Charm, S.E. and Lai, C.J., (1971), "Comparison of Ultrafiltration Systems for Concentration of Biologicals", **Biotech. and Bioeng.**, 13, 185-202

Clifton, M.J., Abidine, N., Aptel, P., and Sanchez, V., (1984), "Growth of the polarisation layer in ultrafiltration with hollow fibre membranes", **J. Memb. Sci.** 21, 233-246

Colman, D.A. and Mitchell, W.S., (1990), "Enhanced Mass Transfer for Membrane Processes", **I. Chem. E. Symp. Ser.**, 118, 87-103

Copas, A.L. and Middleman, S., (1974), "Use of Convection Promotion in the Ultrafiltration of a Gel-Forming Solute", **Ind. Eng. Chem. Process Des. Develop.**, 13(2), 143-145

Dorrington, K.L., Ralph, M.E., Bellhouse, B.J., Gardez, J.P., Sykes, M.K., (1985), "Oxygen and CO_2 Transfer of a Polypropylene Dimpled Membrane Lung with Variable Secondary Flows", **J. Biomed. Eng.**, 7, 89-99

Edwards, M.F. and Wilkinson, W.L., (1971), "Review of Potential Applications of Pulsating Flow in Pipes", **Trans. Instn. Chem. Engrs.**, 49, 85-93

Fairbanks, H.V., (1973), "Use of Ultrasound to Increase Filtration Rate", **Ultrasonics Int. Conf. Proc.**, Imperial College, London, 11-15

Fane, A.G., Fell, C.J.D, Kim, K.J., (1985), "The Effect of Surfactant Pretreatment on the Ultrafiltration of Proteins", **Desal.**, 53, 37-55

Finnigan, S.M. and Howell, J.A., (1989A), "The Effect of Pulsatile Flow on Ultrafiltration Fluxes in a Baffled Tubular Membrane System", **Chem. Eng. Res. Des.**, 67(3), 278-282

Finnigan, S.M. and Howell, J.A., (1989B), "The Effect of Pulsed Flow on Ultrafiltration Fluxes in a Baffled Tubular Membrane System", in **The Membrane Alternative-Energy Implications for Industry Conference**, University of Bath, March 29-30, Elsevier, in press.

Finnigan, S.M. and Howell, J.A, (1989C), "The Effect of Pulsed Flow on Ultrafiltration Fluxes in a Baffled Tubular Membrane System", in **Proceedings 3rd Int. Conf. on Foulng and Cleaning in the Food Industry**, Prien, Germany, June, in press.

Finnigan, S.M. and Howell, J.A., (1989D), "The Effect of Pulsed Flow on Ultrafiltration Fluxes in a Baffled Tubular Membrane System", presented at **6th Int. Symp. on Synthetic membranes in Science and Industry**, Tubingen, Sept 4-8, to be published in J. Memb. Sci.

Finnigan S.M., (1990), "Pulsed Flow Ultrafiltration in Baffled Tubular Membrane Systems, **PhD dissertation**, University of Bath

Goel, V. and McCutchan, J.W., (1976), "Colorado River Desalting by Reverse Osmosis", in **Proceedings. 5th Int. Symp. Fresh Water from the Sea**, Alghero, May 16-20, 4, 143-156

Gregor, H.P. and Gregor, C.D., (1978), "Synthetic Membrane Technology", **Scientific American**, 239, 88-101

Hallström, B. and López-Leiva, M., (1978), "Description of a Rotating Ultrafiltration Module", **Desalination**, 24, 273-9

Hanemaaijer, J.H., (1987), "Fouling of Ultrafiltration Membranes. The Role of Protein Adsorption and Salt Precipitation", in **Proceedings from the workshop on Concentration Polarization and Membrane Fouling**, Twente University, Enschede, May

Harper, W.J., (1980), "Factors Affecting the Application of Ultrafiltration Membranes in the Dairy Industry", in **Ultrafiltration Membranes and Applications**, ed. A.R. Cooper, Plenum Press, New York, 321-342

Hermann, C.C., (1982), "High Frequency Excitation and Vibration Studies on Hyperfiltration Membranes", **Desal.**, 42, 329-338

Hiddink, J., Kloosterboer, D., Bruin, S., (1980), "Evaluation of Static Mixers as Convection Promoters in the Ultrafiltration of Dairy Liquids", **Desal.**, 35, 149-167

Howell, J.A. and Velicangil, O., (1977), "Protease Coupled Membranes for Ultrafiltration", **Biotech. and Bioeng.**, 19, 1891-1894

Howell, J.A. and Velicangil, O., (1981), "Self-Cleaning Membranes for Ultrafiltration", **Biotech. and Bioeng.**, 23, 843-854

Howell, J.A. and Velicangil, O., (1982), "Theoretical Considerations of Membrane Fouling and its Treatment with Immobilized enzymes for Protein Ultrafiltration", **J. Appl. Poly. Sci.**, 27, 21-32

Howes, T., (1988), **On the Dispersion of Unsteady Flow in Baffled Tubes**, PhD thesis, Department of Chemical Engineering, Cambridge University

Kennedy, T.J., Merson, R.L., McCoy, B.J., (1974), "Improved Permeation Flux by Pulsed Reverse Osmosis", **Chem. Eng. Sci.**, 29, 1927-1931

Kroner, K.H. and Nissinen, (1988), "Dynamic Filtration of Microbial Suspensions Using an Axially Rotating Filter", **J. Memb. Sci.**, 36, 85-100

Le, M.S. and Howell, J.A., (1983), "The Fouling of Ultrafiltration Membranes and its Treatment", in **Progress in Food Engineering**, Foster Publ, Switzerland, 321-326

Le, M.S. and Howell, J.A., (1985), "Ultrafiltration", in **Comprehensive Biotechnology**, ed. M. Moo-Young, Ch. 25, 383-409

Leung, W.F., and Probstein, R.F., (1979) **Ind. Eng. Chem.(Fundam.)** 18(3) 274-278

Lowe, E. and Durkee, E.L., (1971), "Dynamic Turbulence Promotion in Reverse Osmosis Processing of Liquid Foods", **J. Food Sci.**, 36, 31-32

Lozier, J.C. and Sierka, R.A., (1985), "Using Ozone and Ultrasound to Reduce Reverse Osmosis Membrane Fouling", **J. AWWA**, Aug, 60-65

Michaels, A.S., (1968), "New Separation Techniques for the CPI", **Chem. Eng. Prog.**, 64, 31-43

Michaels, A.S., Robertson, C.R., Reihanian, H., (1983), "Recent Developments in Ultrafiltration: A Solution to the Polarization/Fouling Problem", **IMTEC Conference Proceedings**, Australia, Nov 8-10, 59-63

Milisic, V. and Bersillon, J.L., (1986), "Anti-fouling Techniques in Cross Flow Microfiltration", **4th World Filtration Congress**, Ostend, April, 11.19-11.23

Murkes, J. and Carlsson, C.G., (1988), **Crossflow Filtration**, John Wiley & Sons Ltd, New York

Nakao, S., Nomura, T., Kimura, S., (1979), "Characteristics of Macromolecular Gel Layer Formed on Ultrafiltration Tubular Membrane", **AIChE J.**, 25(4), 615-622

Peri, C. and Dunkley, W.L., (1971), "Reverse Osmosis of Cottage Cheese Whey. 2. Influence of Flow Conditions", **J. Food Sci.**, 36, 395-396

Poyen, S., Quemeneur, F., Bariou, B., (1987), "Improvement of the Flux of Permeate in Ultrafiltration by Turbulence Promoters", **Int. Chem. Eng.**, 27(3), 441-447

Pritchard, M., Scott, J. A., and Howell, J.A., (1990) D. L. Pyle (ed) "The Concentration of Yeast Suspensions by Crossflow Filtration", **Separations for Biotechnology 2 pt 1**, 65-73

Racz, I.G., Wassink, J.G., Klaasen, R., (1986), "Mass Transfer, Fluid Flow and Membrane Properties in Flat and Corrugated Plate Hyperfiltration Modules", **Desal.**, 60, 213-222

Randerson, D.H., (1983), "Principles Governing Flux Rate in Membranes for Blood-Plasma Separation", **IMTEC Conference Proceedings**, Australia, Nov 8-10, 44-46

Rebsamen, E., (1981), "Fundamentals and Engineering Concept of a Pressure Filter for Dynamic Filtration", **Proceedings Symposium Societe Belge de Filtration**, Louvain-la-Neuve, 247-270

Reed, I.M. and Dudley, L.Y., (1987), **Adsorption of Proteins to Membranes**, BIOSEP report, Chemical Engineering Division, AERE Harwell

Savvides, C.N. and Gerrard, J.H., (1984), "Numerical Analysis of the Flow through a Corrugated Tube with Application to Arterial Prostheses", **J. Fluid. Mech.**, 138, 129-160

Semmelink, A., (1973), "Ultrasonically Enhanced Liquid Filtering", in **Ultrasonics Int. Conf. Proc.**, Imperial College, London, 7-10

Sobey, I.J., (1980), "On Flow Through Furrowed Channels. Part 1. Calculated Flow Patterns", **J. Fluid. Mech.**, 96(1), 1-26

Speaker, L.M., (1985), "Antifouling technology for Membranes and Non-permeable Surfaces", in **Fouling and Cleaning in Food Processing**, Second International Conference on Fouling and Cleaning in Food Processing, ed. D. Lund, E. Plett, C. Sandu, Wisconsin, 454-465

Stephanoff, K.D., Sobey, I.J., Bellhouse, B.J., (1980), "On Flow Through Furrowed Channels. Part 2. Observed Flow Patterns", **J. Fluid Mech.**, 96(1), 27-32

Suki, A., Fane, A.G., Fell C.J.D., (1986), "Modelling Fouling Mechanisms in Protein Ultrafiltration", **J. Memb. Sci.**, 27, 181-193

Thomas, D.G. and Watson, J.S., (1968), "Reduction of Concentration Polarization of Dynamically Formed Hyperfiltration Membranes by Detached Turbulence Promoters", **Ind. Eng. Chem. Process Des. Develop.**, 7(3), 397-401

Van der Waal, M.J., van der Velden, P.M., Koning, J., Smolders, C.A., van Swaay, W.P.M., (1977), "Use of Fluidized Beds as Turbulence Promoters in Tubular Membrane Systems", **Desal.**, 22, 465-483

Van Der Waal, M.J. and Racz, I.G., (1989), "Mass Transfer in Corrugated-Plate Membrane Modules. 1. Hyperfiltration Experiments", **J. Memb. Sci.**, 40, 243-260

Wakeman, R., (1986), "Electrofiltration, Microfiltration and Electrophoressis" June 65-70

Wang, S.S., Davidson, B., Gillespie, C., Harris, L.R., Lent, D.S., (1980), "Dynamics of Enhanced Protein Ultrafiltration Using an Immobilized Protease", **J. Food. Sci.**, 45, 700-702

Winfield, B.A., (1986), "Waste Treatment with Reverse Osmosis Membranes", in **Membrane Separations in Biotechnology**, ed. W.C. McGregor, Marcel Dekker Inc, USA, Ch. 13, 355-373

Wyatt, J.M., Knowles, C.J., Bellhouse, B.J., (1987), "A Novel Membrane Module for Use in Biotechnology that has High Transmembrane Flux Rates and Low Fouling", in **Proceedings of International Conference on Bioreactors and Biotransformations**, ed. G.W. Moody and P.B. Baker, Gleneagles, Scotland, 166-172

Zahka, J. and Leahy, T.J., (1985), "Practical Aspects of Tangential Flow Filtration in Cell Separations", Advances in Chemistry Series, **Amer. Chem. Soc. Symp. Ser.**, 271, 51-69

MEMBRANE FOULING AND CLEANING

A Review

Stephen N Cross (GRACE Dearborn Ltd, UK)

SUMMARY

The use of membrane separation processes in a variety of manufacturing and related industries is growing at a significant rate. The problems of fouling are being studied intensely because they are seen as key factors in the overall economics of the process. The effects of membrane type, foulant and cleaning protocol are discussed. The importance in correctly formulating detergent solutions lies in the effectiveness of the product, its approval status and its membrane compatibility. An alternative approach to the problem of fouling is considered, whereby the use of flocculants as a preventative measure is considered. The economics of this approach is discussed.

1. INTRODUCTION

Various types of membrane filtration have been used in many different process industries since the first practical and efficient membrane was developed in the early 1960's. From those very early days the problems of fouling and cleaning were apparent. In the past decade there have been many studies into the fundamental processes that contribute to fouling, which can be simply defined as an irreversible loss of flux with processing time. Table 1 lists three common factors referred to when describing membrane fouling. As Paulson (1987) stated, the differentiation between these states is not always clear, and feedstream behaviour is notoriously unpredictable.

One of the greatest problems in trying to define mechanisms for fouling is the sheer complexity of the process streams involved. Table 2 presents some of the principle applications for membrane systems in process industries. This variety of situations immediately indicates the scale of the potential for fouling. Most industrial processes involve the handling of immense volumes of fluids, figures of 2 million litres per day are not uncommon (Beaton 1984). All of these process streams consist of both organic and inorganic components. Furthermore, the nature and concentration of the constituents often vary due to them originating from natural (animal, vegetable or microbial) sources. This means that at best there can be seasonal changes in the fluid to be treated (milk for example) and at worst daily or batch-to-batch changes (fermentation/biotechnology).

Typical foulants mentioned in the literature are referred to in Table 3. These cover a range of organic and inorganic materials, colloids and general debris. To complicate matters, molecules such as proteins can and will interact with mineral salts to form complexes requiring special attention when cleaning procedures are considered.

In addition, there are further variables in terms of membrane type and membrane structure. The majority of membranes are based upon organic polymers. These typically vary in hydrophilicity from cellulose acetate (hydrophilic) to polysulphone (hydrophobic). In addition there is the increasing usage of ceramic/mineral membranes which have certain unique properties of their own although they are considered expensive to purchase, not withstanding their long lifetimes. A list of some of the common membrane materials is shown in Table 4, together with comments about their chemical resistance. This is an important factor when considering methods of cleaning and sanitizing.

Effective Industrial Membrane Processes — Benefits and Opportunities, pp.61–90

TABLE 1 FACTORS CONTRIBUTING TO EFFICIENCY OF MEMBRANE SEPARATION

Concentration Polarization	Fouling	Gel Layer
Occurs in retentate	Deposition on/in membranes	Specific type of fouling due to a gelatinous layer
Minimized by crossflow velocity	Results of physical or changes; . precipitation . agglomeration . sieving . adsorption	May not be observed in high solids/high viscosity feed streams
Can reach equilibrium		
Flux loss related to osmotic effects		
Reversible	Rarely reaches equilibrium	Can be a confusing term
	Normally associated with particulate or colloidal process streams.	

TABLE 2 APPLICATIONS FOR MEMBRANES IN PROCESS INDUSTRIES

Industry	Applications
Dairy	- Skimmed Milk Whey Quark
Fruit Juice	
Fermentation/Biotechnology	- Antibiotics Enzymes Proteins Polysaccharides
Beer	
Wine	
Paint	
Water Treatment	

TABLE 3 TYPICAL MEMBRANE FOULANTS

1. Dairy	Proteins Lactose Minerals Fat Micro-organisms
2. Fruit Juice	Pectins Sugars
3. Fermentation Biotechnology	Proteins Polysaccharides Lipids Biomass/Cell Debris Minerals Nucleic Acids Antifoams

TABLE 4 MEMBRANE TYPES

Composition	Chemical Resistance
Cellulose Acetate	pH 2 - 8,
Regenerated Cellulose	pH 2 - 12,
Polyacrylonitrile	pH 2 - 13,
PVDF	pH 1 - 12,
Polyethersulphone	pH 2 - 13,
Polysulphone	pH 2 - 14,
Polypropylene	pH 2 - 14,

Note: This list is not exhaustic. It is simply indicative of some of the more common polymeric membrane types.

Moreover, membrane systems occur in one of five typical configurations;

- . flat sheet
- . spiral
- . hollow fibre
- . tubular
- . plate and frame

Thus it is immediately obvious that the numbers of variables involved makes any study of fouling and cleaning very difficult. This is especially true of realistic industrial situations.

Many people have studied the phenomenon of fouling in great detail. These investigations range from the theoretical works examining such phenomena as gel layers (Issacson 1980), pore blocking (Le 1984) and protein adsorption (Aimer 1986), to more specific investigations of industry-related fouling. The effort of this area of work has been especially concentrated in the Dairy Industry and related sectors (Matthews 1978; Cheryan 1984).

More recently, the impact of Biotechnology has meant that workers are investigating the interactions between antifoams, proteins and various membranes (Cabral 1985; Harris 1988; Kloosterman 1988).

All this effort is targetted towards understanding the mechanisms of fouling since this is considered the biggest single problem preventing acceptance of membranes in chemical and biological processes (Ilias 1987). Fouling impinges upon the economics of a process in terms of productivity, energy, membrane replacement and, of course, cleaning chemicals.

Cleaning of membranes is considered a necessary evil. Nevertheless, this can account for anything from 5% of the operating cost (Beaton 1984), to 20% in extreme cases, when it comes behind only membrane replacement and depreciation in order of importance. Unlike membrane replacement and depreciation, cleaning is something that has to be faced practically, on a day to day basis. Without effective cleaning, most large scale membrane plants will not be operational economically.

Therefore, it is not surprising to find a number of researchers investigating cleaning procedures for membranes used in Dairy (Parkin 1976; Smith 1987), Wastewater (Whittaker 1984) and Biotechnology (Taeymans 1989) industries.

In the earlier days of cellulosic membranes, the study of enzyme detergents was notable, due to the chemical degradation of cellulosics at high pH. Parkin's work (1976) was definitive, also predicting the evolution of non-enzyme detergents in the wake of more chemically resistant membrane types. Whittaker (1984) and others have shown that for most common process streams a formulated product is necessary which combines some or all of the following features;

- . alkalinity
- . chelation ability
- . surface activity
- . dispersing ability
- . enzyme activity
- . chaotropic agents
- . bactericides

These features certainly are all of some importance. However, it is the relationship between the chemistry of the detergent formulation and the nature of the foulant/membrane interactions that still needs to be more clearly understood. Certain surfactants are known to interact with certain membranes (Chong 1985), but this relationship is dependant on certain factors (surfactant charge and critical micelle concentration for example), which can all in turn be affected by the process stream and the hydrophobicity of the membrane surface.

One area of investigation that has received some attention is the possibility of preventing deposits at the membrane surface during the actual process cycle in order to improve the membrane flux and reduce the propensity to foul. This approach has included the use of various types of pre-treatment (pH adjustment, heat, demineralization and pre-filtration) and a small amount of investigation has been carried out into the possibility of utilizing flocculants or filter aids to improve solid liquid separation at the membrane interface. Fane (1983) showed that the addition of filter-aid particulates could affect the flux rate of dextran or gelatin solutions.

However, particle size is very important, as are the conditions of the process stream, the disposal of residual solids (containing potentially cationic flocculants) and the end use of the material being produced. In the biotechnology area, some effort has been placed in the development of naturally flocculating cell strains (Miki et al, 1980) to improve separation. Much work has been done on flocculants in special research programmes by organisations such as the Bioseparations Club. The published work is confidential, but it has involved a detailed study of the interaction of flocculants with cells to improve solid/liquid separation. However, much of this work has been directed towards centrifugation.

The aim of this paper is to highlight some of the problems of membrane fouling and cleaning from selected experiments, to illustrate the potential for the use of flocculants as promoters of increased membrane flux and to indicate areas for future research into the prevention and cure of membrane fouling.

2. METHODS

Three basic types of experiment are reported in this paper. Membrane fouling of different ultrafiltration membranes using E. coli cell broths in order to study cleanability. Fouling and cleaning studies using a model broth to simulate typical microbial broth-fouling on ultrafiltration membranes. And finally the possibility for increasing the performance of microfilters in the concentration of cells, by the addition of flocculating agents.

2.1 E. coli Fouling Studies

This work involved the production of E. coli broth at 20g/litre dry cell weight in a glucose, NZ amine and salts medium. 20 litre Chemap fermenters were used. PPG 2000 was used as antifoam at 500 ppm.

This broth was concentrated by a factor of 5 in a variety of 10,000 MWCO polysulphone membrane systems;

- AMICON PM10 - Flat Sheet
- AMICON PM10 - Hollow Fibre
- AG Technology 10 - Hollow Fibre

The average transmembrane pressure was 2 psi and a linear velocity of 1 m/sec was used.

The cleaning procedures are described in the results section. The aim was to observe differences in membrane configuration and morphology.

2.2 Model Fouling Studies

In order to increase the throughput of the number of fouling and cleaning tests, a model broth has been developed (Table 5) which comprises of yeast extract, glucose and various mineral components. It is not as fouling as typical fermentation broths, but it serves as a base line for development of a more or less reproducible fouling stream to which selected components (protein, lipids, nucleic acids and antifoams) can be added. This model broth has a solids content (after autoclaving) of 1.75 g/litre, and a total protein content of 1.5 g/litre (Sigma Total Protein Determination Procedure, No. 541).

The fouling procedure is outlined in Table 6, and the various cleaning procedures are discussed in the results section.
In certain experiments foulants have been "spiked" into the model broth in order to observe their effects. The experiments discussed herein include the addition of;

. 1.0 g/l BSA ..... see Figure 4
. 0.1 g/l PPG 2000 ..... see Figure 4
. 0.1 g/l Lipids ..... see Figure 5
. 0.1 g/l Nucleic Acids ..... see Figure 5

2.3 Flocculation Studies

Laboratory scale AG Technology hollow fibre cartridges containing 0.002 m^2 of membrane were used to assess the effect of flocculants upon membrane performance. The fibres are 0.22 micron polysulphone units with 1 mm inner diameters. Transmembrane pressure was measured in the range 0-15 psi with gauges fitted to the retentate inlet and outlet, and the permeate outlet. Recirculation and permeate lines consisted of No 15 C-flex tubing (Cole-Parmer). A Cole-Parmer peristaltic pump was used to circulate the retentate at 1 l/min.

The flocculants used were a commercially cationic product (known here as Floc 2), available from GRACE Dearborn Ltd as DARAFLOC 8253, and an experimental cationic product known here as Floc 1. This material is being developed in W.R. GRACE's US laboratories.

Optimal flocculant addition rates were obtained by carrying out initial settling rate tests. These involved adding aliquots of flocculant to 250 ml cell suspensions. The rate of cell settlement was measured over a period of 7 hours at room temperature. Beyond 7 hours cell lysis interferes with the results. The typical, and repeatable, settling rates for Floc 1 are presented in the results section. An optimum dose of 100 ppm, which the settling tests indicated, was used in the microfiltration tests. The optimum dosage for Floc 2 was 25 ppm.

The microfiltration experiments utilize the same E. coli broth as described in section 2.1.

3. RESULTS

3.1 E. Coli Fouling/Cleaning

The experiments in this section were designed to assist in formulating efficient membrane cleaning products for use in biotechnological applications where the foulants are cells, cell debris, cell metabolites, residual nutrient and antifoams.

TABLE 5 MODEL BROTH

Component	Concentration (g/l)
Glucose	5.0
Yeast extract	5.0
$MgSO_4 \cdot 7H_2O$	2.0
K_2HPO_4	3.0
KH_2PO_4	3.0
$(NH_4)_2HPO_4$	1.0
Trace Mineral Soln.	3.0 ml/L

Trace Mineral Soln.	
$FeCl_2 \cdot 6H_2O$	27g/l
$ZnCl_2 \cdot 4H_2O$	2g/l
$CoCl_2 \cdot 6H_2O$	2g/l
$Na_2MoO_4 \cdot 2H_2O$	2g/l
H_3PO_4	0.5 g/l
HCl (conc.)	100ml/l
$CaCl_2 \cdot 2H_2O$	1g/l

TABLE 6 MEMBRANE FOULING EXPERIMENTS

1. Standard Fouling Broth
2. PM 10 (Amicon) Flat Sheet, 0.006 m2
3. Steady state water flux = 157 - 186 LMH
4. TMP = 10 psi, Flow Rate 1 litre/min
5. Fouling cycle is 1 hour at 15 - 18 degrees C
6. Water Flush, 6 litres at 2 litre/min
7. Cleaned with MC-2 for 20 minutes, 65 degrees C, 2 litre/min
8. Water Flush, 6 litres at 2 litre/min
9. Rinse for 40 minutes at 15 - 18 degrees C
10. Efficiency measured as Final Water Flux vs Initial Steady State Water Flux.

Note: All water used is pre-treated through a Millipore RO 6 unit.

By working with a flat sheet PM 10 membrane it was possible to study the effect of varying the composition of key ingredients in order to optimize the cleaning efficiency for removal of the E. coli broth after fouling.

Results for a typical set of experiments are presented in Table 7. They represent the average of three sets of experiments and indicate that there are considerable effects depending upon the components added. The use of a nonionic surfactant does not offer any benefits. However, the selection of the optimum dispersant is critical. Experimental formulation 2/25 is the basis for MC-2, which is discussed in greater detail later in this section. All the formulations in this study are based upon blends of special surfactants, chelants and dispersants. The products do not contain enzymes.

The cleaning regime used in these tests was to initially rinse the membrane with water, then to add the detergent into a 0.4% NaOH solution and circulate for 60 minutes at 60°C. A final rinse was carried out prior to a cleaned water flux being obtained. The percent flux recovery is the ratio of the water flux of the cleaned membrane to that of the new membrane.

Further examples of work in this section are presented in figures 1 and 2. In both cases the process stream is the E. coli broth.

Figure 1 shows the effects of membrane configuration (flat sheet vs hollow fibre) and membrane morphology (anisotropic vs homogeneous) upon the cleanability of the membranes. It is apparent that flat sheet membranes are easier to clean than hollow fibre membranes. This point has been noticed previously and is reported elsewhere (Roe 1989). Moreover, when two hollow fibre systems are compared, it can be deduced that anisotropic (or skinned) membranes are easier to clean than homogeneous (or depth-filter) membranes.

The phenomena are thought to be related to the available surface area in the first case and to the nature of the pore cross-section in the second case.

Figure 2 is the result of further tests upon the MC-1 type of product. In these tests, the basic MC-1 formulation is the control alongside two modifications. The first modification involves the addition of hypochlorite, to give an effective level of 200 ppm of HOCl, at the use concentration of 1% of the MC-1 type product. The second modification shows the effect of just using MC-1 without any surfactants or dispersants, thus relying only upon its chelation ability.

The results indicate that a formulated product such as MC-1 is somewhat more effective than a product based upon hypochlorite (without added surfactants), and a great deal more effective than a product based only upon chelates, in this application. These results are in agreement with recent studies in the area of CIP cleaners formulated for the removal of burnt on deposits in heat-exchangers (Pritchard 1988). In this reference a formulated caustic additive is said to be 6 times more effective than caustic alone, whereas a hypochlorite based product is only 4 times more effective than caustic. The MC-1 EDTA product has very little effect because it lacks both surfactants and dispersants. The inclusion of an amount of the correct surfactant has a dramatic effect.

3.2 Model Fouling/Cleaning

The model broth, described in Table 5, was designed to give a reproducible feed stream that could be "spiked" with known fouling components. The aim of this part of the study was to observe the effects of adding these fouling components, both upon the flux loss and upon the cleanability of the membrane.

Figure 3 represents a comparison of the fouling rates of the model broth with a typical E. coli broth reported in section 3.1. It is apparent that the real broth is considerably more fouling than the model.

TABLE 7 SCREENING TESTS FOR FORMULATION OF CAUSTIC ADDITIVE CLEANER

	Added Component				
Variant	Non-ionic Surfactant	Anionic Dispersant 1	Anionic Dispersant 2	Anionic Dispersant 3	% Flux Recovered
MC-1	-	-	-	-	59%
2/35	+	-	-	-	42%
2/50	+	+	-	-	42%
2/55	+	-	+	-	55%
2/57	+	-	-	+	51%
2/20	-	+	-	-	48%
2/25	-	-	+	-	76%
2/27	-	-	-	+	62%

Note: + indicates component included in formulation

- indicates component excluded

Anionic Dispersants 1, 2 and 3 represent a range of molecular weights.

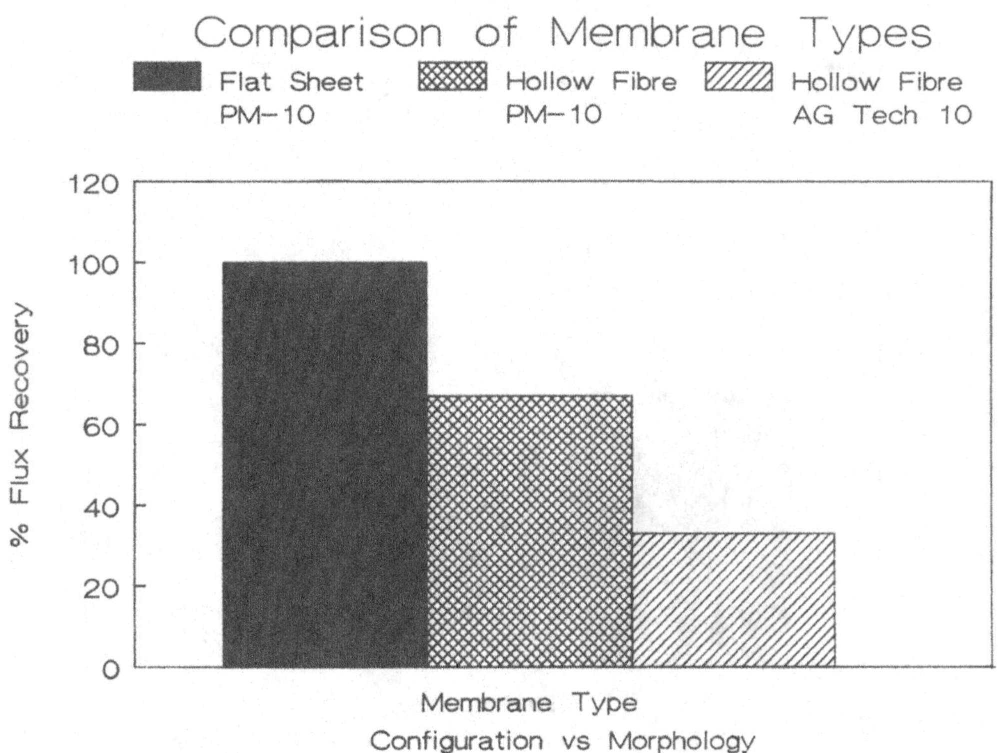
Figure 1
Comparison of Membrane Types
Flat Sheet PM-10
Hollow Fibre PM-10
Hollow Fibre AG Tech 10
120
100
80
60
40
20
0
% Flux Recovery
Membrane Type
Configuration vs Morphology

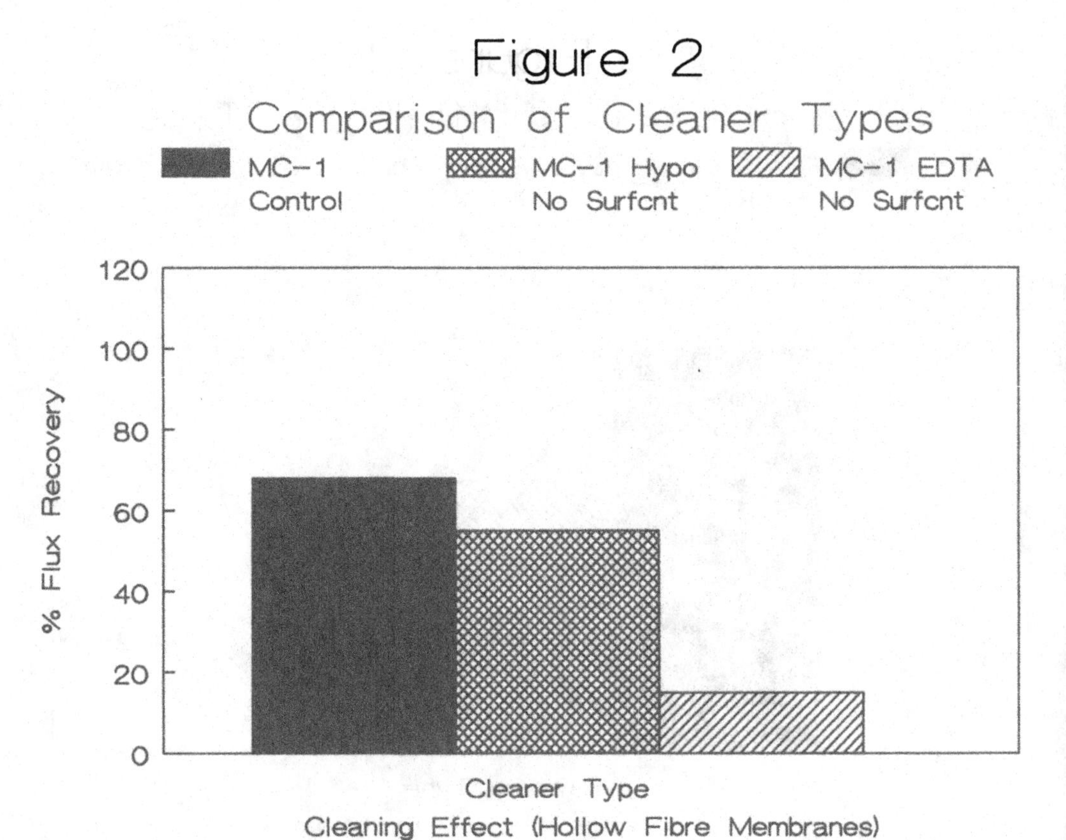
Figure 2
Comparison of Cleaner Types
MC-1
Control
MC-1 Hypo
No Surfcnt
MC-1 EDTA
No Surfcnt
120
100
80
60
40
20
0
% Flux Recovery
Cleaner Type
Cleaning Effect (Hollow Fibre Membranes)

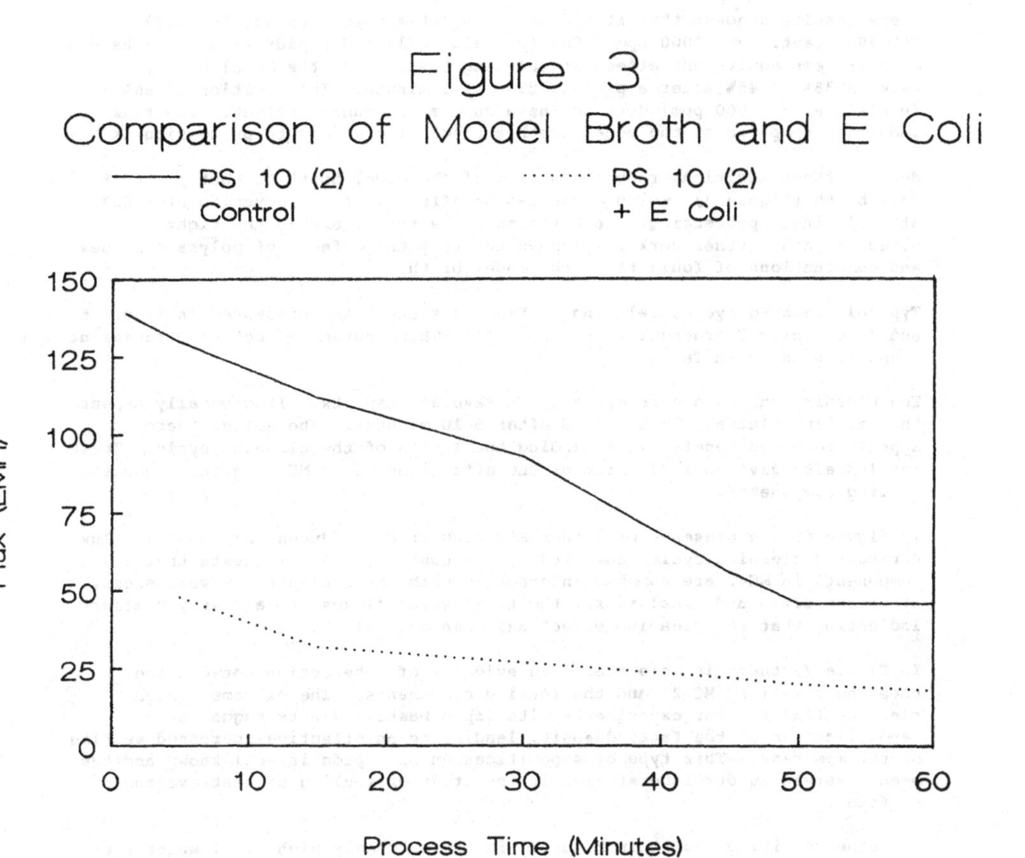
Figure 3
Comparison of Model Broth and E Coli
PS 10 (2) Control
PS 10 (2) + E Coli
Flux (LMH)
150
125
100
75
50
25
0
0
10
20
30
40
50
60
Process Time (Minutes)

However, Figure 4 and 5 (titled Fouling Characteristics I and II) show what happens to the rate of flux loss for the model broth by the incorporation of amounts of suspected foulant.

These results suggest that at the addition rates used (see section 2.2) PPG 2000 (antifoam, 1000 ppm), BSA (protein 1g/l) and Lipids (1000 ppm) have a more or less equivalent effect in reducing the flux of the model by anything between 38% to 46% after a process time of 1 minute. The addition of RNA (nucleic acid, 1000 ppm) does not have such a pronounced effect. The flux reduction compared to the model broth control after 1 minute processing is 32%.

None of these additions reduce the flux of the model broth to that of the E. coli broth (Figure 3), which exhibited an effective flux reduction of ~ 60% after 1 minute processing. Nevertheless, the trends are in the right direction and further work is planned to study the effects of polysaccharides and combinations of foulants in the model broth.

Typical cleaning cycles relating to Figures 4 and 5 and presented in Figure 6 and 7 (Cleaning Characteristics I and II). These curves effectively represent steps 7, 8 and 9 in Table 6.

The cleaning curves are interesting in several respects. It generally appears that a "flux plateau" is attained after 5-10 minutes. Thereafter there appears to be no benefit in extending the length of the cleaning cycle. These results also give an indication of the effectiveness of MC-2 against specific fouling components.

In Figure 6, the presence of either PPG 2000 or BSA reduces the apparent flux during the cleaning cycle, compared to the control. This suggests that the components in MC-2 are somehow interacting with the foulants. Nevertheless, after the clean and final rinse, the final water fluxes are all very similar, indicating that the cleaning effect has been equivalent.

In Figure 7, there is more startling evidence of interaction between the cleaning solution (MC-2) and the fouling components. The extremely high cleaning flux for the experiments with lipid based foulants suggests saponification of the fatty deposit, leading to an effective increased wetting of the membrane. This type of saponification of lipids is well-known and has been observed in our laboratories in the study of fouling of heat-exchanger surfaces.

The other striking result in Figure 7 is the extremely high final water flux for the control. This gives an overall cleaning efficiency (ie. final water flux over original water flux) of 145%. This result is shown graphically in Figure 8 together with those for all the broth variations shown in Figures 6 and 7. Generally, the final water flux falls in the range +/- 10% of the original water flux, except for the odd exception.

This high water flux may be indicative of several occurences, membrane damage being one possibility. It is unlikely that chemical damage has occurred since both MC-1 and MC-2 have passed Amicon membrane compatibility tests. It is more likely to be related to local variations in pore size on the flat sheet sample tested. All the experiments in Figures 4 and 6 were carried out on one piece of membrane, while those in Figures 5 and 7 were carried out on a second piece. The fouling curves for the model broth (Figures 4 and 6) indicate that higher fluxes were observed in the second case, suggesting a larger pore-size distribution, with a more open membrane.

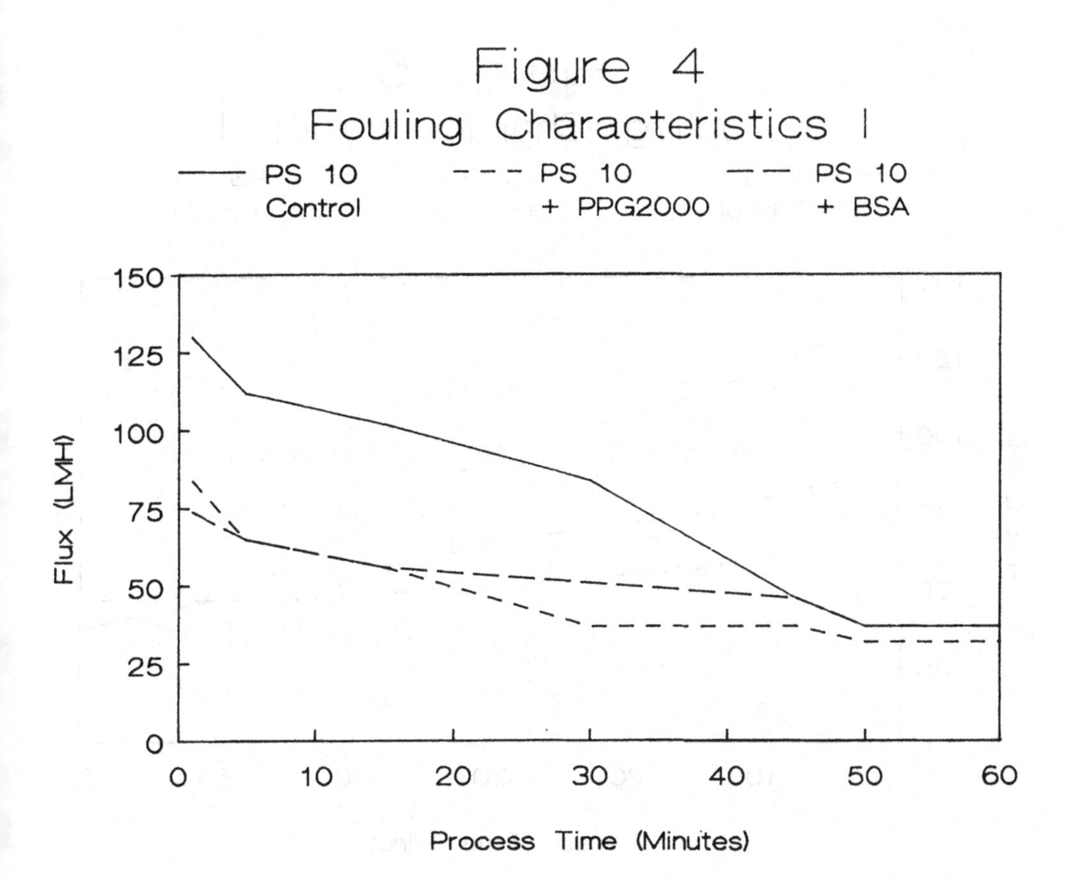
Figure 4
Fouling Characteristics I
PS 10 Control
PS 10 + PPG2000
PS 10 + BSA
150
125
100
75
50
25
0
0
10
20
30
40
50
60
Flux (LMH)
Process Time (Minutes)

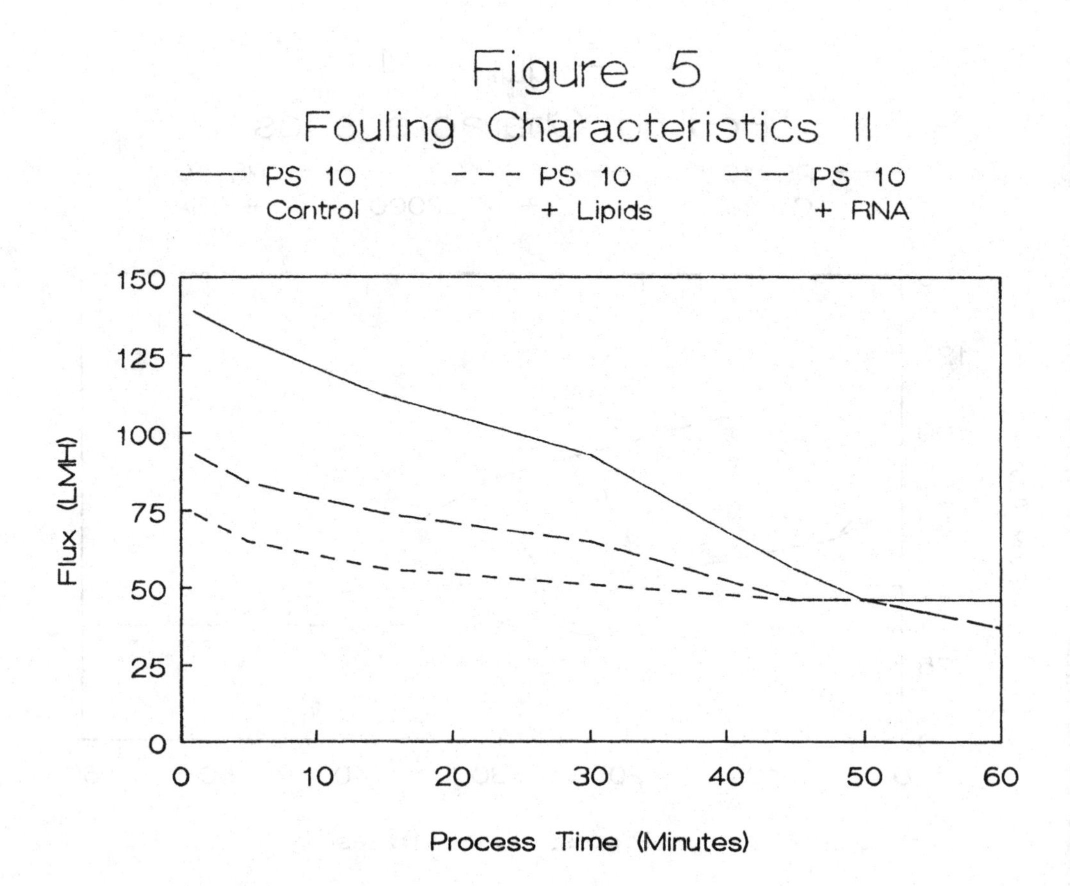
Figure 5
Fouling Characteristics II
PS 10 Control
PS 10 + Lipids
PS 10 + RNA
150
125
100
75
50
25
0
Flux (LMH)
0
10
20
30
40
50
60
Process Time (Minutes)

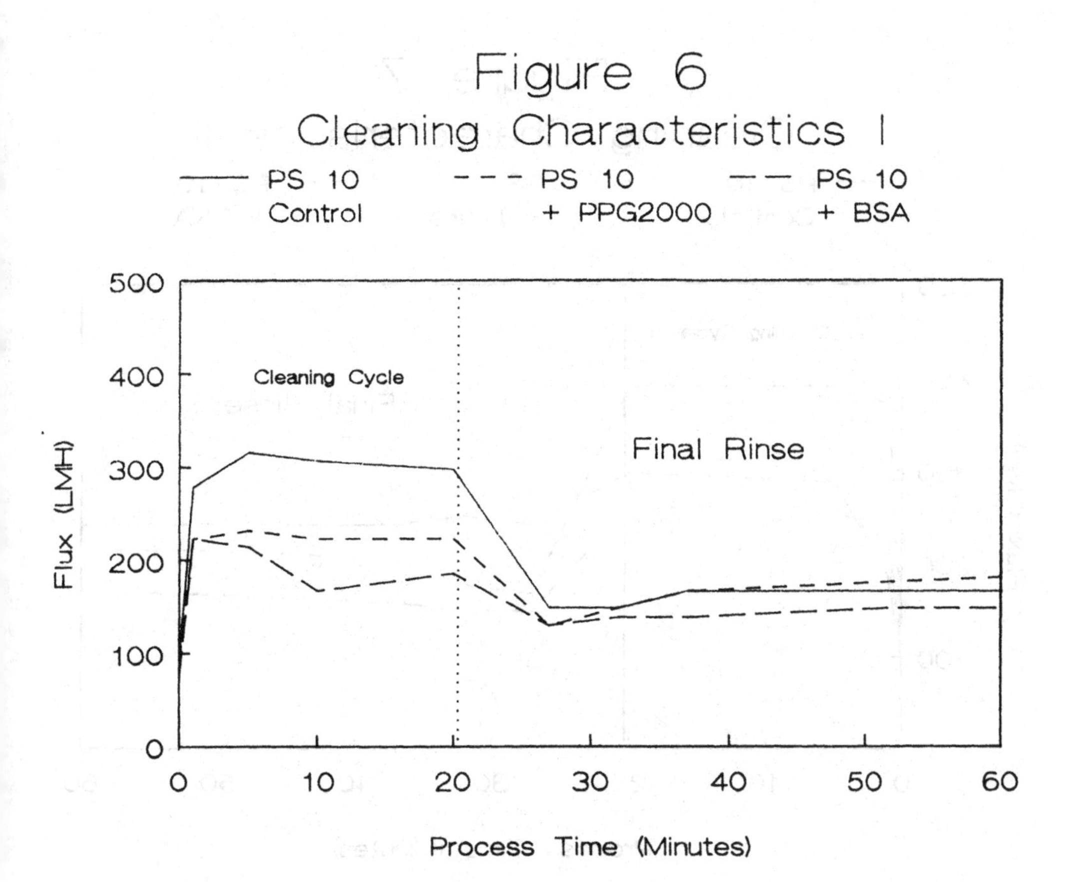
Figure 6
Cleaning Characteristics I
PS 10 Control
PS 10 + PPG2000
PS 10 + BSA
Cleaning Cycle
Final Rinse
Flux (LMH)
0
100
200
300
400
500
0
10
20
30
40
50
60
Process Time (Minutes)

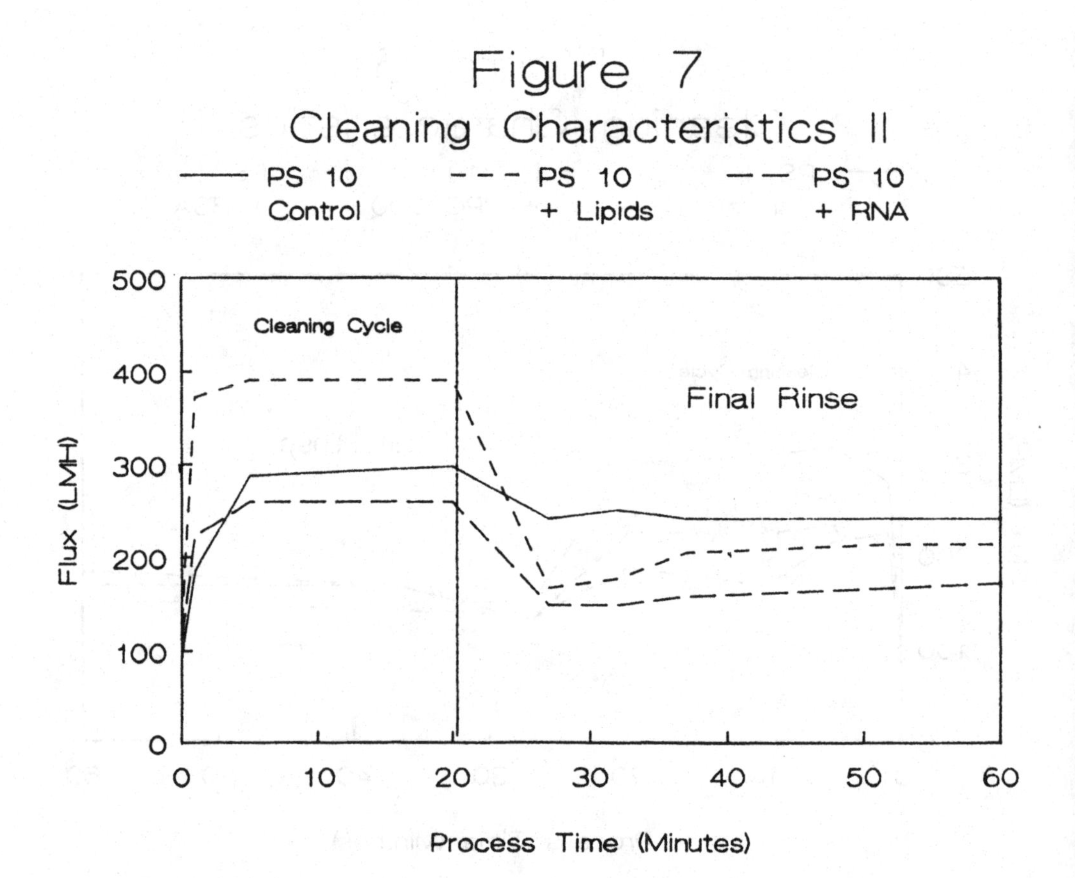
Figure 7
Cleaning Characteristics II
PS 10 Control
PS 10 + Lipids
PS 10 + RNA
Cleaning Cycle
Final Rinse
Flux (LMH)
Process Time (Minutes)
0
100
200
300
400
500
0
10
20
30
40
50
60

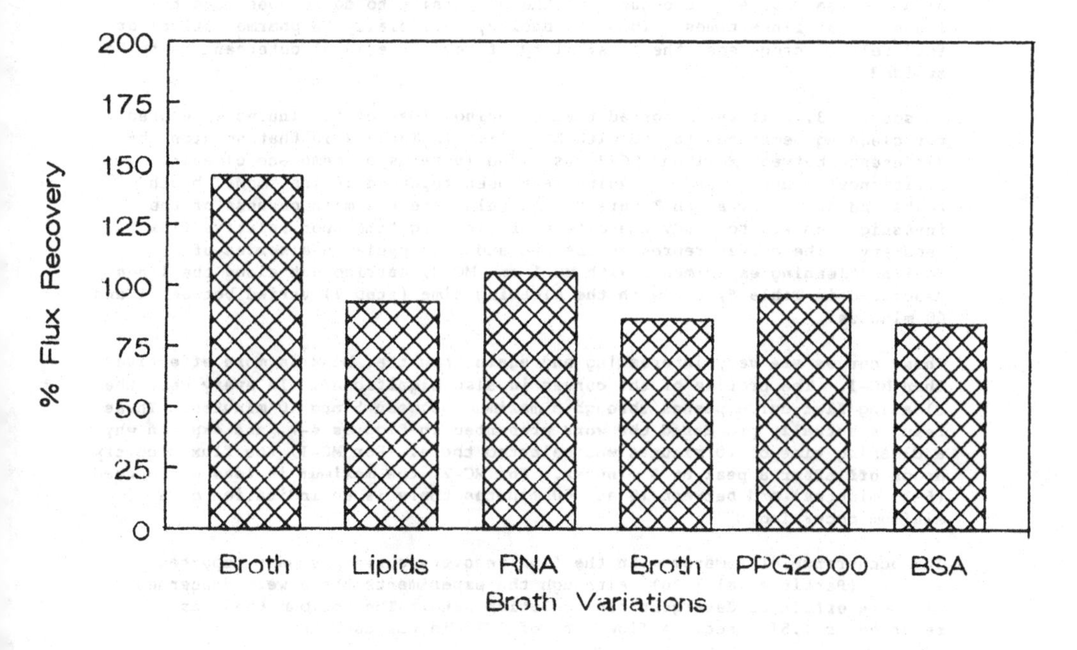
Figure 8
Relative Cleaning Efficiency
Final Water
Flux/Initial
200
175
150
125
100
75
50
25
0
% Flux Recovery
Broth
Lipids
RNA
Broth
PPG2000
BSA
Broth Variations

Finally, it can be seen that the final water rinses are similar in all the cases examined (Figures 6 and 7). There is a flux drop initially as detergent solution is washed away (7 minutes) followed by a re-equilibration period of 10 minutes, where the water flux gradually climbs once more to a plataeu which then remains steady, or increases very slowly for the duration of the final rinse.

The rinsability of the final stage is very important. The use of caustic alone as a membrane cleaner is often not very efficient, particularly on hydrophobic membranes, because it does not wet or disperse the fouling layer. In the rinsing stage, caustic can also be difficult to remove in the final rinse, again because it is, in itself, not very wettable. On the other hand, membrane cleaning detergents which are not optimally formulated can cause foaming during the cleaning cycle due to too much surface activity. This situation may not always occur, but the propensity to do so does lead to longer final rinse times. In biotechnology, especially in pharmaceutical or food related processes, the possibility of having residual detergent is to be avoided.

In section 3.1, it was reported that cleaning times of 60 minutes were used for cleaning membranes fouled with E. coli. In Table 7 in that section the difference between MC-1 and MC-2 was noted in terms of membrane cleaning efficiency. These type of results have been repeated in the "model broth" tests and are reported in Figure 9. In this case the main purpose of the investigation was to study the effect of cleaning time upon relative flux recovery. The curves represent the 5th and 10th cycle in a series of fouling/cleaning experiments with MC-1 and MC-2, carried out along the lines described in Table 6, but with the cleaning time (step 7) varied between 1 and 60 minutes.

These curves are very interesting and again show that MC-2 is more effective than MC-1. The profile of the curves is also significant. In every case the cleaning flux curve passes through a maximum between 5 and 10 minutes. These results actually pre-dated the work presented in Figures 4-7, and explain why a cleaning time of 20 minutes was selected there. For MC-1, the flux recovery drops off after a peak at 5 minutes. For MC-2, the maximum is less pronounced (5-10 minutes) and between 10 and 60 minutes there is an indication of a minimum occurring.

The occurrence of a maximum in the flux recovery curve has been reported earlier (Parkin et al 1976), although the experiments there were concerned with the effect of detergent recirculation rate. The optimum then was reported as 1.51 l/sec. A flow rate of 2 l/min was used here.

3.3 Flocculation Studies

The aim of this work was to investigate the use of flocculants to improve the separation efficiency of an extracellular protease enzyme during microfiltration of a fermentation broth.

The first stage involved assessing the optimum flocculant dosage for the broth of interest. This was done by use of settlement tests as described in section 2.3. Figure 10 shows the typical results obtained for the E. coli broth. The optimum dosage is 100 ppm for Floc 1. Dosages above and below this optimum do not effectively contribute to cell separation in comparison. The doses below 100 ppm do not have sufficient charge to neutralize the anionic surface of the cells. Above 200 ppm the charge of the adsorbed cell/flocculant agglomerates are possibly tending towards a surplus cationic charge and hence increased repulsion and more effective dispersion.

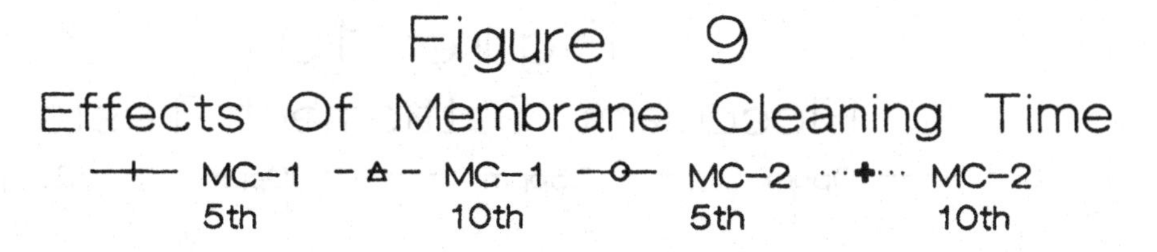
Figure 9
Effects Of Membrane Cleaning Time
MC-1 5th
MC-1 10th
MC-2 5th
MC-2 10th

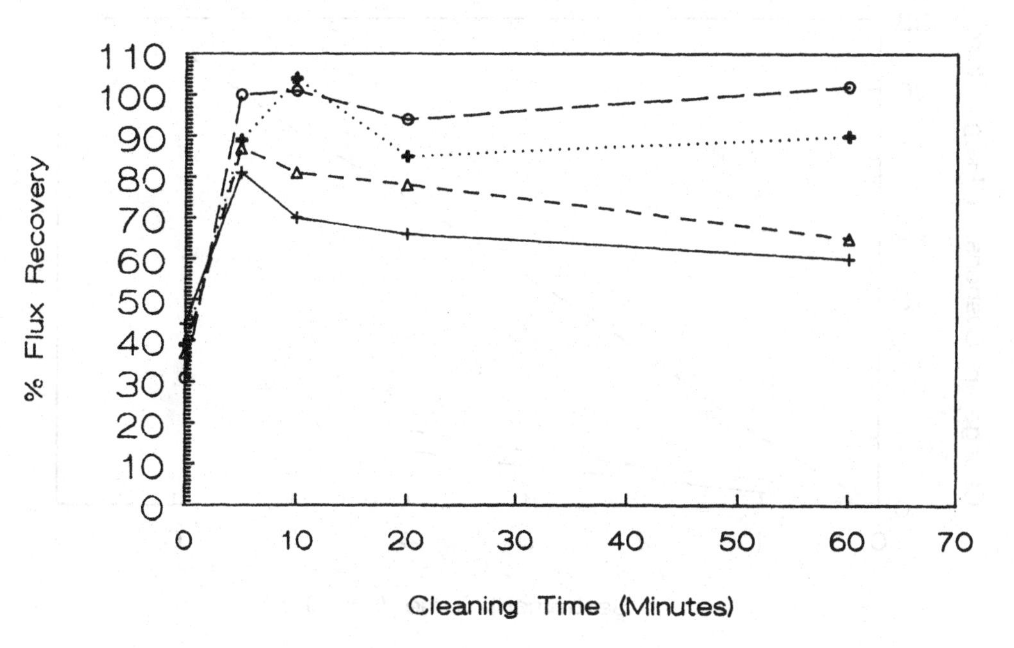
% Flux Recovery
110
100
90
80
70
60
50
40
30
20
10
0
0
10
20
30
40
50
60
70
Cleaning Time (Minutes)

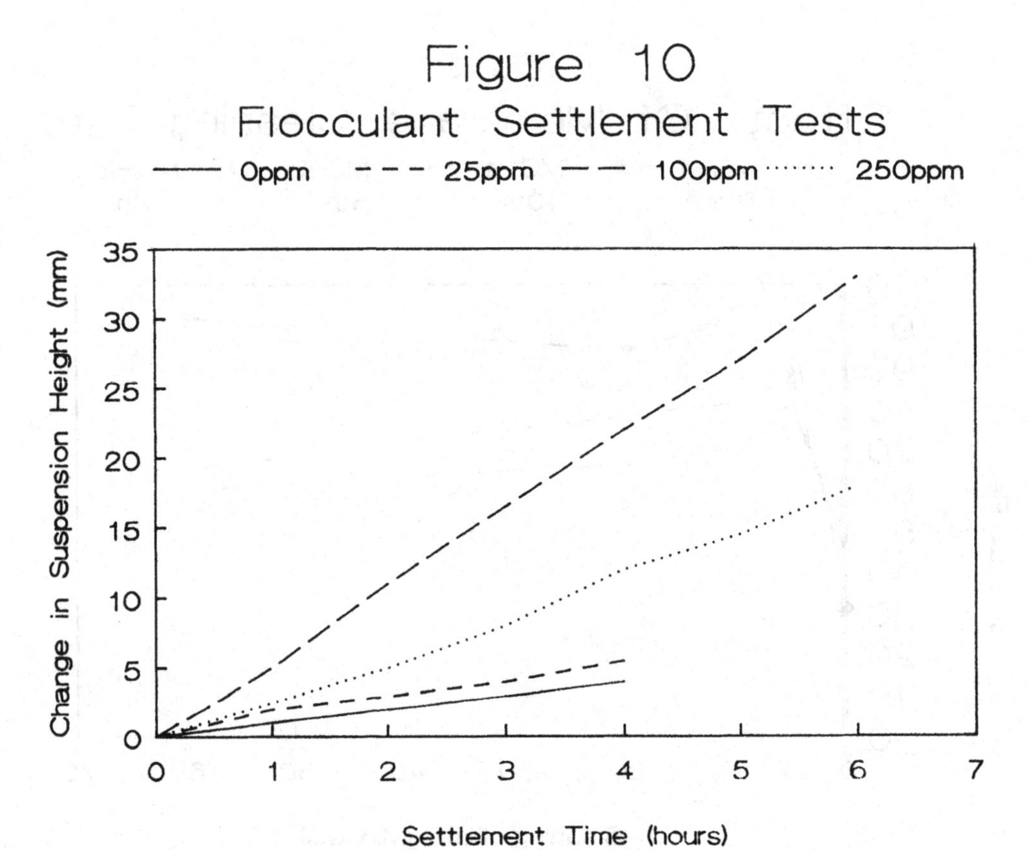
Figure 10
Flocculant Settlement Tests
0ppm
25ppm
100ppm
250ppm
Change in Suspension Height (mm)
35
30
25
20
15
10
5
0
0
1
2
3
4
5
6
7
Settlement Time (hours)

Similar tests gave an optimum dose for Floc 2 of 25 ppm. Floc 2 is known to have a greater cationic charge density than Floc 1, so those results can be interpreted as reflecting that difference.

The second stage was to test the flocculants in several microfiltration experiments. A typical one is reproduced in Figure 11. It is apparent that the addition of 100 ppm of Floc 1 gives a significant improvement to the performance of the hollow fibre microfilter. On the basis of this work, and further studies that indicated that enzyme recovery was always greater than 90%, it was estimated that considerable savings were possible if a new installation was to use a suitable flocculant and thus avoid expenditure on a greater membrane area than would otherwise be necessary.

The potential cost savings (expressed as $/kg enzyme produced) are shown in Figure 12. The annual cost represents the depreciated capital plus the operating cost expressed as an annual average in $/kg of enzyme. In this scenario, both Floc 1 and Floc 2 offer cost savings in excess of 10%. The calculations are based upon the case study and assumptions shown in Table 7.

These are the results of preliminary tests and an internal case study. Nevertheless, they are encouraging. Floc 2 is, as stated, a commercially available product. However, as with many cationic materials there are concerns about residual toxicity. Floc 2, still experimental, is an example of research efforts towards a cost-effective low toxicity alternative.

4. CONCLUSIONS

4.1 E. Coli Fouling/Cleaning

It appears that the relationship between the foulant, the membrane and the cleaning solution is critical. It is certain that the choice of surfactant is important, but decisions also have to be made in terms of dispersant and chelant levels. Several works studying the effectiveness of cleaning more conventional surfaces such as heat-exchangers, have postulated that, despite the combined inorganic/organic nature of the deposit, cleaning with caustic plus additives (chelants, surfactants, dispersants) will have the best overall effect (Pritchard 1988), compared to cleaning with alkali and acid.

Longer term studies in this area should investigate the interaction of different surfactants with the most common types of membrane surfaces (cellulosic, polysulphone, polypropylene and ceramic), to investigate wetting ability and rinsability. A knowledge of the best surfactant/membrane combinations will assist in formulating the best products for cleaning. The final selection of product will depend upon the process stream.

4.2 Model Fouling/Cleaning

It is certain that a model system does not equate to a real process stream. However, as stated in the introduction, there are so many potential process streams that a study has to start somewhere, at a sensible baseline.

In Biotechnology, in particular, the potential for changes in the process stream are vast. Variations not only occur between the different strains of micro-organism but even within one strain. Microbes are known, for example, to excrete more or fewer polysaccharides depending upon the hostility of their environment.

Therefore the case of the model has enabled a study to be carried out of individual, suspected foulants. This has shown a tendency towards fouling on the same scale as a real E. coli broth. Further studies are required to investigate other potential foulants (eg. polysaccharides) and combinations of foulants.

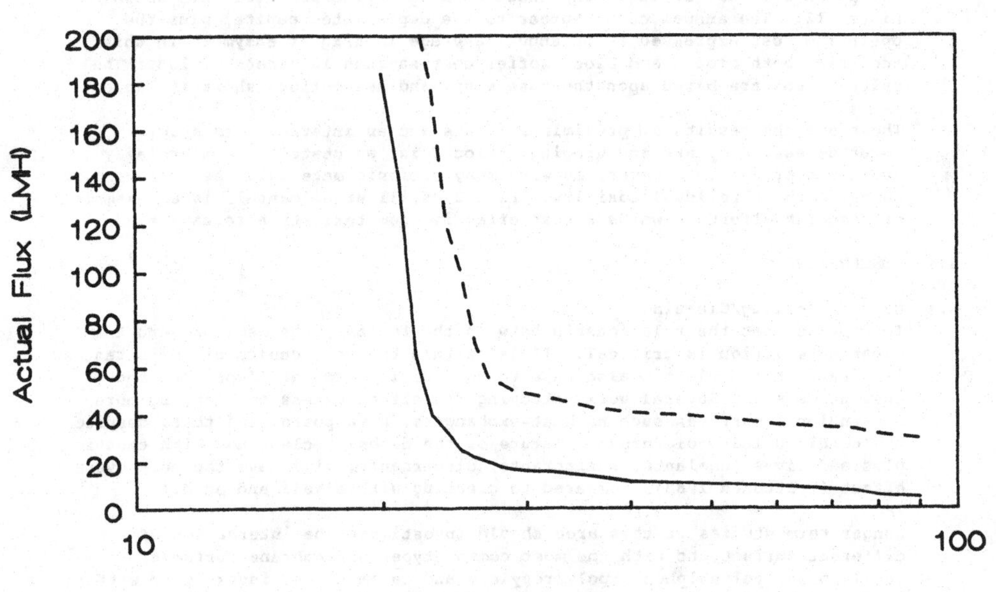
Figure 11
Effect of Flocculant on Microfiltration
Control
100ppm Floc1
Actual Flux (LMH)
200
180
160
140
120
100
80
60
40
20
0
10
100
Cell Concentration (g/l)

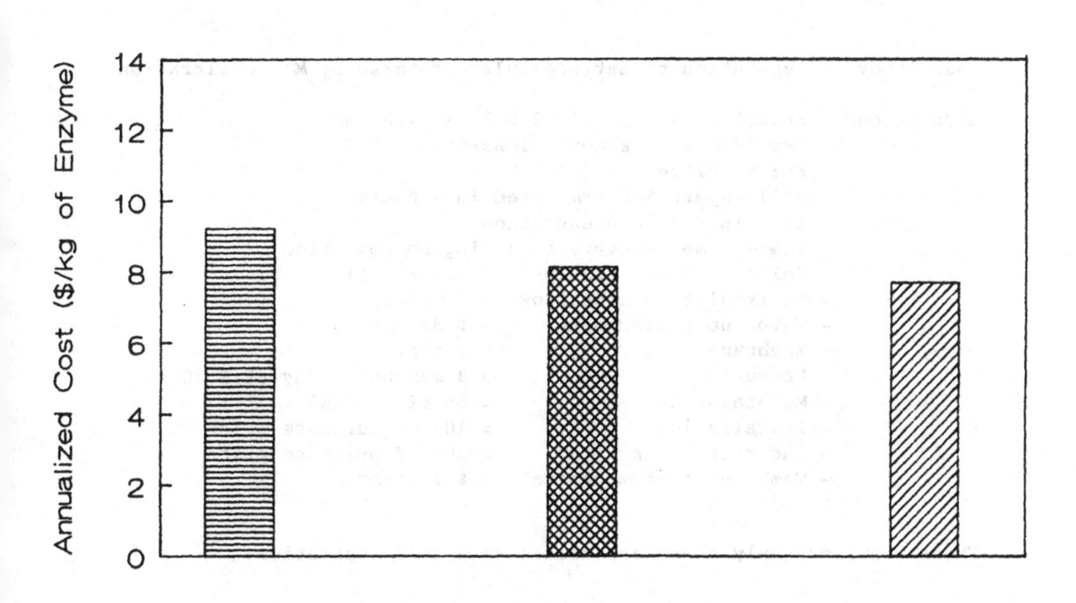
Figure 12
Economics of Cell Separation
Control
Floc1
Floc2
14
12
10
8
6
4
2
0
Annualized Cost ($/kg of Enzyme)
Flocculant Type

TABLE 8 CALCULATION OF THE ECONOMICS OF CELL SEPARATION

These data support the values given Figure 12.

Case Study - Separation of Extracellular Protease by Microfiltration

Assumptions
- Annual capacity = 17,000 kg enzyme/year
- Two 17,000 l fermentations/day
- Enzyme titre = 2 g/l
- Cell separation completed in 4 hours
- 10 fold cell concentration
- 90% enzyme recovery (assuming no rejection)
- Hollow fibre polysulphone membrane (1 mm ID)
- Recirculation crossflow = 1 m/sec
- Membrane replacement = $ 375/m^2
- Membrane life = 1 year
- Labour = 3 man hours/day at $ 20/hr
- Maintainance = 5% of capital
- Installation = 10% of purchase price
- Indirect Costs = 30% of purchase price
- Membrane system capital = $ 1385/m^2

These costs are only those associated with cell separation.

The cleaning study raises questions about optimal cleaning times, redeposition and rinsing. In all this work, cleaning has been carried out in a total recycle mode with the permeate open. This allows intimate contact of the cleaning solution with all the potentially fouled surfaces. However, it also allows for redeposition. Therefore, in the worst cases where a standard cleaning time does not appear to give very good flux recovery, this could be due to redeposition. In these instances, shorter cleaning times, or even possibly lower cleaning additive concentrations, may be the best solution. This apparently illogical proposal is also the conclusion reached by Parkin (1976, p111), where the optimum concentration of an enzymatic cleaner was found to be 0.5 g/litre. Higher or lower concentrations led to less efficient cleaning in terms of the time taken to reach the optimum flux recovery.

In all events, the best cleaning programme for any plant will be best determined on site. A rigorous protocol should be adopted to investigate cleaning time and additive concentration for a particular process stream. It is conceivable that a once through clean for 5 minutes with the solution going to drain may give a better result than a 20 minute recycle, providing such an approach does not cause problems in the effluent discharge, and provided it is cost-effective.

It is certain that closer co-operation between membrane suppliers, membrane users and detergent formulators will solve cleaning problems, provided a flexible approach is taken. In this regard the questions of product approvals (FDA etc) and membrane compatibility need to be addressed.

FDA approvals for new surfactants, dispersants or chelants can take a great deal of time and money. Moreover, there are not necessarily, for example, FDA sections dealing specifically with cleaning chemicals for use in antibiotic manufacture. In these cases, the detergent formulator and the membrane user have to get together to discuss toxicity and residual questions. It needs to be remembered that some standard methods for cleaning membranes range from the use of straight caustic, which may not always be efficient, does not rinse well and can contain low levels of mercury, to the use of hypochlorite which can be corrosive to steel, corrosive to membranes and can react to form undesirable by products (eg. chloramines).

Membrane compatibility is another area where co-operation between membrane suppliers and detergent formulators is and will be important. As stated in the introduction, the variable factors that impinge upon the effectiveness of any given membrane in a certain process situation will often be difficult to predict. This is why membrane suppliers go through lengthy pilot trials prior to selection of the optimal system for scaling up. It is at this stage that a cleaning regime should be selected. If it requires the formulation of new detergents, that can be achieved by competent, professional companies. However, membrane compatibility must also be taken into account. It is conceivable that a more efficient cleaner could be developed for an application which could reduce cleaning down-time so much that a shorter membrane life is not considered a disadvantage. However, the economics of this argument need to be studied more closely.

It is often more difficult to change cleaning programmes in an established process, unless some clear goals for improvement can be set.

In conclusion, the growing number of membrane applications throws up increasing numbers of questions about cleaning. Even ceramic membranes, one of the benefits of which is their sterilizability, need to be cleaned effectively before being subjected to any steam treatment. To solve these new problems will require a greater understanding of the effects of formulated detergents with membrane surfaces and principle foulants. It could be that the cleaning needs for each process stream/membrane combination is almost a type of "lock and key" relationship.

4.3 Flocculation Studies

These preliminary results are encouraging and indicate the potential for the use of flocculating agents to improve the flux performance of membrane systems.

The main areas for further study include;

. toxicity status of residual cationic materials

. effects on enzyme (or other product) performance

. long term effects on membrane lifetime

. approval situation

This study is in a very early stage but the ramifications can be very important.

The main problems with cationic flocculants have been associated with their toxicity to fish and the ability to measure residual amounts of flocculant in a given process stream.

Ashley (1989) in discussing opportunities for the improvement in design of downstream process equipment in Biotechnology, stressed the importance of having greater understanding of flocculation mechanisms.

ACKNOWLEDGEMENTS

My thanks go to W.R. GRACE for allowing me to present this paper, to Dr. J. Sheehan, Dr. G. Thompson and Dr. M. Hughes for carrying out most of the experimental work and to Sue Maxwell for typing and preparing the finished document.

REFERENCES

Aimer, P. and Sanchez, V. (1986) 'Ind. Eng. Chem-Fundam,' Vol. 25, 789-798.

Ashley, M.H.J. (1989) ICHEME - Advances in Biochemical Engineering, 'Development Opportunities for Downstream Bioprocessing Equipment,' 75-82.

Beaton, N.C. (1984) Recent Developments in Separation Science, Ed. N.N. Li, CRC Press, Boca Raton, 1.

Cabral, J.M.S., Casale, B. and Cooney, C.L. (1985) 'Biotechnology Letters,' Vol. 7, No. 10, 749-752.

Cheryan, M. and Merin, U. (1984) 'J. Dairy Sci' Vol. 67, 1406.

Chong, R. Jelan, P. and Wong, W. (1985) 'Separation Sci and Tech,' Vol. 20, Nos. 5 & 6, 393-402.

Fane, A. G. (1983) 'J. Sep. Proc. Technol,' Vol. 4, No. 1, 15-23.

Harris, T. A. J. (1988) 'J. Chem. Tech. Biotechnol,' Vol. 49, 19-30.

Ilias, S. and Govind, R. (1987) A Study on Fouling of Micro/Ultrafiltration Membranes by Dilute Suspensions, N.A.M.S. Annual Meeting, June 3rd-5th.

Issacson, K., Duenas, P., Ford, C. and Lysaght, M. (1980) Ultrafiltration Membranes and Applications, Ed. A. R. Cooper, 507-522.

Kloosterman, J., van Wassenaar, P. D., Slater, N. K. H. and Baksteen, H. (1988) 'Bioprocess Engineering,' vol. 3, 181-185.

Le, M. S. and Howell, J. A. (1984) 'Chem. Eng. Res Des,' vol. 62, 373-380.

Matthews, M. E., Doughty, R. K. and Hughes, I. R. (1978) 'N. Z. J. Dairy Technol,' vol. 13, 307.

Miki, B. L. A. (1980) Advances in Biotechnology, 'Current Developments in Yeast Research,' Proc. Int. Yeast Symp., London (Ontario), p. 193.

Parkin, M. F. and Marshall, K. R. (1976) 'N. Z. J. Dairy Technol,' vol. 11, 107-113.

Paulson, D. J. (1987) 'An Overview of and Definitions of Membrane Fouling,' 5th Membrane Technology Planning Conference, Cambridge (MA), 103-122.

Pritchard, N. J., De Goederen, G. and Hoasting, A. P. M. (1988) 'Fouling in Process Plant,' Inst. of Corrosion Sci and Tech, London, 465-475.

Roe, S. D. (1989) BIOSEP State of the Art Report (SAR 16), 'Selection and Integration of Purification Processes, p. 32 (Commercial in Confidence).

Smith, K. R. and Bradley Jnr, R. L. (1987), 'J. Dairy Sci,' vol. 70, 243-251.

Taeymans, D. and Lenges, J. (1989) 'Belg. J. Food Chem and Biotech,' vol. 44, no. 6, 233-236.

Whittaker, C. Ridgway, H. and Olson, B. H. (1984), 'App. and Environ. Microbiology,' vol. 48, no. 3, 395-403.

Investigation of Protein Fouling Characteristics of Ultrafiltration Membranes

I M Reed, J M Sheldon*
BIOSEP, AEA Environment and Energy, Harwell Laboratory, Oxfordshire
*Department of Biological and Molecular Sciences, Oxford Polytechnic

Abstract

A range of transmission electron microscope (TEM) techniques have been adapted by BIOSEP for the investigation of the protein fouling characteristics of ultrafiltration membranes. Two membrane types, one made from polysulphone and the other from regenerated cellulose, have been studied in some detail. Thin section TEM combined with immunochemical staining can show the distribution of protein within a fouled membrane and replication techniques permit the fine structure of both membranes and proteins to be studied. This work can explain differences in the protein fouling characteristics of the two membranes studied and can greatly assist in selection of membranes to minimise fouling and loss, or denaturation, of valuable products.

Introduction

Although electron microscopes have been widely used to examine membranes most previous studies have employed scanning electron microscopes (SEMs). However, the potential of the wide range of transmission electron microscope (TEM) techniques has yet to be fully realised in this field. This paper describes a number of TEM methods, which have been adapted for use with ultrafiltration membranes for BIOSEP by Oxford Polytechnic. These techniques have been used to investigate the protein fouling characteristics of two different ultrafiltration membranes, one a hydrophilic membrane made of regenerated cellulose and the other a more hydrophobic membrane made from polysulphone. Both membranes had a nominal molecular weight cut-off of 10 000.

Effect of Protein Fouling on Ultrafiltration Membranes

Table 1 shows the quantity of protein, in this case bovine serum albumin (BSA), adsorbed by the two membranes under static and filtration conditions, and shows the resultant changes in water fluxes.

All experiments were carried out using a solution containing 5g/l BSA in 0.05M, pH7 phosphate buffer. In the static tests the membranes were contacted with protein solution for 3 hours with no flow of liquid through the membrane and no applied pressure. Filtration experiments were performed over a period of 2 hours in a small stirred cell operating at a stirrer speed of 500rpm and an applied pressure of 1.7×10^5Pa.

Effective Industrial Membrane Processes — Benefits and Opportunities, pp.91–99

Table 1 Protein fouling characteristics of ultrafiltration membranes

Membrane type	Static conditions		Filtration conditions	
	BSA adsorption (mg/m²)	water flux after contact / water flux before contact (-)	BSA adsorption (mg/m²)	water flux after contact / water flux before contact (-)
Polysulphone	38.2	0.31	185	0.24
Regenerated cellulose	7.3	1.08	66.1	1.0

Under static conditions the polysulphone membrane adsorbed approximately 5 times as much protein as the regenerated cellulose membrane. This difference was reflected in the way the water fluxes of the two membranes responded; protein adsorption resulted in a 3 fold reduction in the water flux of the polysulphone membrane but had almost no effect on the regenerated cellulose membrane. Under filtration conditions there was again no change in the water flux of the regenerated cellulose membrane even though the membrane adsorbed almost twice as much protein as the polysulphone membrane had under static conditions. These results suggest that there is a relationship between protein adsorption and flux decline but that the link is complex. In order to study this link, TEM techniques have been used to investigate the influence of protein fouling on membrane flux.

Advantages of Transmission Electron Microscopy

Although electron microscopy has been used extensively for the examination of membranes, most previous studies have used SEM rather than TEM. The main reasons for this are that SEM is relatively easy to use and produces clear, easily interpreted images with a good depth of focus. Nevertheless, TEM possesses a number of advantages which are particularly useful in the examination of fouled membranes. Firstly, TEM can give greater magnification and can resolve finer details than can SEM. Secondly, a wider range of specimen preparation techniques can be employed with TEM enabling the distribution of particular classes of molecule within a sample to be determined. As a result TEM has the potential to provide more detailed information on the effect of foulants on membrane structure than is generally possible using SEM.

Figure 1 Section through polysulphone membrane using thin section TEM
(bar=10μm V-void S-separating surface)

Figure 2 Section through regenerated cellulose membrane using thin section TEM
(bar=5μm S-separating surface D-downstream surface)

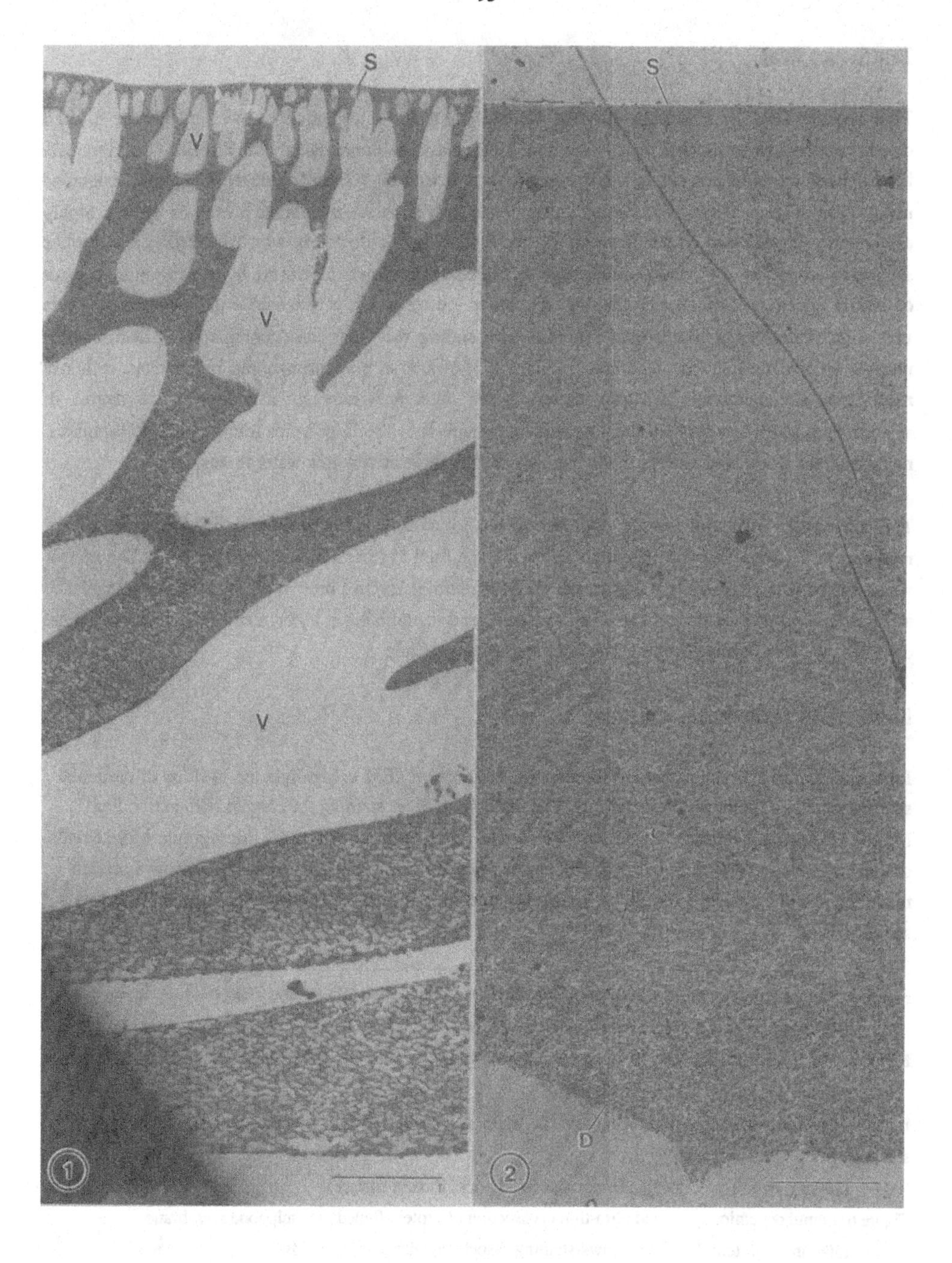
S
V
V
V
1
S
D
2

Thin Section TEM

Thin section TEM is a straight forward sample preparation technique. A wet membrane is first dehydrated through an alcohol series (Sheldon, 1991) and then impregnated with a resin. Once the resin has hardened the specimen can be cut into sections of around 50-100μm in thickness and then examined using TEM. Figure 1 shows a section through the polysulphone membrane which is clearly highly asymmetric. It possesses a thin dense layer about 0.5-1μm in thickness which is responsible for the separating properties of the membrane and is supported against applied pressure by a highly porous layer of 100μm thickness. The large finger shaped voids in the support structure enable the membrane to give very high water fluxes. The polymer matrix surrounding the voids also contains small pores which increase in size through the membrane. Figure 2 shows a section through the regenerated cellulose membrane. The membrane is approximately 50μm thick and, although asymmetric, the degree of asymmetry is much less than for the polysulphone membrane. The large voids found in the polysulphone membrane are absent and as a result the membrane exhibits relatively low water fluxes.

The main reason for using electron microscopy was to investigate the protein fouling characteristics of membranes. Figure 3 shows a polysulphone membrane used to filter a 5g/l solution of BSA. The sample was prepared using the standard thin section TEM method. A layer of protein can be clearly distinguished at the separating surface since it shows up as a pale band in comparison to the membrane. However, there does not appear to be any protein within the membrane structure.

Immunochemical Staining

Immunochemical staining can be combined with thin section TEM to highlight the position of particular species, usually proteins, within a sample. The method involves staining the sample with antibodies linked to a small gold particle which will show up as a distinct black dot on the micrograph. More details of the methods used can be found in Sheldon, Reed and Hawes (1991). Figures 4 and 5 show a protein fouled polysulphone membrane which has been stained in this fashion. In this case the membrane was stained before

---

Figure 3 Surface region of protein fouled polysulphone membrane using thin section TEM
(500nm P-protein S-separating surface)

Figure 4 Surface region of protein fouled polysulphone membrane with pre-embedding immunochemical staining
(200nm P-protein arrows-gold staining)

Figure 5 Downstream surface of protein fouled polysulphone membrane with pre-embedding immunochemical staining
(200nm arrows-gold staining in micropores D-downstream surface)

Figure 6 Immunochemically stained ultra-thin cryosection of protein fouled polysulphone membrane
(500nm P-protein V-voids arrows-staining associated with walls of voids)

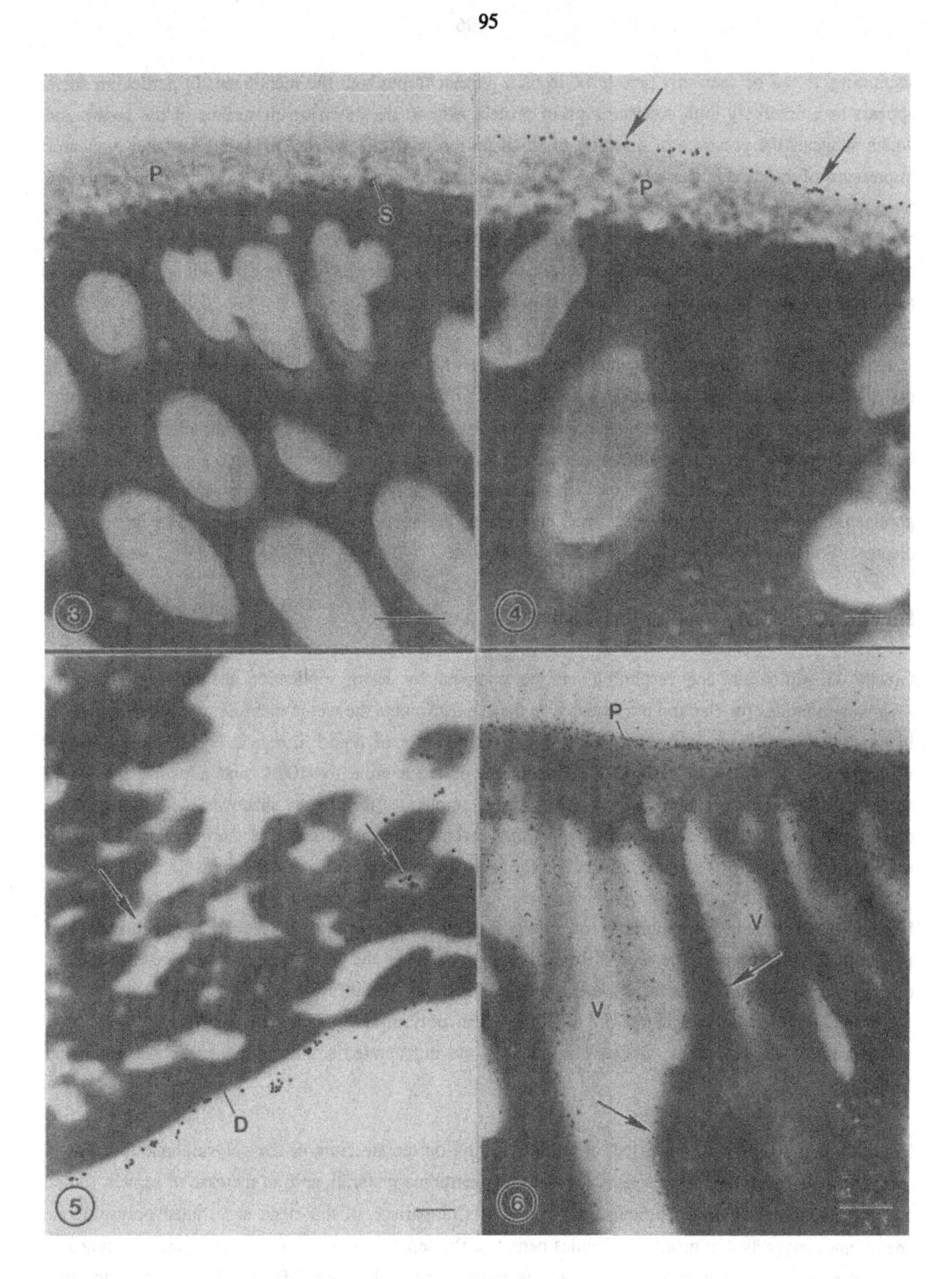
P
S
3
P
4
5
D
P
V
V
6

sectioning. It can be seen that protein is, in fact, present throughout the membrane. In particular, there appears to a relatively high concentration of protein near to the downstream surface of the membrane which suggests the presence of a region of small pore size similar to the separating surface. A clearer impression of the distribution of protein can be obtained by making cryosections of the membrane and then applying the immuno-gold stain. Figure 6 shows a fouled polysulphone membrane stained in this way. The picture is easier to interpret than Figures 4 and 5, and shows clearly that protein is located throughout the fouled membrane. However, this technique is rather difficult to carry out and great care is required to avoid staining of material other than the target protein.

The presence of protein within the membrane is somewhat surprising since BSA has a molecular weight of 68000 and the nominal molecular weight cut-off of the membrane is only 10000. Nevertheless, this micrograph does suggest one possible explanation for the finding that membranes can adsorb a relatively large quantity of protein with little apparent effect on water flux. The membrane support structure is highly porous and can contain adsorb a considerable quantity of protein without affecting membrane permeability. Protein adsorbed to the membrane surface, on the other hand, would have a disproportionately high effect on membrane performance.

Use of Replication Techniques to Prepare Membrane Samples

Greater magnification and resolution can be obtained by using replication techniques to prepare membrane samples for electron microscopy. In these experiments the wet membrane was frozen rapidly in liquid propane which had been cooled to the temperature of liquid nitrogen. The surface layers of water were then etched by raising the temperature of the sample to -100°C and allowing the ice to sublime for between 15 and 60 minutes. The frozen sample was then rotary shadowed with a mixture of carbon and platinum and a replica built up using carbon. The membrane was then removed from the replica by dissolving in a suitable solvent and the replica was examined using TEM.

Figures 7 and 8 show micrographs of replicas of the surfaces of the regenerated cellulose and polysulphone membranes. The level of detail is very great since the dark areas show the location of pore openings in the membrane surfaces.These appear to be about 4-7nm in diameter which is consistent with the nominal molecular weight cut-off of 10000. The polysulphone membrane appears to have quite regular almost circular pores whereas the pores in the regenerated cellulose membrane are somewhat irregular.

Figures 9 and 10 illustrate the effect of protein fouling on the structure of the polysulphone membrane. The micrographs reveal the presence of strands of membrane material, and, in the case of protein fouled membranes, the structure of protein fouling layers. Comparison of the clean and fouled polysulphone membranes suggests that protein molecules penetrate the separating layer of the membrane resulting in a marked reduction in porosity. In addition, the protein molecules in the fouling layer appear to form

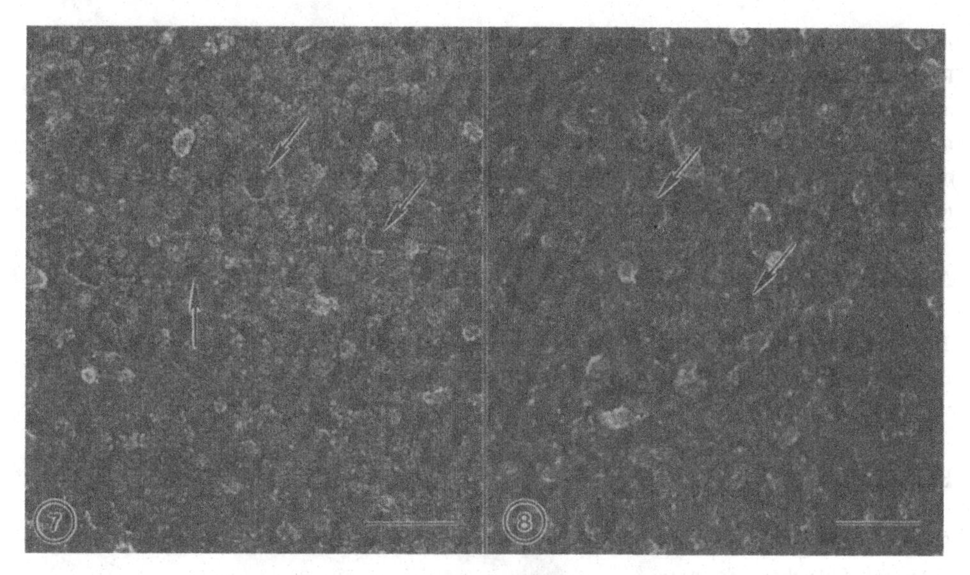

Figure 7 Replica of surface of polysulphone membrane
Figure 8 Replica of surface of regenerated cellulose membrane
(bars=50 nm arrows indicate pore openings)

strands rather than to be globular in shape as in the free solution (see Fig. 13). There are also indications that strands of protein are passing into the membrane structure. This suggests that the polysulphone membrane is altering the tertiary structure of the protein molecules. By contrast the structure of protein molecules in fouling layers on the regenerated cellulose membrane (see Figures 11 and 12) appears to be very similar to that found in the pure solution. It is not possible to determine whether protein is passing into this membrane since protein and membrane have a similar appearance in the microscope image.

From these micrographs it appears that, on the polysulphone membrane, BSA unfolds to some extent forming compact fouling layers with a low hydraulic permeability. In addition, this unfolding seems to permit the protein to enter the membrane pores and to cause substantial blocking of the separating layer of the membrane. Both of these phenomena can lead to a substantial reduction in membrane permeability and a consequent loss of flux. The regenerated cellulose membrane, on the other hand, appears to have little affect on protein structure, which therefore forms open porous fouling layers. As a result there is relatively little reduction in membrane flux due to protein binding.

Conclusions

Transmission electron microscopy can reveal much information on protein fouling that is not usually possible using the, more common, scanning electron microscopy. Thin section TEM combined with immunochemical staining enables the distribution of protein within a fouled membrane to be determined. Replication techniques permit details of the structure of both membranes and proteins to be visualised and suggest that the structure of protein molecules in fouling layers depends on the chemical nature of the membrane material. This work can explain differences in the protein fouling characteristics of the two membranes and can greatly assist in the diagnosis of fouling problems and in the selection of membranes to minimise membrane fouling and denaturation of valuable products.

ACKNOWLEDGEMENT

This work was funded by BIOSEP, the Biotechnology Separations Club, which is supported by the Biotechnology Unit of H.M. Governments Department of Trade and Industry.

REFERENCES

Sheldon J M. (1991). "The fine structure of ultrafiltration membranes: (i) clean membranes". Submitted to Journal of Membrane Science.

Sheldon J M, Reed I M, Hawes C. (1991). "The fine structure of ultrafiltration membranes: (ii) protein fouled membranes". Submitted to Journal of Membrane Science.

Figure 9 Replica of surface region of polysulphone membrane in cross-section
(bar=100nm S-separating surface arrows-micropores)

Figure 10 Replica of surface region of protein fouled polysulphone membrane
(bar=100nm P-protein S-separating surface arrowheads-protein penetrating membrane surface arrows-protein in matrix)

Figure 11 Replica of surface region of regenerated cellulose membrane in cross-section
(bar=100nm S-separating surface arrows-micropores)

Figure 12 Replica of surface region of protein fouled regenerated cellulose membrane
(bar=100nm arrows-separating surface P-protein M-membrane matrix)

Figure 13 Replica of BSA molecules in solution
(bar=20nm arrows show where protein molecules are associated)

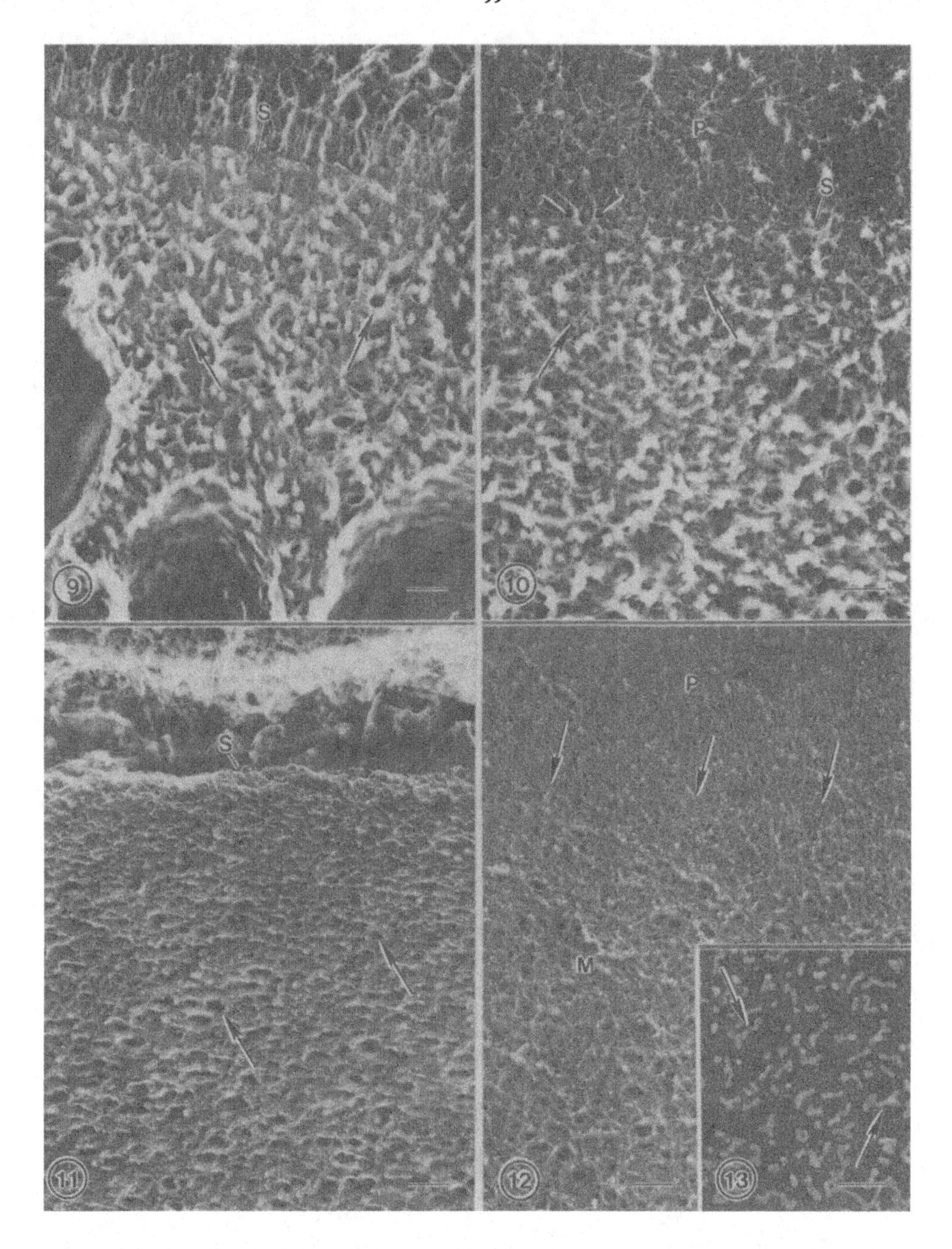
S
9
P
S
10
S
11
P
M
12
13

SESSION C:

WASTE WATER TREATMENT I

A COMPARATIVE EVALUATION OF CROSSFLOW MICROFILTRATION MEMBRANES FOR RADWASTE DEWATERING

R.G. Brown[1], M.H. Crowe[1], D.J. Hebditch[1], R.N. Newman[1] and K.L. Smith[2]

1. Berkeley Nuclear Laboratories, Nuclear Electric plc
2. University of Bath

SUMMARY

Pilot testing of five diverse types of dewatering unit has established crossflow microfiltration as the most appropriate for thickening of mineral waste slurries. A comparative testing programme has been carried out using stainless steel and ceramic membranes. The main experimental measurements made were permeate flux and clarity and retentate concentration. The experimental systems have been examined for fouling, wear, ease of control and suitability for use with radioactive material.

Membranes used were in tubular form ie Pall PSS (2.5 μm limit of separation), Fairey Microfiltrex FM4 (1μm) and APV Ceraver (1.4 μm). Feed slurries were generally at 5 w/o solids concentration with particle size distributions in the range 1-100 μm and comprised mixtures of magnesium hydroxide, slate, filter precoat (Dicalite Speedplus), graphite, oil and water. Slurries were dewatered to 30-50 w/o depending on the rheological properties. On-line backwash and off-line chemical cleaning were employed. Flux decline by irreversible fouling did not occur during 80 h operation of one of the membrane modules. Operation has been batch or semi-batch using 0.2-1 m^3 initial inventory and approximately 0.1 m^2 membrane area. Average velocities in the crossflow tubes were in the range 3-5 $m\ s^{-1}$. Average transmembrane pressures were in the range 0.8-1.9 bar (g). With partially thickened material permeate fluxes were found to be of magnitude $100 \times 10^{-6}\ m^3\ m^{-2}\ s^{-1}$.

Crossflow microfiltration has been found to provide effective dewatering of a range of simulant radioactive wastes. Performance data and various process constraints have enabled selection of an optimal system for the particular application. Pressure drop data has been analysed to predict slurry viscosity for comparison with other measurements and with a view to use as a parameter for control of thickening.

INTRODUCTION

During 1988 the then Central Electricity Generating Board recognised the need to develop a dewatering plant for use with its proposed mobile plant for the encapsulation of wet intermediate level radioactive waste (ILW) sludges and resins. A shortlist of five types of dewatering plant was selected by CEGB staff at Barnwood using a procedure derived from decision tables devised by Warren Spring Laboratory. The selection criteria included feed characteristics, product parameters and suitability for use in a radwaste process system. Berkeley Nuclear Laboratories (BNL) undertook pilot testing of three of these dewatering devices and the overall assessment of suitability of the five. Two other devices were tested concurrently by separate contractors.

Effective Industrial Membrane Processes — Benefits and Opportunities, pp.103–113

The dewatering pilots tested at BNL were a scroll decanter centrifuge, a Lamella settler with thickener and a rotary drum vacuum filter (RDVF). A crossflow filter and a Delta Stak settler were tested respectively by AEA Technology at Winfrith and Heriot Watt University at Edinburgh. Work was performed with non-radioactive simulants of radwastes. The main objectives of this programme were:

(1) to perform comparative testwork on selected types of dewatering plant at pilot scale
(2) to assess the performance, operational features and design of the dewatering plants
(3) to identify the most suitable device(s) for process design, and for further testing and development as necessary

Performance tests were carried out at various slurry concentrations and flowrates. The product dryness and filtrate clarity were determined, see Figure 1.

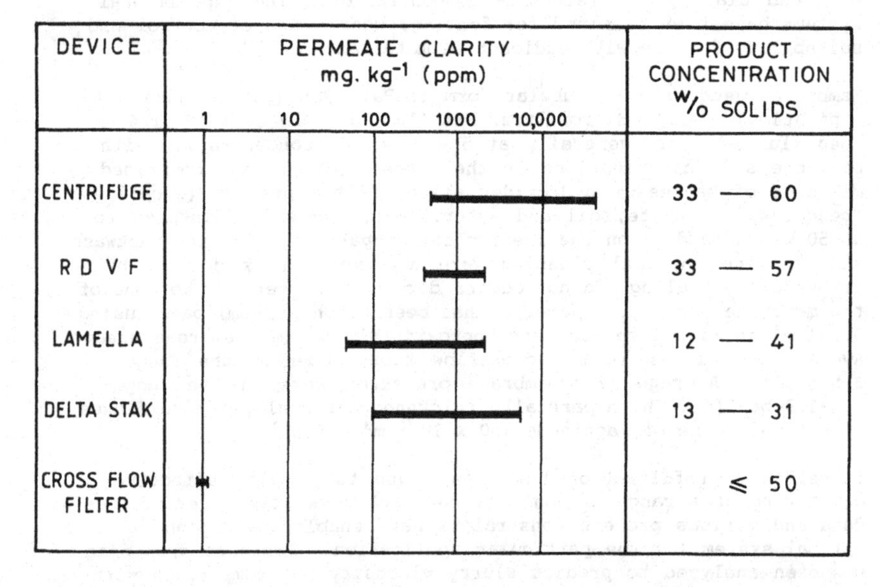

Figure 1: Scheme of Dewatering Performances

Observations were made of plant operability and product characteristics. The plant assessment included consideration of maintenance, complexity, cleaning/decontamination, fault conditions and process compatability.

The conclusions from this work were as follows:

(1) Of the five plants considered, the crossflow filter and the Lamella settler appeared to be the more effective dewatering devices for the conceptual Nuclear Electric mobile radwaste encapsulation plant.

(2) The crossflow filter had the better performance and operating characteristics. Given suitable instrumentation, the control of the extent of dewatering is much more positive than for other devices. However it is subject to potential adverse factors in relation to blockage, fouling, wear and overall

cost. If these can be cleared it is the first choice depending on overall process compatibility.

(3) The Lamella settler had a lesser performance but is a simple, established technology. The extent of dewatering and the clarity of overflow are inferior to several other devices. This and difficulties of control may cause sub-optimal utilisation of waste matrices and greater difficulty of integration within the overall process.

It was recommended that:

(1) Crossflow filtration using tubular membranes should be adopted as the provisional choice of optimal dewatering technology.

(2) An experimental study be made to compare the performance of several proprietary membranes and to further confirm the suitability of the technique.

This study is reported here. Particular objectives of the study were to:

(1) Measure and compare permeate fluxes and dewatering cycle times. Determine permeate clarities.

(2) Evaluate membrane blockage and techniques for unblocking and cleaning.

(3) Examine the extent of membrane wear.

(4) Investigate the reproducibility of rate data and the potential occurence of membrane fouling over moderate operating periods.

The thickening of mineral slurries using crossflow microfiltration, based on tubular inorganic membranes, is a modern technique. Recent improvements in fabrication of ceramic and metallic porous materials have enabled this progress.

A process design study for the dewatering module has also been completed which shows crossflow technology to be compatible with the mobile plant concept. The plant duty has provisionally been defined as a daily capability to dewater a 10 m^3 batch of 5 w/o Magnox slurry to a concentration of 25 w/o within a period of 8 h.

EXPERIMENTAL PROCEDURES

Crossflow Dewatering Test Rig

Crossflow microfiltration of liquids employs high velocity flow, e.g. 3-5 m s^{-1}, across a membrane surface through which permeation occurs under the influence of a pressure difference, e.g. 0.5 - 2 bar. In the present case tubular inorganic membranes of diameter 5 - 10 mm were used to resist abrasion and to prevent gross flow blockage. The high shear which occurred in the slurry flowing within these narrow tubes, length around 1 m, is considered to minimise deposition and cake formation at the wall. This helped maintain the permeate flux. The ratings of the filters were around 1 μm. Modules containing many tubes in parallel were assembled like shell and tube heat exchangers. Slurries were concentrated by repetitive passes through the tube bundles.

A simplified schematic of the BNL Crossflow Dewatering Test Rig is shown in Figure 2.

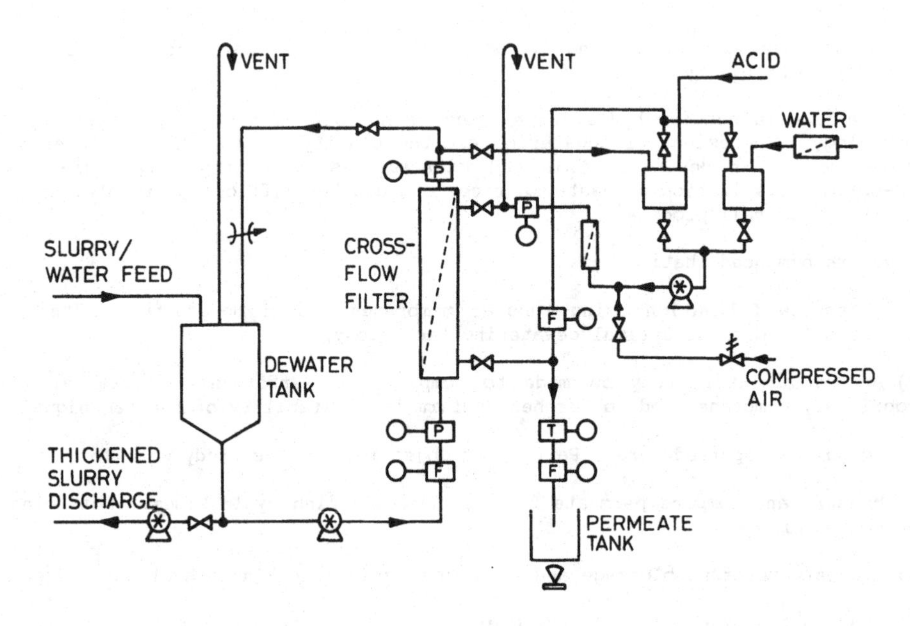

Figure 2: Simplified Schematic of Crossflow Dewatering Rig

The main components of the system were as follows: the crossflow circuit including filter modules and hold-up tank, the backwash and chemical cleaning circuit, the permeate collection system, the slurry feed circuit and the thickened product collection system. The Rig is shown in Figure 3. It was operated in batch or semi-batch mode.

Some details of the Rig were as follows. The crossflow module was mounted vertically with slurry flow in an upward direction. The crossflow circuit was constructed of 28 mm OD copper piping and the dewater batch tank of aluminium alloy. The tank capacity was 0.27 m^3. The permeate lines were connected to a collection tank or drain and to a 10 μm filtered water supply for backwashing or to a chemical cleaning tank normally containing 10% nitric acid or sodium hydroxide. Sections of Durapipe plastic were built into the circuit adjacent to the crossflow filter both to prevent the formation of a galvanic couple between the copper pipe and stainless steel of the filters and to protect that part of the circuit coming into contact with acids. Permeate could be passed through flow and turbidity meters for rate and solids content measurements.

The slurry feed circuit included two 1 m^3 make-up tanks containing large agitators. The circuit was designed such that these tanks could be substituted for the dewater tank when larger scale operation was required. A small Mono pump was used in the product collection circuit to empty thickened material to a waste skip for disposal. The main crossflow pump was a Mono type with a maximum flow rate of 4 $m^3\ h^{-1}$. This initially necessitated operating with four tubes only when using the Pall and Microfiltrex filters, see Figure 4a where three of the Microfiltrex tubes are blanked off. This was to maintain the desired in tube velocity. In later modifications, a large capacity Wemco recessed impellor centrifugal slurry pump and a replacement agitator capable of suspending 1 mm sand particles in one of the two large feed tanks were fitted. Full use of the Pall and APV modules was then made.

Figure 3: View of Crossflow Dewatering Rig

Filter Modules

These are shown in end and side views in Figure 4.

The Pall module was supplied in seven tube form, tube designation PSS grade S025. The membrane was fabricated from sintered 316L stainless steel powder and was seamless. The tubes were welded to a tube sheet at each end. The maximum working pressure was 6 bar (abs). The tube outer and inner diameters were 11 and 9 mm respectively. The tube length was 1.1 m. This provided a tube inner area of 31.1×10^{-3} m^2. The tubes had a particle rating of 2.5 μm.

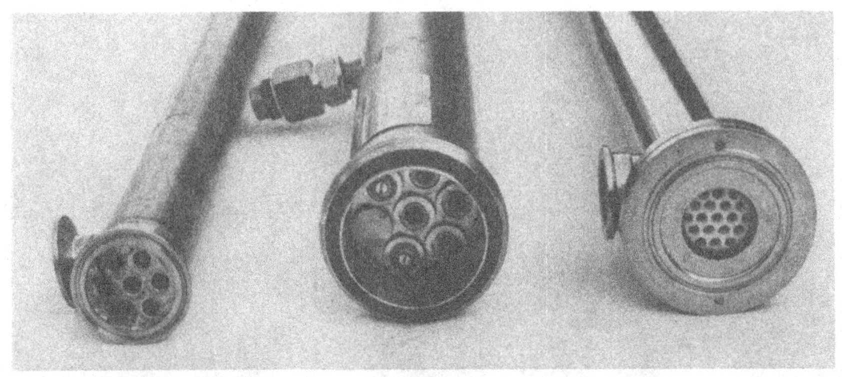

(a)

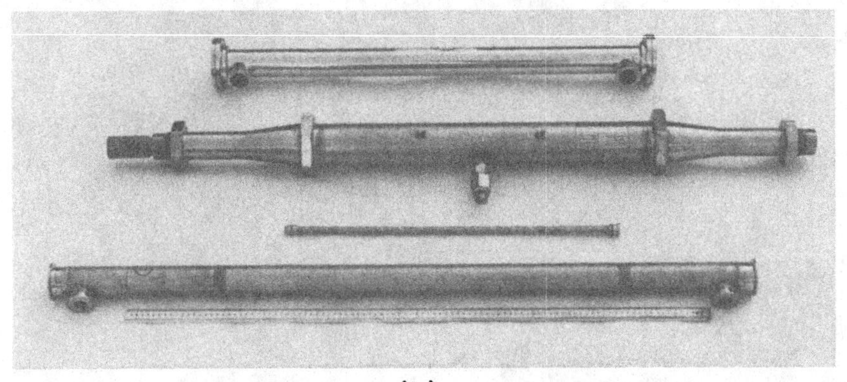

(b)

Figure 4: Crossflow Filter Modules
a) Pall, Microfiltrex, APV - left to right
b) Pall with metre rule, Microfiltrex with tube, APV - front to back

The Microfiltrex module was also provided in seven tube form, tube designation FM4. The membrane consisted of a layer of fine 316L stainless steel fibres sintered onto a porous substrate and welded into tubular form. The tubes were removable and employed rubber 'O' ring seals. The maximum working pressures were 3 and 10 bar (abs) on the outside and inside of the tubes respectively. The tube outer and inner diameters were 11 and 9.5 mm respectively. The tube length was 554 mm. This provided a tube inner area of 17.4×10^{-3} m^2. The tubes had a particle rating of 1 μm.

The APV Baker Ceraver module consisted of 19 tubular holes in a hexagonal block of alumina. The membrane was formed by sintering a layer of fine ceramic particles within the bores. The block was fitted within a stainless steel housing and sealed to it with an elastomer gasket. The maximum working pressure was 15 bar (abs). The tube inner diameter was 6 mm. The tube length was 850 mm. This provided a tube inner area of 16.0×10^{-3} m^2. The membrane had a particle rating of 1.4 μm.

Feed Slurries

These were chosen as simulants of wet radioactive wastes. The two main solid materials used were Dicalite Speedplus (a diatomaceous earth used as a filter precoat) and a mixture of 50 w/o magnesium hydroxide with 50 w/o slate dust (Magnox sludge waste). The magnesium hydroxide was a 50 w/o mixture of fine and coarse grades, as supplied by Pennine Darlington Magnesia Ltd, see Table 1. Use was also made of fine graphite powder, coarse sand and oil.

Name	Composition	Particle Size Range µm	Particle Size Median µm
Dicalite Speedplus	Diatomaceous earth	5-80	22
PDM/H3/200	$Mg(OH)_2$	1-10	3.5
PH10/200	$Mg(OH)_2$	1-50	-
Delafila 90/200	Slate powder	5-150	-

Table 1: Selected Properties of Main Solid Materials

Rig Operation

Feed slurries were made up to 5 w/o and circulated through the filter module at the desired flow rate with permeate valves shut. A throttling valve downstream of the module was adjusted to provide the desired pressure differential. This was quantified as the average transmembrane pressure, TMP, and defined as half the sum of the tube inlet and outlet pressures (gauge). Some readjustment of the restrictor was required during the run. The standard conditions for the study were a slurry velocity of 3.5 m s^{-1} and an average transmembrane pressure of 0.9 bar. Readings were made of the permeate flow rate, permeate turbidity, accumulated permeate volume, circuit pressures, slurry flow rate and slurry volume in the batch tank at various chosen time intervals in the range 10-30 minutes. The slurry concentration was determined by volumetric means during runs but periodic checks were made by testing samples by gravimetric and densitometric methods. Some dewatering runs lasted more than a working day in which case the crossflow circuit was drained down and left overnight in flushed condition. Runs were terminated either when target conditions were reached or when gross circuit blockage appeared imminent. Slurries were not re-used following dewatering runs due to potential particle attrition effects.

After operations were finished, the crossflow circuit was washed out and a clean water permeate rate measured over a period of about two minutes. The filter was then backwashed with either water or air/water and a further measurement of permeate rate made. If the rate was not restored to the clear condition, then the module was cleaned in 10% nitric acid, generally using recirculation for 0.5 - 1 h. Backwashing was also carried out during runs. The pressure in the crossflow circuit was reduced to near zero gauge and that in the backwash circuit raised briefly to about 0.5 bar(g). Backwashing lasted about 10 s.

RESULTS

Some 40 dewatering test runs, all of 2 - 20 h, were performed in this comparative evaluation. The major tests for each filtration module were the two series where 1 m^3 each of Dicalite Speedplus and Magnox slurries were dewatered.

In these runs the dewatering durations for both slurry types were some 10% less for the Microfiltrex as compared to Pall membranes. The programme commenced with measurement of clean water permeation rates in order to provide a performance baseline. Magnox sludge simulant and Dicalite Speedplus were thickened up to 50 w/o and 30 w/o respectively without difficulty. These levels are in excess of targets for further treatment. The general pattern for the permeate rate was an initial steep fall from the clean water rate over the first 10 - 20 minutes, followed by a near plateau region of gentle decline over many hours. A more rapid reduction was sometimes noted at high solids concentrations where significant viscosity increases occurred. The plateau rate value was independent of run length. Each run was characterised by a mean permeate rate derived from the plateau region. An example of data from a dewatering run is shown in Figure 5. Average transmembrane pressures were in the range 0.8 - 1.9 bar(g). Over all slurry materials and operating conditions, permeate fluxes were found to be in the range 30 - 400 x 10^{-6} m^3 m^{-2} s^{-1}. Averaged permeation fluxes for the three membranes and two slurry types are given in Table 2. The spread of rates within an average was up to -50% +100% but for one case (Microfiltrex-Magnox) was within +/- 22%. For the 1 m^3 runs only, the fluxes of the Microfiltrex membrane were found to be 10% greater than those of Pall for both slurry types.

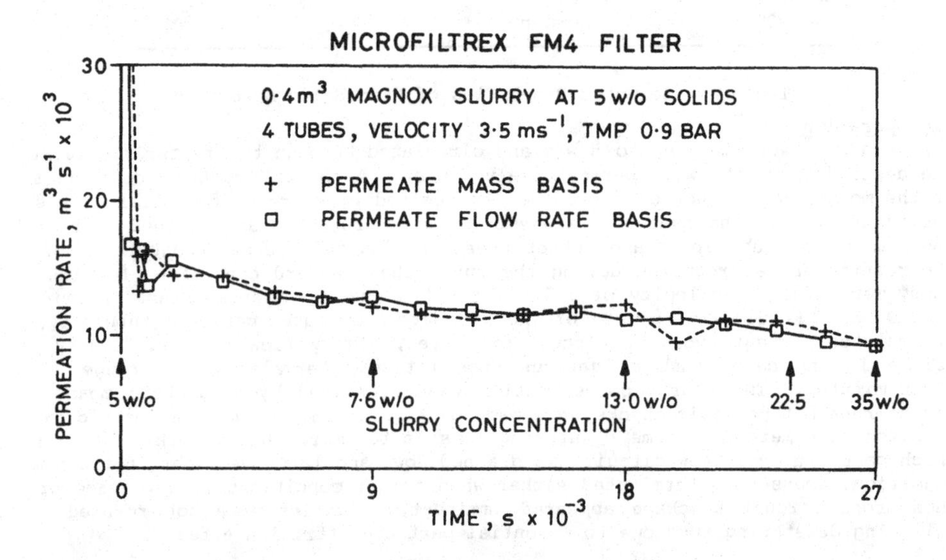

Figure 5: Permeate Rate Measurement

Some experimentation was made concerning the effects of water and air backwash during runs. Minor additions of oil and graphite powder had no significant effect on dewatering performance. For all three membrane types, permeate clarities of approximately 1 NTU (nephelometric turbidity unit) were generally observed, which equated to a solids content of magnitude 1 mg kg^{-1} (ppm).

Pall Tests

Clear water permeation rates were found to decrease within tens of minutes when using waters filtered to 10 μm and 0.2 μm. The decline was slower with unfiltered water. Chemical cleaning was required to restore the flux. A mean initial clean water permeation rate of 2.1 x 10^{-3} m^3 m^{-2} s^{-1} ± 15% at 2 bar(g) inlet pressure was measured.

Runs were generally performed at the standard condition (3.5 m s^{-1} and 0.9 bar(g)) and at 5 m s^{-1} and 1.8 bar(g). Within experimental variation there was no systematic difference of permeate rates between these two sets of conditions for a given slurry type. However, short term increases of rate were noted in standard runs following an increase of TMP. Comparative Magnox dewatering runs were performed both with and without water backwash during the test. Backwashing was for 10 s every 1800 s. The durations of dewatering to 35 w/o were very similar. The permeate flux with concentrated slurry was typically 10% of the clear water value. Backwashing after a run often restored the water flux to the 50% level. Acid recirculation was required to fully restore the clear water flux. No trend to permanent blocking (fouling) was detected after 80 h cumulative dewatering operation despite exposure to insoluble solids. No membrane wear was detectable, using a micrometer, after slurry circulation, despite 8 h flow of 1 mm particle size sand during dewatering tests.

Fairey Microfiltrex

A mean initial clear water permeation rate of 5.4 x 10^{-3} m^3 m^{-2} s^{-1} at 2 bar(g) inlet pressure was measured.

All dewatering runs were performed at the standard conditions. Permeate flux values are given in Table 2. Comparative Dicalite and Magnox slurry dewatering runs were performed both with and without water backwash during the tests. Backwashing was for 10 s every 600 s. The durations of dewatering to given concentrations were reduced by around 30% when using backwash. Compressed air was also used to effect backwash by both water and air/water mixtures. A smaller improvement in dewatering times was observed. Given the variability of dewatering rates, these observations require further confirmation for design validity. Following the completion of dewatering runs, water flushing and backwashing tended to restore the clear water flux but some acid cleaning was still required.

APV Baker

A mean initial clear water permeation rate of 550 x 10^{-6} m^3 m^{-2} s^{-1} at 2 bar(g) inlet pressure was measured.

Only five dewatering runs were performed. All were at the standard slurry velocity of 3.5 m s^{-1} which in the modified circuit effected a TMP of 1.1 bar(g). Permeate flux values are given in Table 2. Comparative Dicalite and Magnox slurry dewatering runs were performed with and without water backwash but results were not conclusive. Acid recirculation was required to restore clear water flux after dewatering runs. Cleaning was more difficult than with the other membranes, and stronger acid solutions were used (20% HNO_3).

	PALL	MICROFILTREX	APV
Magnox	170	180	150
(No Runs)	(11)	(8)	(3)
Dicalite	110	200	60
(No Runs)	(9)	(10)	(2)

Table 2: Average Dewatering Permeation Rates, m^3 m^{-2} s^{-1} x 10^6

THEORY

It has been argued that the marked reduction of pure solvent permeate flux by particulate suspensions during microfiltration is due to concentration polarisation at the membrane surface (Porter 1979, pp. 2.32-2.53; Zydney and Colton 1986). In such models, flux is related to shear rate at the wall which is in turn influenced by fluid rheology and flow type. To assist the quantification of operating conditions, the shear rate, the Reynolds number and the flow type have been evaluated. Using pressure drop, the shear viscosity has been calculated and compared to rheometric data. The further purpose of this work is to develop criteria for dewatering control.

For Magnox slurry and using measured viscosities, tube Reynolds numbers of 3600 and 260 were calculated for concentrations of 5 w/o and 25 w/o respectively at a slurry velocity of 3.5 m s^{-1} and tube diameter of 9 mm. The former indicated transitional/turbulent flow, for which there is no simple quantitative model for pressure drop. As thickening occurred, the flow became laminar. The wall shear rate was calculated as 3100 employing the conventional Newtonian and laminar flow model (Coulson and Richardson 1982, pp. 38-55)

Using measured pressure drops and the pipe friction chart method (Coulson and Richardson 1982, pp. 38-55), the following data were calculated, see Table 3.

Magnox Slurry Concentration w/o	25	35
Viscosity mPa s	66	100
Reynolds Number	780	540
Shear Rate s^{-1}	4400	4700

Table 3: Predicted Viscosity and Flow Data

These viscosity values are of similar magnitude to but rather lower than those measured using commercial viscometers. With improved instrumentation, tube length variation and calibration with known fluids it is likely that the crossflow rig could provide accurate viscosity determinations on simulants. The Reynolds number values confirm laminar flow for thickened material and the shear rate values are in good agreement with the theoretical value derived above. Due to the expectation of transitional flow, values were not predicted for the 5 w/o case.

Similarily, for viscous materials in streamline crossflow conditions, the Poiseuille equation (Coulson and Richardson 1982, pp. 38-55) is applicable and predicts a viscosity of 50 mPa s for a pressure drop of 1 bar. However there is a poor correspondence between the gradient of the Poiseuille equation, 2×10^6 s m^{-1}, and that of an experimental plot of pressure drop taken to low viscosities. This is considered to reflect the change of flow type as well as the possible influence of entry and exit effects.

CONCLUSIONS

(1) Using crossflow microfiltration, the two main slurry types were dewatered to values of 50 w/o and 30 w/o which was in excess of target. Permeate solids content were of magnitude 1 mg kg^{-1} and were considered good.

(2) Average permeate fluxes for a given slurry with the three membrane types were very similar, particularly so for Magnox sludge simulant. At 1 bar TMP the magnitude was $100 \times 10^{-6}\ m^3\ m^{-2}\ s^{-1}$. This may be governed by slurry properties since differences were noted in the clear water permeation rates of the membranes.

(3) No dependence of permeate flux on TMP and slurry velocity was detected between dewatering runs at the two sets of conditions used and within the experimental variability found. Permeate fluxes were considered to be close to the asymptotic value.

(4) Recirculatory acid cleaning was required for all three membranes to maintain clear water flux values. No trend to irreversible fouling was detected after operational periods up to the order of 10^2 h. No membrane wear was detectable after slurry circulation.

(5) Backwashing during dewatering runs may have some limited scope for reducing batch times but was not necessary for runs of several tens of hours duration.

(6) Predictions of shear viscosity from pressure drop data indicate potential application for control and instrumentation purposes.

REFERENCES

Coulson, J. M. and Richardson, J. F. (1982) Chemical Engineering, volume 1, third edition, Pergamon Press, Oxford, pp. 38-55

Porter, M. C. (1979) Membrane Filtration, Section 2.1, Handbook of Separation Techniques for Chemical Engineers, Schweitzer P. A. (editor), McGraw-Hill, New York, pp. 2.32-2.53

Zydney, A. L. and Colton, C. K. (1986) 'A Concentration Polarization Model for the Filtrate Flux in Crossflow Microfiltration of Particulate Suspensions', Chemical Engineering Communications, vol. 47, pp. 1-21. A

ACKNOWLEDGEMENTS

This paper is published by permission of Nuclear Electric plc. We gratefully acknowledge assistance from APV Baker, Fairey Microfiltrex Ltd. and Pall Process Filtration Ltd. during the tests. We thank Messrs A. S. D. Willis and R. D. Empsall of Nuclear Electric, Barnwood for many helpful discussions during the course of this work.

THE USE OF REVERSE OSMOSIS FOR THE PURIFICATION OF COAL GASIFICATION LIQUORS.

A.R. Williams, British Gas plc., Westfield Development Centre, Fife.

SUMMARY

Laboratory trials have been conducted at the Westfield Development Centre to assess the potential of reverse osmosis as a stage in the treatment of coal gasification waste liquors arising from operation of the BGL Slagging Gasifier. The experiments were performed using feed streams from a pilot plant. This plant comprises solvent extraction, dissolved gas stripping, biological oxidation and activated carbon treatment. This process renders the effluent suitable for sewer or estuary discharge. Reverse osmosis was identified as a means of lowering the chloride and trace metal levels in this effluent so making it suitable for river discharge or re-use as process water.

Experiments have taken feeds from various points of the treatment process and the quality and quantity of permeate obtained at a range of temperatures, operating pressures, pH and volumetric concentration factors observed. The trials utilised a tubular laboratory reverse osmosis unit and thin film composite membranes. Trials were conducted on both a batch basis, to observe the effect of process parameters, and on a long term basis, to identify operational problems such as membrane fouling.

Results have established the feasibility of the use of reverse osmosis in the treatment process and have established operating conditions for further pilot plant scale work. Optimal positioning of the reverse osmosis stage within the treatment process has also been determined. Component rejections have been consistently good and a final effluent of river discharge standard produced.

1.0 INTRODUCTION

The Westfield Development Centre at Cardenden, Fife, is currently used by British Gas plc for the commercial scale development of the British Gas / Lurgi (BGL) Slagging Gasifier. Work at Westfield is aimed at demonstrating a process for the production of Substitute Natural Gas (SNG) from coal but also includes work on the application of the BGL Gasifier to Integrated Gasification Combined Cycle (IGCC) power plants.

A by-product of the BGL gasification process is an ammoniacal liquor. This liquor is a combination of the aqueous condensates collected during cooling of the product gases and arises from moisture present in and on the feed coal and excess steam added to drive the gasification reaction. In addition to ammonia, the liquor contains sulphides, cyanides, a wide range of phenols, chloride and various other organic and inorganic pollutants. Table 1 gives analyses of liquor resulting from British and American coals. Liquor production is equivalent to 20 tonnes per 100 tonnes of coal gasified. It is required to process this liquor to a standard that makes it suitable for discharge or for reuse in the gasification plant.

1.1 Treatment of the Liquor

The liquor is stripped of phenols by solvent extraction. The stripped phenols are recovered and the solvent regenerated by distillation. The dephenolated liquor then undergoes two stage steam stripping. The first stripper removes the dissolved gases ("free" ammonia, hydrogen sulphide, hydrogen cyanide and carbon dioxide). sodium hydroxide is then added to the liquor to "free" fixed ammonia which is then removed in the second stripper. The gases from both strippers are passed to a Claus plant for oxidation and sulphur recovery. The deammoniated liquor is cooled, neutralised with sulphuric acid, and passed to a biological oxidation vessel. After clarification the biologically treated liquor passes through a bed of activated carbon which reduces the remaining Chemical oxygen demand (COD) and trace phenols. This process route has been widely used for treating coal and other phenolic liquors.

Effective Industrial Membrane Processes — Benefits and Opportunities, pp.115–126

TABLE 1 Typical Liquor Compositions

All units (except pH) are in mg/l.

Component	Markham coal liquor	Pittsburgh coal liquor	Illinois coal liquor
Free ammonia	710	590	2500
Fixed Ammonia	4000	7400	4700
Nitrate	92	67	220
Cyanide	3	60	<1
Thiocyanate	340	870	780
Sulphide	270	3200	1100
Sulphate	36	130	55
Total phenol	5000	5500	7600
TOC	5500	5600	N/A
COD	20000	21000	28000
Suspended solids	74	100	36
Chloride	4200	1400	1900
pH	9.1	9.0	9.0

2.0 THE EFFLUENT TREATMENT PILOT PLANT

A simplified flow diagram of the effluent treatment pilot plant is presented as figure 1. Liquor collected during BGL Gasifier trials is stored in the buffer storage tank under a nitrogen blanket to avoid oxidation of certain components. The plant is a single process stream capable of treating up to 60 l/hr of effluent. The solvent used for dephenolation is methyl iso-butyl ketone. Typical effluent compositions through the plant are detailed in table 2. This is based on liquor arising from Pittsburgh coal.

The high level of COD present immediately following solvent extraction is due to the liquors being saturated with solvent. This solvent is stripped off in the first stripping column, condensed out of the vapours and returned to the solvent distillation column. The final effluent is suitable for sewer or estuary discharge, but unsuited for river discharge owing to the high levels of chloride, sulphate (arising from the neutralisation with sulphuric acid prior to biological oxidation) and total organic carbon (TOC) present. In the future it is likely that discharge limits will tighten and water costs increase. Reverse osmosis was identified as a means by which the final effluent could be rendered suitable for river discharge or on site reuse as process water.

3.0 EXPERIMENTAL PROCEDURE

3.1 Reverse Osmosis

A semi-permeable membrane is one through which only the smallest molecules in a solution, usually the solvent molecules, will pass. When solutions of different concentrations are separated by such a membrane, solvent (usually water), flows from the more dilute solution to the more concentrated side. If allowed to reach equilibrium the flow will eventually produce a head of liquid equal to the osmotic pressure difference between the solutions. This effect can be reversed by the application of a physical pressure exceeding the osmotic pressure difference to the more concentrated solution, further concentrating it and producing a very dilute permeate.

Figure 1 The Effluent Treatment Pilot Plant

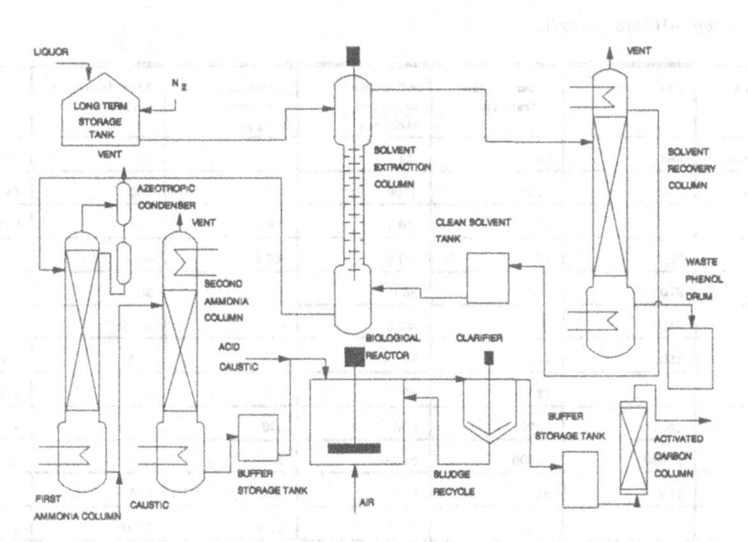

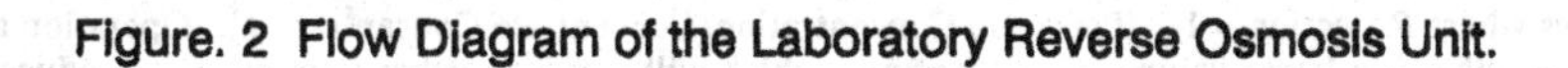

Figure. 2 Flow Diagram of the Laboratory Reverse Osmosis Unit.

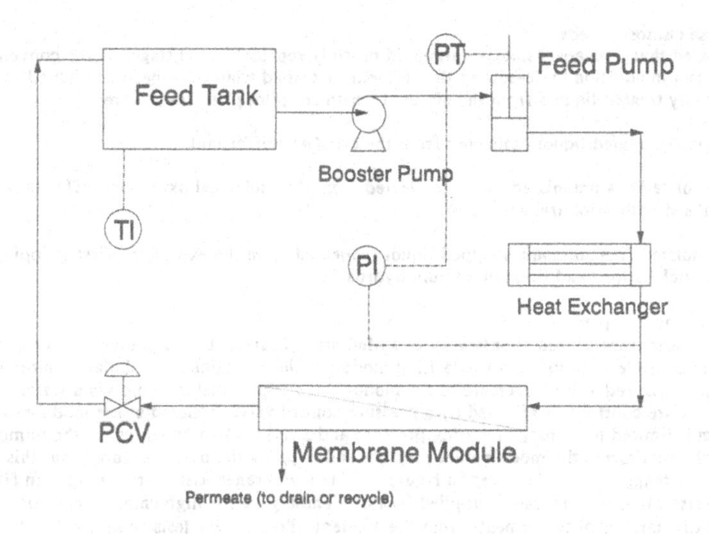

TABLE 2 Liquor Compositions During the Effluent Treatment Process

All units (except pH) are in mg/l.

Component	Feed liquor	After solvent extraction	After free ammonia stripping	After fixed ammonia stripping	After Biological oxidation	Final effluent
Free ammonia	590	650	<1	5	<1	<1
Fixed ammonia	7400	6800	1200	<1	100	54
Nitrate	67	72	<0.5	<0.5	760	820
Cyanide	75	70	<0.5	<0.5	<0.5	<0.5
Thiocyanate	870	840	880	740	23	<2.5
Sulphide	3200	2600	<0.5	8	<0.5	<0.5
Sulphate	130	130	52	190	7600	7200
Total phenol	5500	180	49	12	3	<0.5
TOC	5600	2200	990	880	220	40
COD	21000	48000	3500	2500	760	200
Susp. solids	100	33	980	380	470	170
Chloride	1400	1300	1300	1200	1100	1200
pH	9.0	9.1	4.8	13.2	6.8	7.8

Water removal performance of an reverse osmosis unit is often described in terms of volumetric concentration factors (VCFs), defined as the ratio of the feed flowrate to the concentrate flowrate (or feed volume : concentrate volume for a batch operation). The efficiency of separation is expressed as rejections of species where Rejection = 1 - (Permeate Concentration/Concentrate Concentration). Operation at high VCF is desirable as a lower volume of concentrate, which will require further treatment, is produced.

3.2 Reverse Osmosis Feeds

It was anticipated that a reverse osmosis unit could possibly replace several stages of the conventional treatment process in addition to de-salting the effluent. Detailed trials have been conducted with three different partially treated liquors from the effluent treatment pilot plant these were:

(i) Biologically treated liquor (collected from the clarifier buffer tank).

(ii) Dephenolated deammoniated liquor (collected from the biological oxidation buffer tank and neutralised with sulphuric acid to pH 7).

(iii) Dephenolated free ammonia stripped liquor (collected from the exit of the first stripping column, immediately prior to addition of sodium hydroxide).

3.3 Experimental Apparatus

A laboratory reverse osmosis unit supplied by PCI Membrane Systems Ltd. was used for the experiments. This was a tubular system mounting a single 1.2m module which contained 18 tubular membranes arranged in series. Liquor was fed at high pressure to the module by a triplex piston pump via a water cooled heat exchanger. Pressure control was effected by a pressure control valve mounted at the module exit. The PCI tubular system is limited to 70 barg operating pressure and a relief valve mounted on the pump and a pressure switch, monitoring the module inlet pressure and tripping the pump, ensured that this was not exceeded. The arrangement is illustrated in figure 2. The membranes used were a tight thin film composite reverse osmosis membranes supplied by PCI. These gave the high chloride rejection necessary to produce river discharge quality permeate from the effluent. Preliminary tests conducted by PCI on samples of liquor from the BGL Gasifier had shown these to be suitable for such work.

3.4 Membrane Cleaning

Cleaning of reverse osmosis membranes is necessary to maintain the permeate flux and to avoid membrane damage caused by the build up of foulants on and in the membrane. The cleaning cycle used in the experimental programme was as follows:

(i) Flush with process water at minimum operating pressure.

(ii) Measure the permeate flux obtained operating with process water at an operating pressure of 40 barg and a temperature of 15•C.

(iii) Rinse for half an hour with a dilute solution of a proprietary alkali detergent mixture at minimum pressure at 50•C.

(iv) Flush the membranes with water.

(v) Rinse for half an hour with a dilute solution of nitric acid at minimum pressure and 50•C.

(vi) Flush the membranes with process water and measure the permeate flux under the same conditions as (ii). The two measurements of permeate flux give an indication of the efficacy of the cleaning cycle. Comparison with previous water flux measurements indicates any decline in membrane performance during the trials.

This cleaning cycle was conducted at the conclusion of each batch concentration trial, after each cycle in the recycle trials and every two days during the continuous operation trials.

3.5 Objectives and Tests Conducted

The aims of the trials were to ascertain the usefulness of reverse osmosis as a treatment method for coal liquor and to identify the best source of feed liquor (within the effluent treatment pilot plant) for the reverse osmosis unit. Further tests were conducted to discover the optimum reverse osmosis operating conditions, to gain the experience necessary for design of a commercial scale reverse osmosis plant and to identify any problems that might arise in long term operation of treatment plant.

3.5.1 Batch Concentration Trials To observe the effects of increasing feed concentration upon permeate flux and permeate quality, batch concentration tests were carried out. The feed tank was filled with approximately 200 litres of the liquor and the reverse osmosis unit operated in a total recycle fashion to allow the membrane flux to steady and the liquor to reach the desired test temperature. The latter was achieved by balancing the heating action of the pump and the water flow (hot or cold) to the heat exchanger. Concentration of the liquor then commenced, the permeate being diverted to drain. The liquor volume in the tank, permeate flux and permeate quality were measured hourly and samples of concentrate and permeate taken at suitable intervals.

3.5.2 Continuous Operation Trials These experiments were conducted whilst the effluent treatment pilot plant was operating, providing a steady supply of liquor. They were intended to simulate continuous operation of a reverse osmosis pilot plant and to provide data on long term operation at high VCF.

The unit was started in batch concentration mode to produce a VCF 6 (or in one case VCF 12) concentrate. A cleaning cycle was then conducted before continuous operation commenced. Fresh liquor was fed, using a peristaltic pump, to the feed tank, the permeate sent to drain and concentrate was allowed to overflow out of the feed tank. The liquor feed was initially set at VCF/(VCF−1) times the permeate flow to maintain the VCF of the concentrate. Operational parameters were monitored on an hourly basis and samples of feed liquor, permeate and concentrate taken every twelve hours. The trials were conducted for as long as liquor supplies allowed with membrane cleaning cycles being conducted as appropriate. Two modes of reverse osmosis operation were used. In the first the feed liquor flow was decreased as the permeate flux declined so as to maintain the VCF of the concentrate. In the second the feed liquor flow was left at the initial value throughout each two day cycle. This meant that the concentration factor decreased as the permeate flux declined (since VCF = Feed/(Feed − Permeate Flux)), but maintained permeate quality and is more likely to be the operating procedure on a full scale plant. Before starting the next cycle it was necessary to concentrate the feed liquor to re-establish the initial VCF.

3.6 Analysis

The following parameters were monitored during the trials, usually at hourly periods, but more frequently at the start of the trial:

Operating pressure
Feed temperature
Permeate flow
Feed flow (for continuous operation trials)
Level in feed tank (for concentration trials)
Total dissolved solids (TDS) concentration in permeate
TDS concentration in feed

Samples of the permeate, concentrate and feed (for continuous trials) were taken at suitable times during trials and analysed for total ammonia, sulphate, chloride, thiocyanate, TDS and chemical oxygen demand (COD). During the continuous trials composite samples were made up during each operating cycle and underwent a more detailed analysis covering 27 species. The results of these analyses were used to calculate rejections of species and to monitor performance of the reverse osmosis unit during operation.

4.0 RESULTS

4.1 Batch Concentration Tests

The results from a selection of the batch concentration trials conducted upon liquors arising from Pittsburgh and Markham coals are summarised in table 3.

The best liquor concentration results were obtained treating the low chloride coal liquors, and in particular the dephenolated free ammonia stripped liquor. The higher dissolved solids content of the liquors collected further down the pilot plant treatment route is caused by the addition of sodium hydroxide solution to "free" the fixed ammonia and the later addition of sulphuric acid to neutralise this solution. This lower dissolved solids concentration means that the solution has a lower osmotic pressure allowing it to be concentrated further and giving higher permeate fluxes. The dephenolated free ammonia stripped liquor was concentrated up to VCF 10 whilst similar biologically oxidised and dephenolated deammoniated liquors were only concentrated up to VCF 7.

The liquors arising from high chloride coal showed less variation and all could be concentrated to at most VCF 7. The high concentration of chloride swamped the differences in concentration of other species.

TABLE 3 Typical Permeate Fluxes Observed During Batch Reverse Osmosis Trials

Liquor type	Component concentrations in feed liquor (mg/l)			Permeate flux (l/m^{2}hr)		
	Chloride	Sulphate	TDS	At VCF 1	At VCF 3	At VCF 6
Low chloride coal						
Biologically oxidised	1600	11000	12000	35	21	10
Dephenolated deammoniated	1100	4100	11000	33	18	9
Dephenolated free ammonia stripped	1300	59	3700	35	22	14
High chloride coal						
Biologically oxidised	4900	7400	21000	40	12	0.9
Dephenolated free ammonia stripped	7700	69	21000	35	8	0.8

The quality of the permeates produced during typical batch trials is summarised in table 4. The dephenolated free ammonia stripped liquor permeate can be seen to have lower levels of sulphate than the other liquors and a lower overall concentration of dissolved solids (TDS) but to have higher levels of thiocyanate, ammonia and total organic carbon (TOC) and a higher overall chemical oxygen demand (COD). The permeates produced from the biologically oxidised liquors showed the lowest levels of thiocyanate and COD. The permeate produced from high chloride coal liquor is markedly inferior to that arising from low chloride liquors even when operating at a lower VCF.

Further batch trials were conducted investigating the effects upon permeate flux and permeate quality of the feed pH and operating temperature. Increasing temperature was found to increase the permeate flux in accordance with the temperature correction data supplied by PCI for the membranes. Tests varying the feed pH by adding alkali or acid to the feed liquor revealed that optimum permeate quality occurred at pH 5. This is close to the normal pH of the dephenolated free ammonia stripped liquor.

TABLE 4 Typical Permeate Quality Observed During Batch Reverse Osmosis Trials

Liquor type	Component concentration in permeate (mg/l)					
	Ammonia	Thiocyanate	Sulphate	Chloride	TDS	COD
Low chloride coal (permeate at VCF 6)						
Biologically oxidised	13	0	360	50	1300	15
Dephenolated deammoniated	4	30	13	11	190	50
Dephenolated free ammonia stripped	80	70	8	20	470	350
High chloride coal (permeate at VCF 3)						
Biologically oxidised	30	5	16	120	1130	7
Dephenolated free ammonia stripped	190	110	1	140	920	1198

4.2 Continuous Operation

Continuous operation trials were conducted with biologically oxidised and dephenolated free ammonia stripped liquors, these having been identified in the batch trials as producing the best permeate and being most easily concentrated respectively. The trials were conducted with liquors from a low chloride coal (Illinois 6), at an operating pressure of 60 barg and an operating temperature of 20°C. The low temperature used avoided the reverse osmosis units demand for liquor exceeding the supplies available. Figures 3 and 4 are plots of the permeate flux versus VCF for dephenolated free ammonia stripped liquor and biologically oxidised liquor respectively during the batch concentration of the liquors prior to continuous operation. It can be seen that significantly higher permeate flowrates are achieved for the free ammonia stripped liquor and that a far higher VCF was reached.

Continuous trials for the free ammonia stripped liquor were initially conducted at VCF 12 but the permeate produced contained high levels of ammonia and thiocyanate. Trials were continued at VCF 6, where a permeate of far higher quality was produced. Figure 5 plots the permeate flux during the two week trial period and figure 6 the levels of various species in the permeate. The permeate flux was between 15 and 25 l/m^2hr throughout and the species rejections were nearly all above 90% when based upon the feed concentration and nearly all above 95% when based upon the concentrate present in the system. The initial cycle was conducted at a constant VCF, whilst subsequent cycles were conducted with a constant feed rate.

The continuous treatment of biologically oxidised liquor was also conducted at a VCF of 6. The first cycle produced unusual results; the flux did not appear to decline, in spite of the fouling nature of the feed and the permeate produced was of low quality. Replacing the membranes solved the problem and the second period of operation produced a high quality permeate at a flowrate only slightly lower than that seen in the treatment of dephenolated free ammonia stripped liquor. The permeate flux and the concentrations of various species in the permeate throughout the whole period of operation are presented in figures 7 and 8 respectively.

Figure 3.

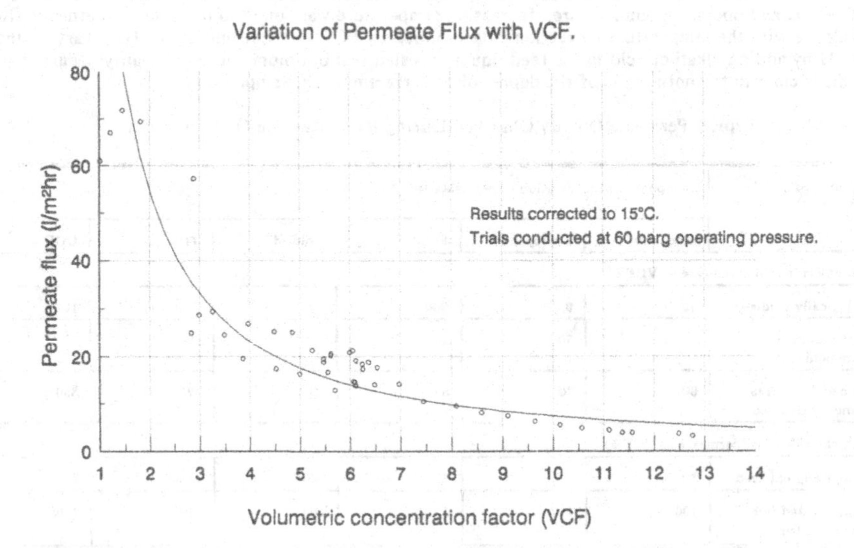
Batch Concentration of Dephenolated Free Ammonia
Stripped Illinois Coal Wastewater.
Variation of Permeate Flux with VCF.
Permeate flux (l/m²hr)
Results corrected to 15°C.
Trials conducted at 60 barg operating pressure.
Volumetric concentration factor (VCF)

Figure 4.

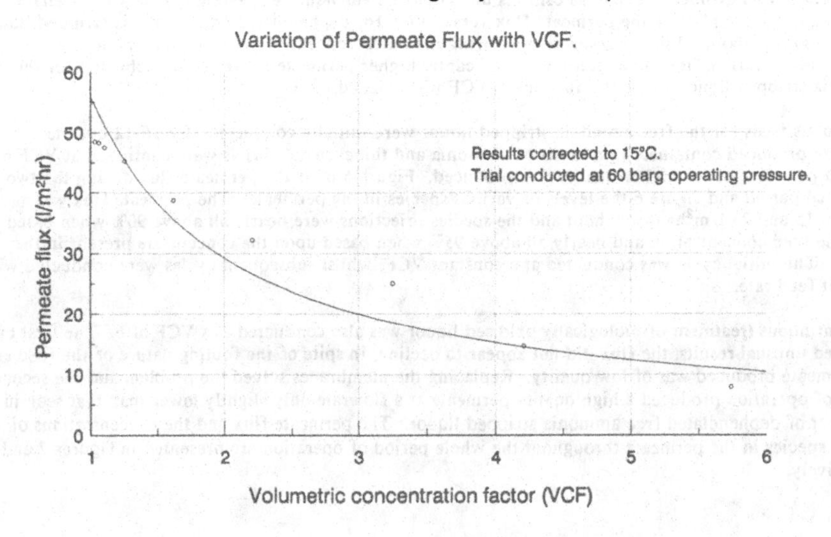
Batch Concentration of Biologically Oxidised Liquor.
Variation of Permeate Flux with VCF.
Permeate flux (l/m²hr)
Results corrected to 15°C.
Trial conducted at 60 barg operating pressure.
Volumetric concentration factor (VCF)

Figure 5 **Variation of Permeate Flux with Time during Continuous RO Treatment of De-phenolated Free Ammonia Stripped Illinois Coal Wastewater.**

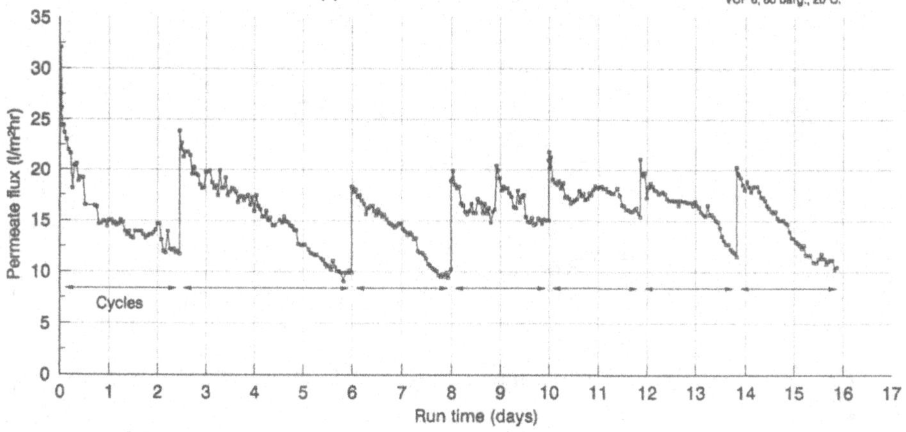

VCF 6, 60 barg., 20°C.

Permeate flux (l/m²hr)

35 30 25 20 15 10 5 0

Cycles

0 1 2 3 4 5 6 7 8 9 10 11 12 13 14 15 16 17

Run time (days)

Figure 6. **Permeate Quality during Continuous Concentration of Dephenolated Free Ammonia Stripped Liquor.**

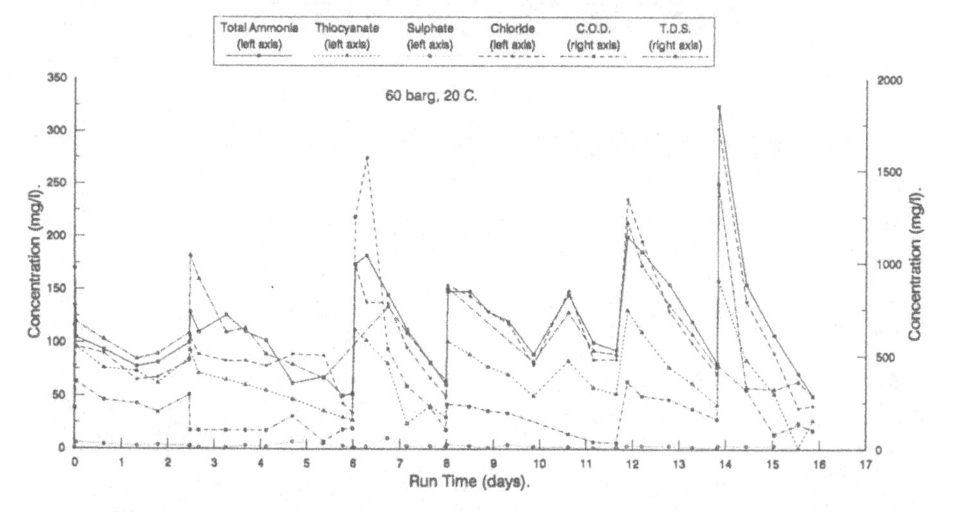

Figure 7 **Variation of Permeate Flux with Time during Continuous RO Treatment of Biologically Oxidised Illinois Coal Liquor.**

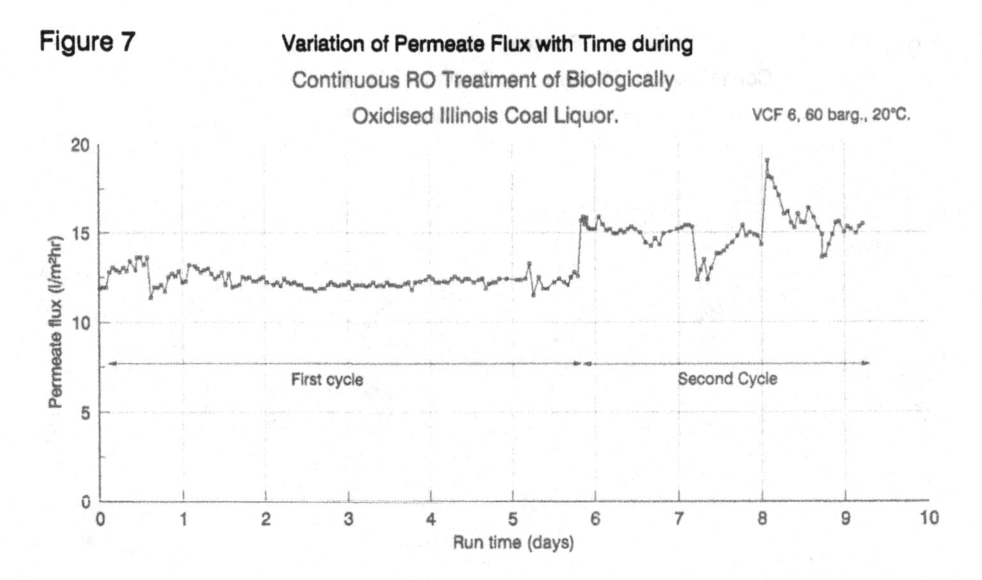

Figure 8. **Permeate Quality during Continuous Concentration of Biologically Oxidised Liquor.**

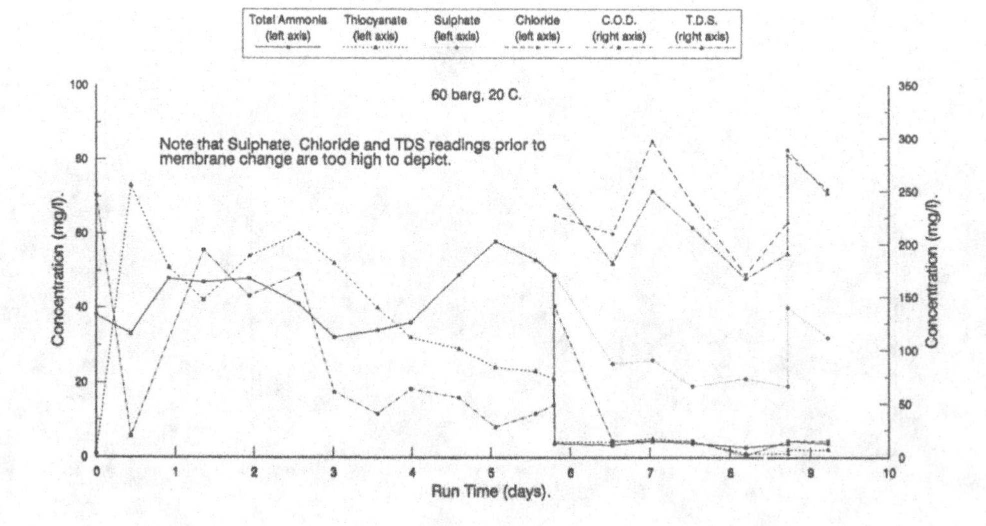

5.0 DISCUSSION

5.1 Comparison with Discharge Criteria

Typical discharge consent limits (Ebbins & Ruhl 1988) are listed in table 5 along with typical levels of pollutants present in the permeates produced during the continuous treatment trials. The permeate produced from biologically oxidised liquor can be seen to meet all the river discharge limits whilst the permeate arising from free ammonia stripped liquor has overly high levels of ammonia and thiocyanate and as a result also fails the criteria for COD.

It is perhaps unwise to compare the permeate with a particular set of existing discharge limits as limits vary from place to place and are almost certain to become more stringent in the future. A more useful measure of permeate quality is whether the permeate could be re-used on site as cooling or process water. Since the requirements for process water might well be less stringent this could save on effluent treatment as well as water costs. In the case of these trials, no problems are envisaged in the use of either the biologically oxidised or free ammonia stripped permeates as process water.

5.2 Membrane Fouling and Damage

The high levels of suspended solids present in the feeds were not found to cause any problems in operation of the reverse osmosis unit, a two day cleaning cycle being typical for tubular reverse osmosis systems. Water flux measurements before and after membrane cleaning gave no indication of any permanent membrane fouling.

The failure of one set of membranes during the continuous treatment of the biologically oxidised liquor is thought to have been caused by presence of free chlorine in the supply water used to make up the cleaning solutions. In a full size plant permeate would be used for this purpose and such a problem would not occur.

TABLE 5 Comparison of Permeates Produced with Discharge Criteria

Component	Discharge consent limits (mg/l except pH)			Level in permeate (mg/l except pH)	
	Sewer	Estuary	River	Biologically oxidised	Dephenolated free ammonia stripped
Total ammonia	1500	100	5	4	150
Sulphide	10	-	0.5	<1	50
Sulphate	1200		-	28	2
Cyanide	5	0.5	0.1	<1	<1
Thiocyanate	30	5	1	<1	84
Total phenols	50	20	0.5	<1	<1
Chloride	-	-	200	67	150
COD	600	500	20	<15	260
pH	6-10	5-9	5-9	7.6	8

5.3 The Overall Process

The addition of a final reverse osmosis stage to the conventional treatment process for Illinois coal liquor upgrades the effluent to river discharge quality. Such a process would be extremely robust to process upsets: Partial failure of the biological treatment would be recovered by the reverse osmosis unit. In a similar manner the presence of the reverse osmosis unit would add flexibility to start up of the biological reactors.

Replacement of the second stripping stage, biological oxidation vessel and activated carbon treatment stages of the effluent plant with a reverse osmosis plant produces an effluent which although not of river discharge quality is well suited to reuse as process or cooling water. To produce a river quality effluent the permeate could be subjected to a second stage of reverse osmosis treatment. reverse osmosis treatment of permeate has not been tested experimentally but conservative estimates indicate that a two stage reverse osmosis plant to treat dephenolated free ammonia stripped liquor would be smaller than the single stage plant required to treat a similar volume of biologically oxidised liquor. A further advantage of using the feed directly after free ammonia stripping is that it can be supplied to the reverse osmosis unit at high temperature from the outlet of the stripping column. This would increase the permeate flux. Further investigation is required in this respect since higher operating temperatures can reduce membrane life. The elimination of the biological oxidation unit results in considerable cost savings and such a plant would probably be cheaper to operate than a conventional plant without a reverse osmosis stage. Due to the lower power requirements of reverse osmosis, both treatment routes have considerably lower running costs than comparable incineration plants.

Results of trials treating high chloride coal liquors are less encouraging. Because of the lower permeate fluxes observed a reverse osmosis plant to treat the liquor produced from Markham coal would need to be twice the size to achieve a similar water recovery and would produce a much lower quality effluent. To produce a permeate of high quality from such a liquor in a single stage reverse osmosis plant would require operation at 50%-65% water recovery. This would result in a large volume of concentrate requiring further treatment. Alternatively a two stage reverse osmosis plant, as described above for the low chloride dephenolated free ammonia stripped liquor, could be used.

6.0 CONCLUSIONS

Liquors produced during gasification of coal in the BGL Gasifier can be treated with a combination of conventional methods and reverse osmosis to produce a final effluent which is suitable for re-use within the plant or discharge to an inland waterway. A simplified treatment route comprising dephenolation, free ammonia stripping and reverse osmosis produces a permeate which is suitable for re-use within the gasification complex.

The chloride content of the coal and hence the concentration in the liquor has an important bearing upon the performance of the reverse osmosis unit. 80% water recovery can be achieved when treating liquors arising from the gasification of low chloride coals, such as Pittsburgh 8 or Illinois 6. Operation using a high chloride Markham coal would necessitate either a larger, two stage, reverse osmosis plant producing a high quality effluent at 80% water recovery or a similar sized plant producing a good quality effluent at only 65% water recovery.

7.0 REFERENCES

Ahrabi, F. (1988) "Purification of Aqueous Liquors from the BGL Slagging Gasifier", Paper presented at the Institute of Gas Engineers Joint Meeting with the Pipeline Industries Guild, 20th September, 1988, Solihull.

Ebbins, J.R. & Ruhl, E. (1988) "The BGL Gasifier: Recent Environmental Results", Paper presented at the Eighth Annual EPRI Conference on Coal Gasification, October 19-20th, 1988, Palo Alto, California.

8.0 ACKNOWLEDGMENT

The author would like to thank the directors of British Gas plc for their permission to publish this paper. The views expressed are my own and do not necessarily reflect the views of British Gas plc.

SESSION D:

WASTE WATER TREATMENT II

A fresh approach to sludge dewatering using tubular membrane pressing technology

S B Tuckwell and B A Carroll[1], (Wessex Water Business Services Ltd.)
and
J B Joseph, (Exxpress Technology Ltd.)

SUMMARY

The traditional disposal routes for sludges from water and sewage works are becoming expensive or unavailable and the satisfactory disposal of sludges is an increasing problem. A woven polyester textile tube forms the heart of a fresh approach to dewatering sludges, the Exxpress process. It has successfully dewatered sludges from dissolved air flotation and from sedimentation processes used for drinking water treatment, producing a firm cake with 10 to 17% w/w dry solids which lost further water by air drying. Sludges from ferric chloride coagulant were less robust than from ferric or aluminium sulphates, leading to poorer quality permeates from the dewatering process. The process performed well on a variety of sludge feed concentrations, at operating pressures around 250 kPa and with cycle times of less than 25 minutes. It represents an economic alternative to the traditional methods of dewatering by pressing, centrifuging or lagooning.

INTRODUCTION

The treatment of sewage and the chemical coagulation of surface waters for the production of drinking water both produce sludges for disposal. The satisfactory disposal of sludges is a problem experienced in several countries. Concern about disposal of water treatment sludge in Belgium led to a survey which found that the largest water suppliers annually produced sludge with more than 10,000 tonnes dry solids (Aerts and Maes, 1986). Half of this was dumped at public waste disposal sites; the other half remained at the waterworks premises. A study of the production and disposal of sludge in the Netherlands (Koppers, 1985) estimated that for half the 30,000 tonnes per year of dry solids produced there were no treatment methods other than drying beds or lagoons at the waterworks. In particular smaller treatment works were identified as needing more efficient treatment methods.

In the U.K. there is an estimated 30 million tonnes of sewage sludge (1.2 million tonnes of dry solids) for disposal each year (IWEM, 1989). At present about 30% of sewage sludge is disposed of at sea but the Secretary of State for the Environment has given an undertaking that this will cease by 1998, possibly resulting in additional demand for disposal by landfill.

U.K. water supply works were estimated by Warden and Craft (1980) to produce liquid sludge effluents containing 182,000 tonnes of dry solids per year. Although 37% of this was disposed of as solid waste, another 36% was apparently discharged to lagoons and left to accumulate - merely postponing the question of its disposal. At many works supernatant from the dewatering of lagoons is discharged to adjacent water courses and simple settlement is often inadequate to achieve the conditions for the discharge consent. Tightening of the control of these discharges by the regulators has led to a re-examination of sludge disposal practices.

---

1 Now the Technical Director of Biotreatment Ltd.

Effective Industrial Membrane Processes — Benefits and Opportunities, pp.129–141

Warden and Craft found that 48% of the supernatant water recovered from the treatment of sludges was returned to works inlets and 15% was returned to raw water sources. The polyelectrolytes commonly used in sludge treatment are not approved for drinking water treatment, making the recovered water unacceptable for recycling. The recycling of water also needs to be reconsidered in view of the concern expressed by the Expert Group on *Cryptosporidium* in Water Supplies (Badenoch, 1990) about the recycling of *Cryptosporidium* oocysts. If these are present in water recovered from filter washing and sludge treatment, such waters will require adequate treatment to prevent the recycling of viable oocysts or they will have to be disposed of safely elsewhere. The recovery and re-use of coagulants has been carried out at full scale (Bishop *et al.*, 1987), but there is the potential problem of the interference of coagulant aids in the recovery.

Changes in the organisation and control of hazardous waste disposal in England and Wales will take place when the Environmental Protection Act is implemented. The disposal of wastes will be carried out by companies owned by local authorities but regulated by the county authorities. A consequence of the increased demand on disposal sites and their altered management is expected to be an increase in the cost of disposal. One expert has suggested that the cost of landfill disposal may treble in the next two to three years (Hawkins, 1990). Although sewage and waterworks sludges are not classified as hazardous wastes for landfill purposes, it is likely there will be a consequential increase in their disposal costs.

These points all lead to the conclusion that an efficient method of dewatering waterworks or sewage sludges would find ready application. The traditional techniques for sludge dewatering rely on the use of barrier filtration involving, for example, large plate and frame presses. More recently centrifugation has been introduced, requiring plant which is expensive to buy and install, is demanding of energy and can be complex to run. This paper describes the use of the *Exxpress* process - a tubular woven membrane press capable of dewatering sludges at low cost, without the use of polyelectrolytes or other chemical aids.

DESCRIPTION OF THE *EXXPRESS* PROCESS

The essential part of the Exxpress process is the array of flexible woven polyester tubes, typically of 25mm diameter, which forms the heart of the dewatering unit. Up to thirty tubes can be woven in parallel into a single piece of fabric, with the array suspended vertically from the top tube, the axis of each tube being horizontal. The ends of the individual tubes are connected to inlet and outlet manifolds (Figure 1). Sludge from the feed tank is pumped to the inlet manifold. The release of sludge from the outlet manifold into the discharge system is controlled by a valve. A full scale plant uses several arrays of tubes in parallel, operated so that the cleaning cycle is sequential. This enables continuous operation of the unit.

At the start of the dewatering operation, sludge is pumped into the tubular array with the outlet valve open, allowing crossflow filtration to take place. In this mode the pressure differential across the walls of the tube is sufficient to allow some sludge solids to form a layer on the inside of the membrane tube. Most of the flow of sludge is returned to the feed tank. After a short time the outlet valve is closed and the tubes are then operating in the barrier, or dead-end, mode. Further sludge is pumped into the tubes, causing permeate to be squeezed out through the filtration membrane. Solids accumulate on the inside of the membrane, forming an annular cake which increases in thickness with time (Figure 2). As the thickness builds up there is an increase in headloss across the membrane. The permeate is collected underneath the tubular array and is returned to the works inlet or disposed of by other means.

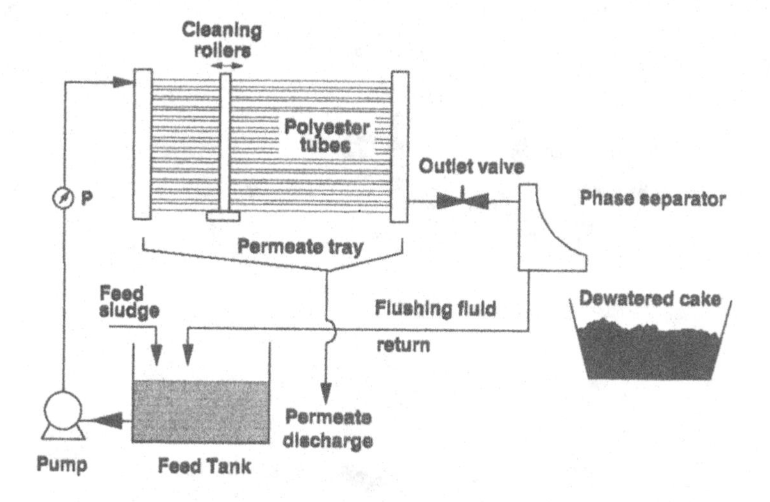

Figure 1 Schematic diagram of the Exxpress process.

After a suitable time when the thickness of sludge on the inside of the tube is adequate, the decaking cycle begins. The outlet valve is opened and flushing fluid (usually the untreated sludge itself) is pumped through the tubes. At the same time a carriage traverses along the array of tubes, with pairs of cleaning rollers mounted externally on each side of the tubes. The rollers are set so that the tubes are gently squeezed. This squeezing action breaks the sludge cake away from the walls of the tube (Figure 2). It also creates a Venturi effect at the point of compression and the increased velocity scours the flakes from the tube wall; the flushing fluid removes them from the tube through the outlet manifold. The mixture of sludge and flushing fluid is deposited onto a phase separator consisting of a wedge wire screen or open-texture conveyor. The excess fluid drains away to be returned to the sludge feed tank and the flakes of dewatered sludge are collected for material recovery or disposal.

Whereas mining effluents with large proportions of clay minerals have been dewatered successfully from feed concentrations exceeding 20% w/w, waterworks coagulation sludges tend to have optimum feed concentrations in the range 1 to 2.5% w/w solids. The process can be designed to dewater satisfactorily feed sludges with a wide range of solids concentrations which may vary rapidly, although the process efficiency may suffer. With a thin sludge large volumes of water have to be pumped; with a viscous sludge the rate of dewatering declines because of the viscosity. Nevertheless, at both extremes the process can achieve satisfactory separation of solid and liquid phases.

The fundamental measure of the performance of any pressure-driven dewatering process is the specific cake yield, C_s (kg dry weight/m^2/h). It is a measure of the rate of production in terms of dry solids and is a function of time, the sludge cake solids content and the cake mass produced during an operating cycle. The quality of the permeate may also be important in many applications. Generally the best cake yields from waterworks sludges are achieved with a dewatering cycle time of 20 to 40

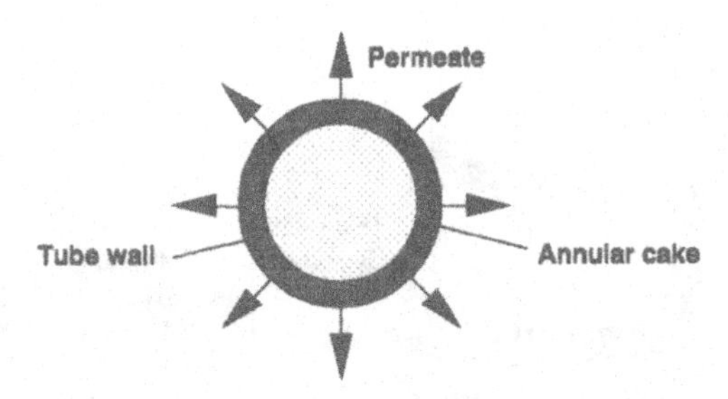

Build-up of dewatered cake in the tube.

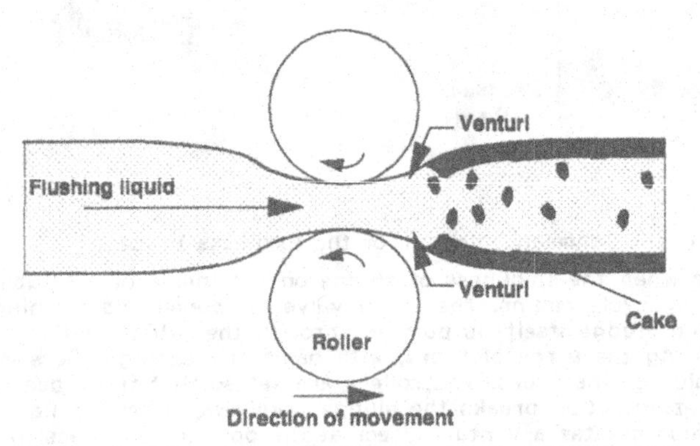

Cleaning the tube using a venturi effect created by external rollers.

Figure 2 The formation and removal of dewatered cake in the tube.

minutes, although optimum efficiencies have been recorded with 5 minutes or 60 minutes in different circumstances. The operating pressure range is usually between 200 and 400kPa, with the optimum for many sludges around 250 to 300kPa.

EXPERIMENTATION

Test sites.

At Site A an extensive pilot plant investigation was carried out between 1987 and 1989 to determine the appropriate treatment processes for a proposed 60 Ml/d water supply scheme. River water from a rural lowland catchment was pumped to shallow storage lakes, forming a source of eutrophic raw water for treatment. The water was moderately hard (alkalinity 280 mg/l $CaCO_3$), with low colour (20 - 30 °Hazen) and turbidity (1 - 5 NTU) but it supported a considerable algal growth throughout the year with chlorophyll a concentrations ranging from 20 to 300 mg/l. Dissolved air flotation (DAF) was confirmed as the most suitable clarification process.

At Site B an existing sedimentation/rapid gravity sand filtration works was scheduled for refurbishment and pilot scale DAF trials were carried out in 1989 and 1990 to confirm the optimum design parameters for the proposed change of process. The raw water supplying the works was similar to that of Site A, but although significant blooms of diatoms occurred in spring and of blue-green algae in summer, algal production was less intense overall.

Pilot plant evaluation.
At both sites aluminium and iron coagulants were evaluated in two stage flocculation DAF plants, rated at 10 m^3/h. The Exxpress process was investigated as an alternative to conventional sludge dewatering options at both sites, using a pilot plant which embodied the principles described previously. The tubular array consisted of a single curtain of six parallel tubes of length 1.2m with an effective filtration area of 0.5 m^2. The plant was capable of operating automatically, but for experimental purposes it was generally operated manually with batched sludge feeds. Its performance was investigated by varying the operating pressure and the duration of the dewatering cycle. Evaluations were made of the production rates of sludge and permeate, as well as their chemical qualities.

Samples of the feed sludge, permeate and final cake were taken from each trial run for analysis as follows:

feed sludge:	pH, total dried solids (105°C)(%w/w);
permeate:	turbidity (NTU), colour (° Hazen);
sludge cake:	total dried solids (105°C)(% w/w).

Analysis was carried out in on-site laboratories using standard methods (HMSO). Occasional random samples were sent for confirmatory analysis to the Wessex Water regional laboratory.

RESULTS

The results of processing sludges from the two sites are presented in Tables 1 to 5. At site B the raw water quality in March and April was affected by a bloom of diatoms (*Stephanodiscus hantzschii*). Ferric chloride coagulant was used in the pilot DAF plant. The batches of sludges were made up with different solids concentrations (2.1% and 1.8% w/w) and dewatering trials were carried out at two different operating pressures for varying cycle times (Table 1).

Table 1 Dewatering of ferric chloride flotation sludge, Site B, March - April 1989.

Feed solids concentration (% w/w)	Operating pressure (kPa)	Cycle time (min)	Cake Dry Solids (% w/w)	Cake Yield Wet weight (kg)	Specific cake yield Dry weight ($kg/m^2/h$)
2.1	200	20	14.6	0.52	0.43
		25	15.1	0.63	0.43
		30	14.1	0.45	0.24
		35	14.4	0.35	0.16
1.8	300	20	15.4	0.67	0.59
		25	16.8	0.51	0.39
		30	15.9	0.60	0.36
		35	16.4	0.71	0.38

In some trials the dewatered cake was washed with water immediately after it was discharged. This was to assess the negative contribution on performance which any residual feed sludge (used to remove the cake) might have on the mean dry solids content of the cake (Table 1a).

Table 1a The effect of washing cake with water, Site B, March - April 1989.

Feed solids concentration (% w/w)	Operating pressure (kPa)	Cycle time (min)	Cake Dry Solids (% w/w)	Cake Yield Wet weight (kg)	Specific cake yield Dry weight (kg/m²/h)
Unwashed cake					
2.0	200	30	15.5	0.56	0.33
	250		15.7	0.72	0.43
	300		16.0	0.68	0.41
Washed cake					
2.0	200	30	15.8	0.47	0.28
	250		16.0	0.59	0.36
	300		16.0	0.59	0.36

Tables 2 and 3 compare the dewatering of ferric sulphate and ferric chloride coagulants used in DAF treatment of water from Site A in August and September 1989. The quality of the permeate from the dewatered sludges was also measured.

Table 2 Dewatering of ferric sulphate flotation sludge, Site A, Aug - Sept 1989.

Feed solids concentration (% w/w)	Operating pressure (kPa)	Cycle time (min)	Cake Dry Solids (% w/w)	Cake Yield Wet weight (kg)	Specific cake yield Dry weight (kg/m²/h)	Permeate Turbidity (NTU)	Permeate Colour (°Hazen)
1.4	100	10	12.2	0.11	0.15	1.2	50
		15	11.6	0.23	0.20	2.3	52
1.6 - 1.8	100	20	12.2	0.27	0.19	2.3	50
		25	12.2	0.48	0.27	0.9	58
	200	10	12.9	0.33	0.48	1.0	64
		15	13.3	0.44	0.44	1.0	66
		20	13.1	0.55	0.41	0.9	65
		25	12.3	0.58	0.33	0.9	68
	300	10	12.5	0.28	0.40	0.8	70
		15	12.7	0.41	0.40	0.8	70
		20	14.8	0.54	0.46	2.0	75
		25	14.4	0.54	0.35	2.0	73

The aluminium sulphate sludge from the sedimentation plant at Site B was tested in August and September 1988 (Table 4). Extensive trials were carried out at this time, with duplicate testing in a random manner and excellent agreement between duplicates was recorded.

Table 3 Dewatering of ferric chloride flotation sludge, Site A, Aug - Sept 1989.

Feed solids concentration (% w/w)	Operating pressure (kPa)	Cycle time (min)	Cake Dry Solids (% w/w)	Cake Yield Wet weight (kg)	Specific cake yield Dry weight (kg/m²/h)	Permeate Turbidity (NTU)	Permeate Colour (°Hazen)
0.80 - 0.86	200	15	10.1	0.31	0.24	13	10
		20	9.6	0.44	0.32	12	15
		25	10.2	0.58	0.27	16	10
		30	10.7	0.28	0.11	16	15
1.3	250	15	11	0.23	0.19	58	21
		20	11	0.30	0.16	40	23
		25	10	0.35	0.16	34	23
		30	11	0.24	0.10	30	28

Table 4 Dewatering of aluminium sulphate sedimentation sludge, Site B, Aug - Sept 1988.

Feed solids concentration (% w/w)	Operating pressure (kPa)	Cycle time (min)	Cake Dry Solids (% w/w)	Cake Yield Wet weight (kg)	Specific cake yield Dry weight (kg/m²/h)	Permeate Turbidity (NTU)	Permeate Colour (°Hazen)
0.4 - 0.5	200	15	10.2	0.32	0.25	1.5	17
		20	9.4	0.25	0.14	1.5	15
		25	10.4	0.42	0.20	1.3	17
	250	15	10.2	0.35	0.27	2.0	16
		20	10.8	0.25	0.15	1.5	17
		25	10.8	0.35	0.17	2.0	17
	300	10	11.1	0.19	0.25	-	-
		15	10.8	0.33	0.27	2.3	18
		20	10.9	0.30	0.19	2.6	18
		25	11.8	0.43	0.23	1.5	18
		30	11.6	0.38	0.17	-	-

The effect of polyelectrolyte on sludge dewatering was assessed on ferric chloride dissolved air flotation sludge from Site A in August 1988 (Table 5). Trials were carried out with and without a cationic polyelectrolyte (Allied Colloids LT31) at a dose of 1 mg/l in a sludge solids feed of 1.1 to 1.3% w/w.

DISCUSSION

Sludges produced from similar waters can vary considerably in their characteristics and comparisons should be treated with caution. However, in these trials, in addition to being able to make the above cautious comparison, it was possible to compare the dewatering of sludges produced from the same water by the same process using different coagulants. The principal parameters for comparison are the specific cake yield, cake dryness and permeate quality. For process design purposes when sizing plant for a particular site the specific cake yield is the most important parameter.

Table 5 Effect of polyelectrolyte preconditioning of ferric chloride flotation sludge, Site B, Aug - Sept, 1988.

Feed solids concentration (% w/w)	Operating pressure (kPa)	Cycle time (min)	Cake Dry Solids (% w/w)	Cake Yield Wet weight (kg)	Specific cake yield Dry weight ($kg/m^2/h$)
Without polyelectrolyte					
1.1 - 1.3	200	15	11.7	-	-
		20	11.3	0.41	0.26
		25	11.5	0.45	0.24
		30	9.9	-	-
With cationic polyelectrolyte - 1 mg/l (LT31, Allied Colloids Ltd)					
1.1 - 1.3	200	15	10.8	0.35	0.29
		20	11.5	0.45	0.29
		25	11.2	0.48	0.24

Specific cake yield.
Specific cake yield (C_s, kg dry weight/m^2/h) is of value in comparison of the process performance on different sludges.

Cycle time. In general there was a decrease in specific cake yield with increasing cycle time. This suggested that the cake accumulated rapidly at first after which the rate declined. All the different sludges treated demonstrated a similar trend. The optimum time varied slightly from sludge to sludge, but frequently it was between ten and twenty minutes. There was no benefit from increasing the cycle time to more than 25 minutes (Figure 3b). Selecting too short a cycle time would reduce the overall efficiency of operation by increasing to an unacceptable proportion the time needed for cleaning the dewatered cake from the tubes.

Operating pressure. In most of the trials increasing the operating pressure had little effect on the yield of wet cake and hence on C_s, once a lower threshold value of about 100 kPa had been exceeded. This was most clearly seen with the ferric sulphate DAF sludge from Site A (Table 2) where, if the small variation in feed solids concentration was ignored, the wet yields produced at each cycle time were very similar at both 200 and 300 kPa operating pressure and were two to three times those at 100 kPa.

The exception to the above was with the ferric chloride DAF sludge at Site A. Two sets of tests were carried out at different feed concentrations (Table 3) and increasing the operating pressure and the feed concentration gave a significant reduction in C_s. An increase in feed concentration is not likely to result in a decrease in C_s. It is probable that the ferric chloride floc was less robust than its ferric sulphate equivalent and that the deposited cake began to deform and break down as the pressure was increased. The permeate quality data supports this hypothesis. There was a 300% increase in turbidity and a 30% increase in colour of the permeate from the ferric chloride trial at the higher pressure.

Sludge feed solids concentration. The effect of sludge feed solids concentration on C_s was not specifically studied but can be demonstrated from a comparison of ferric chloride DAF sludges produced at different times at Site A (Tables 2 & 3). With cycle times between 20 and 30 minutes at 200 kPa operating pressure, reducing the feed from 2.1% to 0.86% solids decreased the wet weight of cake produced, with a reduction in C_s of between a quarter and a half. Subsequently, results from many other sites have shown a similar relationship, although there have been exceptions. The relationship appears to be a function of individual sludges rather than the effect of the operating conditions.

Comparison of sludges from different coagulants. The specific cake yield from ferric chloride sludge was approximately half that of ferric sulphate sludge produced from the same process and using the same water (Tables 2 & 3). At similar feed solids concentrations and operating pressures of over 250 kPa, the wet cake yield and the dry solids contents were lower for the chloride-derived sludge than for the sulphate-derived one. This was most probably due to the loss of fine material in the permeate, as shown by the permeate turbidity data. The difference in wet weights produced suggests that the chloride cake allowed less free passage of water through it than the sulphate cake. The aluminium sulphate sludge solids concentration from the sedimentation process was considerably lower than the DAF sludges (Table 4), as expected due to the nature of the processes. However, the wet weight of cake produced was similar to that from the ferric chloride DAF trial (Table 3) and specific cake yields were similar.

Polyelectrolyte conditioning. The data (Table 5) on the effects of preconditioning the sludge with a cationic polyelectrolyte (Allied Colloids LT31) was too limited to draw any firm conclusions, but it indicated that there was little effect on specific cake yield. This has been confirmed subsequently from other studies.

Cake dryness

Dryness is a major consideration in cake management. Drier cake means smaller volumes to handle and less water to cart to the disposal site, or to be removed by secondary drying processes if the cake is to be reused. Cake dryness is often less important than specific cake yield and its maximum may not coincide with maximum specific cake yield. In most cases the balance between the two is chosen as an economic compromise and as such is site or process related.

There is no obvious relationship between the length of dewatering time and the cake dryness (Figure 3a). The data obtained with ferric chloride flotation sludge indicated that the sludge solids are not readily deformed at pressures up to 300 kPa (Table 1). Increased pressure does produce drier cake and it is possible that this is because the solids break down into smaller, denser particles which pack more tightly together, reducing the volume of liquid which can be held between them. Feed solids concentration was changed concurrently with operating pressure in this trial. In other trials it has been found that feed concentration can affect final cake dryness, with thinner feeds yielding slightly drier cake, above a threshold of 0.2 to 0.3% w/w solids for water works coagulation sludges. The effect is always very much less than that from increasing operating pressures.

The solids contents of the cakes were characteristic of each particular sludge and different dry solids contents would be expected with other sludges. Differences were apparent in the dryness of cake produced from ferric chloride and ferric sulphate coagulants, using the flotation process at Site B (Tables 2 and 3). Average dry solids content of the chloride cake was very much less than that of the sulphate cake, with mean chloride cake dryness of 10.15% w/w compared to 12.9% w/w for the sulphate cake, a difference of up to 25% at 200 kPa. After discharge from the plant, the flakes of sludge continued to lose water by air drying, achieving a solids content in excess of 50% w/w within two days. In other studies solids contents have tripled in four days.

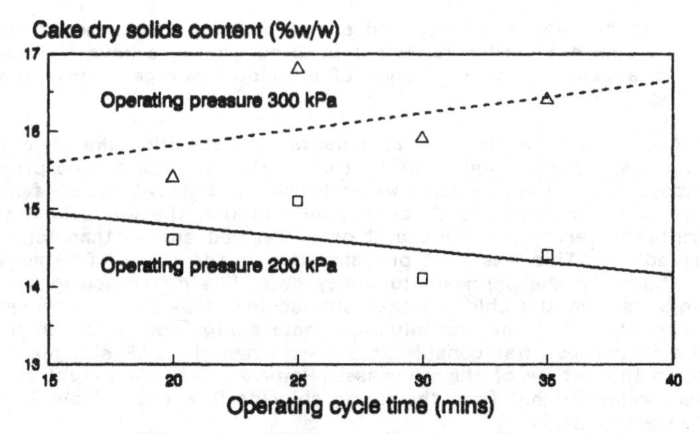

Figure 3a Effect on cake dry solids content.

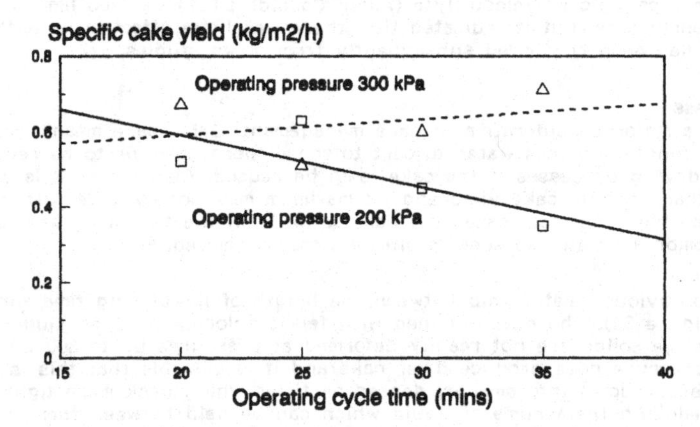

Figure 3b Effect on specific cake yield.

Figure 3 Effect of operating cycle time and pressure on cake dry solids and specific cake yield.

Feed sludge was used to flush the dewatered cake from the tubes during desludging. It was thought that the more viscous sludges might not drain very readily from the cake after flushing, leading to reduced average cake dryness while increasing the wet weight of cake. A test was made of the effect of brief washing of ferric chloride flotation sludge cake with water, typically for five seconds at very low pressure as soon as it had been discharged. The results (Table 1a) showed that washing actually reduced the specific cake yield, because it had no effect on measured dry solids content, but did reduce the wet weight of cake produced. It seems likely that washing removed the smaller flakes of dewatered material and that the feed sludge used for dewatering was sufficiently thin to drain anyway.

There was no significant difference in the dryness of the cakes from sludges with and without polyelectrolyte conditioning (Table 5).

Permeate quality

The quality of permeates from the aluminium and ferric sulphate sludges was good under all operating conditions (Tables 2 and 4). The slight increase in colour released from the sludges as operating pressures were increased indicated limited tendency of the sludge to break down. In contrast turbidity in the permeate from the ferric chloride sludge was already greater at 200 kPa than those from the sulphate sludges and it deteriorated markedly as operating pressure increased to 250 kPa (Figure 4). This indicates that the ferric chloride sludge is weak and some breakdown occurs readily under pressure.

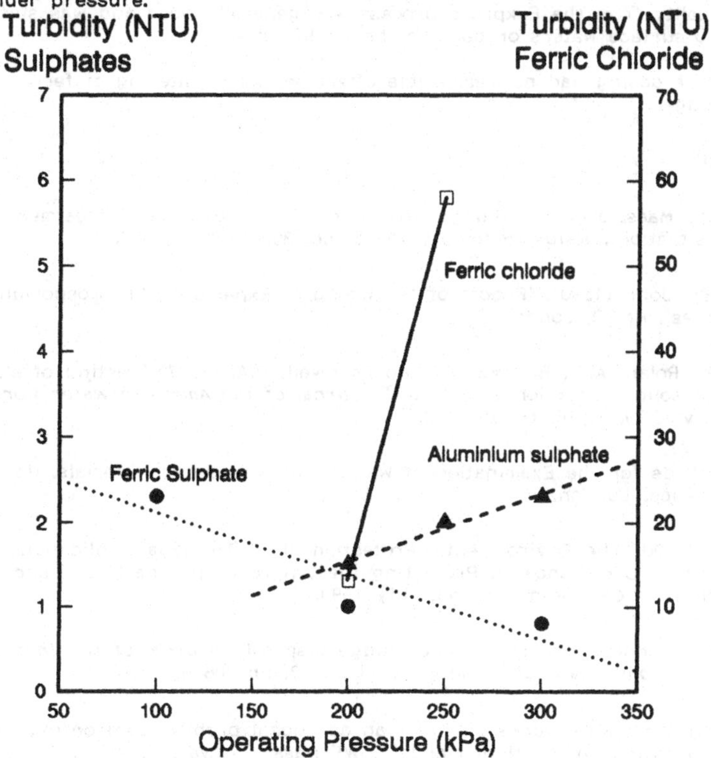

Note: The turbidity scale for ferric chloride sludges is ten times that of the other coagulants.

Figure 4 Effect of operating pressure on permeate turbidity from sludges of various coagulants.

CONCLUSIONS

The Exxpress process successfully dewatered coagulation sludges, formed by aluminium or ferric sulphate and ferric chloride, derived from dissolved air flotation or upflow sedimentation clarification. It operated at low pressures and over a short operating cycle. The cake was produced as solid, drip free flakes.

Ferric chloride sludge was not as robust as ferric sulphate sludge and did not produce such dry cake. Aluminium sulphate sludge dewatered similarly to ferric sulphate sludge.

The dewatered cakes, produced from a variety of feed sludges with widely varying concentrations, regularly contained 10 to 17% w/w dry solids when discharged. The structure of the flakes of cake was such that upon storage and air drying for 48 hours the dry solids content tripled.

Permeate quality from the Exxpress process was generally good and was suitable for discharge to surface waters or back to the works' inlet.

Polyelectrolyte dosing had no measurable effect on the dewatering of ferric chloride flotation sludge.

REFERENCES

Aerts, M. and Maes, J (1986) 'Sludge production in drinking water treatment: part 1. The actual situation', *Water (Belgium)*, vol. 5, no. 30, pp 130 - 133.

Badenoch, Sir John (1990). 'Report of the Group of Experts on Cryptosporidium in Water Supplies'. HMSO, London.

Bishop, M.M., Rolan, A.T., Bailey, T.L. and Cornwell, D.A. (1987) 'Testing of alum recovery for solids reduction and reuse,' *Journal of the American Water Works Association*, vol. 79, no. 6, pp 76 - 83.

H.M.S.O. Methods for the Examination of Waters and Associated Materials. Her Majesty's Stationery Office. London.

Hawkins, R. (1990) The Environmental Protection Bill - The legal implications. Proceedings of a Conference on Protecting the Environment. The Public and Local Service Efficiency Campaign. London, July 1990.

IWEM, (1990) Technical paper: 'Sewage Sludge Disposal', *Journal of the Institution of Water and Environmental Management*, vol 3, no. 2, pp 208 - 211.

Koppers, H. (1985) 'Waterworks sludge - an additional problem confronting water supply undertakings in the Netherlands', *GWF-Wasser/Abwasser*, vol. 126, no. 10, pp 513 - 518.

Warden, J.H. and Craft, D.G. (1980) Technical Report TR150 'Waterworks sludge - Production and disposal in the U.K.' Water Research Centre, Medmenham.

ACKNOWLEDGEMENTS

The authors gratefully acknowledge the contribution to the project from Ruth Freer and George Bancroft in installing and operating the pilot plant, the expert laboratory analysis from the staff of the Wessex Water regional laboratory and the help given by the operational staff of the water treatment works at Site B. They are also indebted to the owners of the Exxpress process for permission to publish this paper.

Landfill Leachate Treatment by Reverse Osmosis

B Weber and F Holz

(Haase Energietechnik GmbH, Neumünster, Germany)

SUMMARY: Leachate from landfill sites represents a highly polluted waste water. It containes biodegradable compounds but also inorganic salts and trace recalcitrant pollutants. The reverse osmosis process with or without biological pretreatment has found to be most effective in the treatment of this special waste water.

Biological pretreatment of raw leachate enhances the effectiveness of the reverse osmosis process. The task of biological pretreatment is the removal of biodegradable compounds, especially if rather highly concentrated leachate occurs. The task of the reverse osmosis process is the rejection of all compounds or, as a 2nd treatment stage, the rejection of recalcitrant pollutants and inorganic salts. Operational data for a semi-technical treatment plant and for full-scale plants are presented.

1. Introduction

The state of the art for the treatment of landfill leachate is represented by the following process combinations:

- biological pretreatment including nitrification and denitrification; chemical/physical treatment with flocculants and activated carbon; residue treatment by sludge dewatering and disposal as hazardous waste.
- 2-stage reverse osmosis (RO) process (1st stage tubular modules, 2nd stage spiralwounded modules); residue treatment by concentrate evaporation, drying and disposal as hazardous waste.
- biological pretreatment including nitrification and denitrification; reverse osmosis process (tubular modules); residue treatment by concentrate evaporation, drying and disposal as hazardous waste.

Effective Industrial Membrane Processes — Benefits and Opportunities, pp.143–154

The aim of scientific research activities is the improvement of chemical oxidation processes in combination with biological pre- and posttreatment including nitrification/denitrification, sludge dewatering and residue disposal as hazardous waste.

2. Quality and quantity of leachate

Leachate from sanitary landfill sites can be divided into four categories:

- leachate from the acidic phase (acidic leachate)
- leachate from the methanogenic phase (stabilized leachate)
- diluted leachate
- leachate from the disposal of residues

On old landfill sites started up in the seventies the leachate often is diluted by surface water. With the improvement in techniques for sealing over naturally impervious sites many operators are observing an increase in leachate concentrations. The high organic concentrations of leachate from the acidic phase, occuring only during the start up, should be diminished by anaerobic pretreatment (e.g. anaerobic waste fixed bed). The quality of anaerobically pretreated acidic leachate is comparable to leachate of the methanogenic phase. Not only the age of the site plays a role but also extensive recycling actvities in the future, which will leave only residues for disposal. The contents listed for such future landfills can only represent rough guesses, with a trend towards leachates higher in inorganic and lower in organic contaminants. Because nobody knows what the future really will bring, in this paper only results of the treatment of diluted and stabilized leachate are presented. Nevertheless the RO-process is predestined for the treatment of the future's inorganic polluted leachate.

kind of leachate	COD mg/l	NH_4-N mg/l	AOX mg/l	Cl^- mg/l	conduc. mS/cm	amount $m^3/(ha \cdot d)$
acidic phase	>15,000	1,200	2 to 5	3,000	25	2 to 5
methanogenic phase	<5,000	1,800	4	3,000	18	5
diluted leachate	1,500	500	1.2	1,000	10	7.5
future leachate	1,500	700	2	3,000	15	5

table 2-1: average concentrations and amount of different kinds of leachate

3. Leachate discharge limits

Discharge limits for the effluent from leachate treatment plants are given in table 3-1.

		Fed. Rep. of Germ.		Switzerland		Italy	Netherlands		Austria	
		indirect discussed at pres. (1989)	direct discussed at pres. (1989)	indirect (1988)	direct (1988)	direct	indirect	direct	indirect (1981)	direct (1981)
COD	mg/l	400[2,3]	(150) 200[2]	–[4]	–[4]	160	-	-	-	75,90 max
BOD_5	mg/l	-	20[2]	–[4]	20	40	300-400[4]	7-20[4]	-	20,25 max
SS	mg/l	20[1]	20[1]	–[4]	20	80	-	-	-	30, (50)
Cl^-	mg/l	-	-	–[4]	–[4]	1200	-	-	-	–[4]
SO_4^{2-}	mg/l	-	-	–[4]	300	1000	300	500	–[4]	–[4]
N-Kj.	mg/l	-	-	-	-	-	300	8-15[4]	-	-
NH_4-N	mg/l	-	(10), 50[2]	–[4]	–[4]	12	200	4-8[4]	-	–[4]
NO_3-N	mg/l	-	-	–[4]	–[4]	20	-	-	-	–[4]
NO_2-N	mg/l	-	(10)[2]	3	0,3	0,6	-	1-4[4]	9	1,5
Pb	µg/l	500[1]	500[1]	500	500	200	200	50	1000	1000
Cd	µg/l	100[1]	100[1]	100	100	20	10	2,5	100	100
Cr(IV)	µg/l	-	-	500	100	200	375	75	100	100
tot-Cr	µg/l	500[1]	500[1]	2500	2100	2200	-	-	2100	2100
Cu	µg/l	500[1]	500[1]	1000	500	100	250	50	1000	1000
Ni	µg/l	500[1]	500[1]	2000	2000	2000	170	100	2000	2000
Hg	µg/l	50[1]	50[1]	10	10	5	2	0,5	10	10
Se	µg/l	-	-	-	-	30	-	-	-	-
Zn	µg/l	2000[1]	2000[1]	2000	2000	500	1000	200	3000	3000
Sn	µg/l	-	-	2000	2000	10000	-	-	2000	2000
Fe	mg/l	-	-	20	2	2	-	-	–[4]	2000
AOX	µg/l	500[1]	500	-	-	-	-	-	-	-
EOX	µg/l	-	-	-	-	5	5	-	-	-
phenol	µg/l	-	-	5000	50	500	-	-	20000	100
tot-HC	mg/l	-	-	20	10	-	-	-	-	-
fishtox.class		(2)[1]	2	-	0-5[4]	2	-	-	-	-

table 3-1: Discharge limits for the effluent from leachate treatment plants in Europe

4. The reverse osmosis process

Reverse osmosis is a pressure driven membrane process. During the last years this technique has been proved to be useful in treating leachate from landfill sites (e.g. JANS et al., 1988).For the lay out of a full scale reverse osmosis plant the following basic data should be evaluated very carefully:

- quantity of the leachate
- quality of the leachate influenced by the manner of primary purification

- possible concentration factor to keep the required quality of the permeate
- kind of modules, membrane material and cleaning procedures
- pressure, temperature and velocity influencing the permeate flux and the permeate quality
- treatment of the concentrate.

Generally physical pretreatment by sand or drum filters is recommended to preserve the pumps, membranes and measuring instruments. The quality of the permeate is given by the minimum requirements or even lower values. The quality of the permeate can be controlled by the volumetric concentration factor C_{FV}, which is the leachate flow rate divided by the concentrate flow rate. If the concentrations of process limiting compounds like calcium (Ca) and iron (Fe) or of critical parameters like NH_4-N and AOX are high, the physical and economic range of the RO-process can be exceeded. The parameter C_{FV} is very important, because it determines the quantity of concentrate produced for a given quantity of leachate. A C_{FV} of 5 may be sufficient for a process to be economic, provided the concentrate can be recirculated back to the landfill site. If the concentrate must be evaporated, the C_{FV} must be greater than 5 to minimize the treatment costs of the concentrate.

There is a strong interrelationship between the other basic process data. Modules and membranes control the permeate quality. The possibility of cleaning and the lifetime of the membranes are very important in leachate treatment. Due to the possible damage of the membrane surface by circulating sponge balls, chemical cleaning was found to be more suitable. The size of the plant, operating pressure, temperature and recirculation rates determine both the quality of the permeate and the plant's power consumption.

These basic process data and the suitability of the modules and membranes should be evaluated during tests. The tests should also answer the question, of whether advanced biological pretreatment is necessary or whether the purification of raw leachate is sufficient and economic.

Test runs have been carried out with a semi-technical RO-plant for different kinds of leachate and for the effluent of different treatment stages of a semi-technical and a full-scale biological purification plant. Figure 3-1 shows the flow scheme of the multi-stage pilot plant. The RO-plant was

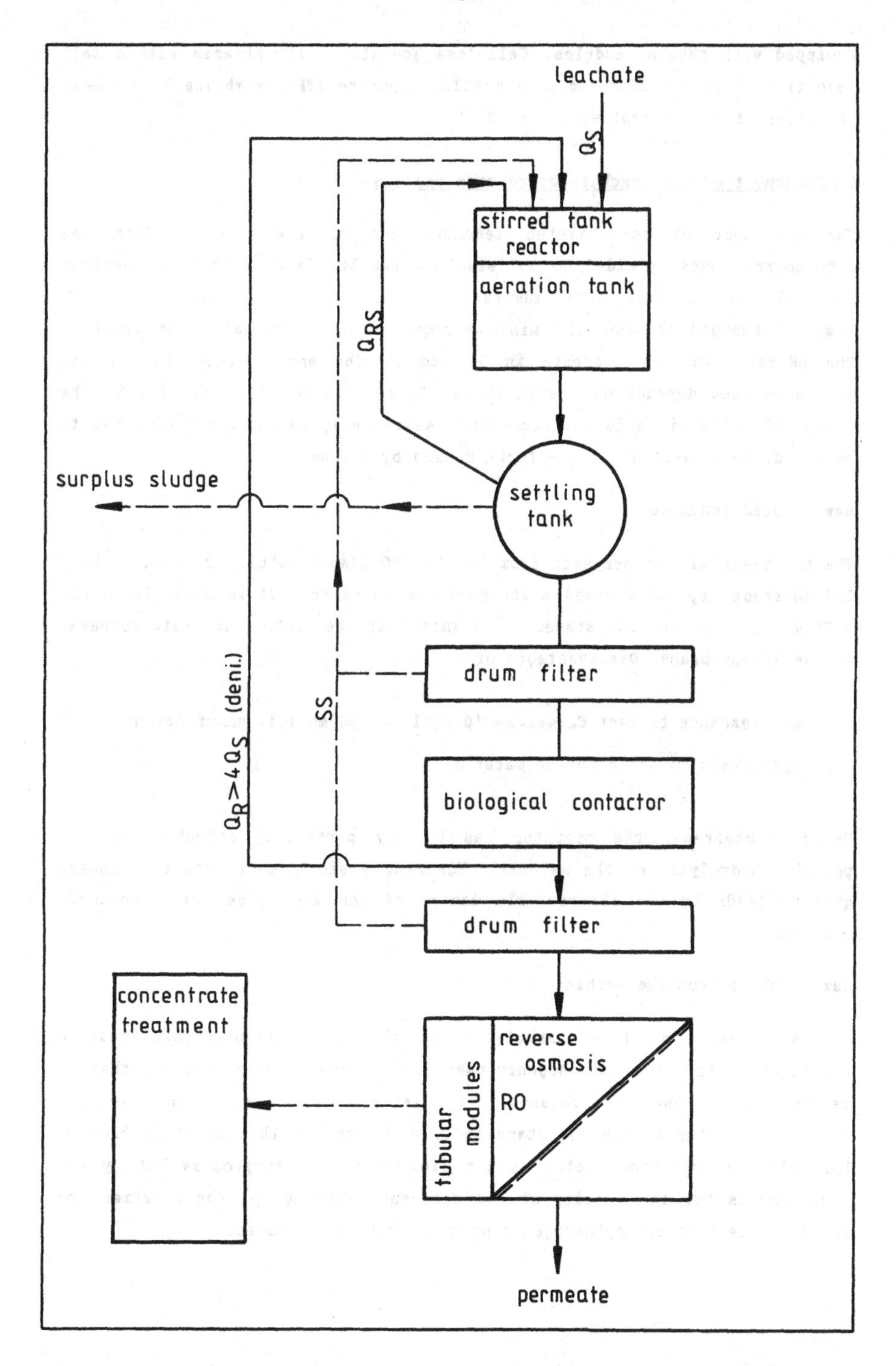

fig. 3-1: flow scheme of the multi-stage leachate treatment plant

equipped with tubular modules. Cellulose acetate (CA) membranes with a salt rejection of 95 per cent and a thin film composite (TFC) membrane with a salt rejection of 99 per cent were tested.

5. Treatment of raw leachate by reverse osmosis

The treatment of raw diluted leachate and of raw leachate from the methanogenic phase yielded no satisfactory results (see table 7-1). Neither the cellulose acetate (CA) membrane nor the thin film composite (TFC) membrane managed to meet the minimum requirements, especially for ammonia. The pH-value of raw leachate is 7.5 to 8. The ammonia-rejection of the membranes used depends on the pH-value. To reach a suitable pH of 6.5, the dosage of sulphuric acid was necessary. Also precipitation-inhibitors had to be added, to prevent scaling effects caused by gypsum.

Raw diluted leachate

The treatment of the permeate from the 1st RO-stage (tubular membranes) by a 2nd RO-stage (spiral wounded membranes) was necessary. It is possible to use a CA-membrane in the 1st stage. An advantage is the higher flux rate compared to the TFC-membrane. Disadvantages are

- no clearance to meet $C_{e,NH4-N}$= 10 mg/l, if 50 mg/l is unsufficient
- hydrolysis of the membrane material

Using CA-membranes the operator has to pay particular attention to the possible hydrolysis of the material. Egnoring a significant drop in permeate quality leads to the irreversible damage of the 2nd stage spiral-wounded-membranes.

Raw leachate from the methanogenic phase

On easy terms it might be possible to meet the minimum requirements treating raw leachate from the methanogenic phase by reverse osmosis. But for that it is necessary to use TFC-membranes in the 1st and 2nd stage. The maximum concentration factor in the 1st stage is C_{FV}= 3.5 and in the 2nd stage C_{FV}= 5. Here the concentration factor is not limited by the RO-process but by the requirements for the quality of the effluent. Summing up, the treatment of raw leachate from the methanogenic phase cannot be recommended.

Cleaning procedures

Chemical cleaning in place was carried out every 75 hours of operation using the detergent Ultrasil and every 200 hours of operation using citric acid.

6. Treatment of ASP-effluent by reverse osmosis

The biological treatment of raw leachate by the activated sludge process (ASP) and a sludge loading of $B_{TS,BOD}$ >0.25 kg/(kg·d) diminishes only the TOC (total organic carbon) and the AOX (adsorbable organic halogens). The reduction of the ammonia by incorporation in the sludge totals up to only 10 per cent. Because ammonia is the bottle-neck for the rejection by reverse osmosis but not the TOC, COD, BOD or the AOX there are no great benefits of applying a relative high loaded activated sludge process (see table 7-1). Subordinated advantages are:

- buffer for COD/BOD-peaks
- removal of scaling and fouling causing compounds
- enhancement of flux rate and concentration factor.

7. Advanced biological pretreatment of leachate and final purification by RO

Advanced biological treatment of raw leachate including nitrification and denitrification can be carried out by the following processes and treatment stages:

1. Activated sludge process and biological contactor (ASP/BC-system)

a) stirred tank reactor	denitrification, TOC-removal
b) aeration tank	TOC-removal
sludge loading of a) plus b): $B_{TS,BOD}$ = 0.2 kg/(kg·d)	
c) settling tank surface overflow rate q_A= 0.6 m/h	
d) drum filter surface overflow rate q_A= 5.0 m/h	separation of solids
e) rotating biological contactor surface loading rate $B_{A,N}$ = 5 gN//m²·d)	nitrification
f) recirculation pump to a with Q_R= 4·Q	
g) drum filter surface overflow rate q_A= 5.0 m/h	separation of solids

2. Activated sludge process (ASP-system)

a) stirred tank reactor — denitirification, TOC-removal
   connected in plug flow with
b) aeration tank — TOC-removal, nitrification
   nitrogen sludge loading of b):
   $B_{TS,N}$ = 30 gN/(kg·d)
   volume of a) is 25 per cent of b)
c) recirculation pump to a) $Q_R = 5 \cdot Q$
d) settling tank
e) drum filter — separation of solids

The task of the fabric-drum filters is the enhancement of the nitrification process and the protection of the reverse osmosis.

Advantages of system 1. (ASP/BC-system), which is part of the used semi-technical treatment plant, are

- fixed film of nitrifying bacteria
- enhanced N-loading rate due to optimum BOD/N-ratios of less than 0.1

An additional disadvantage of system 2. (ASP-system) is the possible stripping of organic halogen compounds caused by the long term aeration. A full-scale system 2. treatment plant is operating at the Venneberg-landfill site near Lingen, FRG.

Removal efficiencies and effluent values of advanced biological pretreatment are listed in table 7-1.

The effluent of the two advanced biological purification plants has been treated by reverse osmosis. The minimum requirements were met by using a one stage RO-plant. In addition the operation of the RO-plant was more economical compared to the treatment of raw leachate, because the flux rate or the possible concentration factor were enhanced (see table 7-1). The nitrification process diminishes the buffer capacity and the pH-value of the pretreated leachate. So a pH-control to reach a value of 6.5 by adding sulphuric acid is not necessary or achieved by dosing with less acid.

Cleaning procedures using Ultrasil were carried out every 150 hours of operation and every 500 hours using citric acid. The longer cleaning interval

quality	diluted leachate				leachate methanogenic phase					
quantity	7.5 m³/(ha·d)x20ha				5 m³/(ha·d)x20 ha					
manner of pretreatment	raw		nitri/deni B_{TS}=0.2 $B_{A.N}$=5		raw		ASP B_{TS} > 0.2		nitri/deni B_{TS}= 0.2 $B_{A.N}$= 5	
effluent pretreatment										
COD mg/l	1,500		1,000		5,000		2,000		1,500	
TKN mg/l	600		50		2,000		1,700		100	
NH_4-N mg/l	500		<10		1,800		1,600		<10	
NO_x-N mg/l	0		250		0		0		400	
AOX µg/l	1,200		1,000		4,000		2,500		2,000	
RO 1st stage	tubular membranes									
membrane	CA	TFC	CA	TFC	CA	TFC	CA	TFC	CA	TFC
effluent										
COD mg/l	90	50	50	30	300	175	100	60	75	40
TKN mg/l	140	80	10	5	400	260	340	235	20	10
NH_4-N mg/l	125	70	<2	<1	375	250	325	225	<2	<1
NO_x-N mg/l	0	0	25	15	0	0	0	0	40	20
AOX µg/l	160	75	125	50	600	275	300	125	250	100
flux for C_{FV} = 5 (l/m²·h)	19	17	32	29	15	13	28	25	31	27
membrane area (m²)	300	330	175	190	250	280	135	150	120	135
possible C_{FV}	5	5	7.5	7.5	4	4	6	6	7	7
RO 2nd stage	spiral-wounded TFC-membranes									
effluent										
COD mg/l	all values less than 15 mg/l									
TKN mg/l	15	8	<2	<1	45	30	35	25	<2	<1
NH_4-N mg/l	10	5	<1	<1	40	25	30	20	<1	<1
NO_x-N mg/l	0	0	2	1	0	0	0	0	3	
AOX µg/l	<50	<50	<50	<50	100	50	55	<50	<50	<50
recommended treatment processes	ASP + nitri/deni + 1st stage TFC-RO or 1st CA-RO + 2nd TFC-Ro					ASP + nitri/deni + 1st stage TFC-RO				
minimum requirements										
COD	less than 150 mg/l									
TKN/NO_x-N	no values up to now									
NH_4-N	less than 50 (10) mg/l									
AOX	less than 500 µg/l									

table 7-1: Results of leachate treatment by reverse osmosis

suggests a lower fouling and scaling potential of the aerobic treated leachate compared to raw leachate.

8. Results and special aspects of multi-stage treatment

In addition to the removal of biodegradable compounds and the improvement of the permeate quality the advantages of primary biological purification can be specified as follows:

- only recalcitrant pollutants remain in the leachate
- enhancement of the flux rate and/or enhancement of the possible concentration factor
- diminished treatment costs
- prevention of biofouling caused by high concentrations of organics
- removal of ferrous and calcic compounds by aeration and precipitation to prevent scaling of the membrane surface
- nitrogen removal to prevent problems caused by ammonia during the evaporation of the concentrate.

Summing up biological pretreatment has a great influence on the reverse osmosis feed. Figure 8-1 shows the flux decline Φ_P/Φ_0 dependence on the conductivity C_{LF} for raw and biologically pretreated leachate of a certain landfill site. This graph was obtained by doing flux measurements at various concentrations and fitting a linear regression.

Biological pretreatment diminished the conductivity of every leachate, that has been tested. This offers the possibilty of evaluating the effect of primary purification on the RO-process. Figure 8-2 shows the permeate flux versus the volumetric concentration factor C_{FV}. The more the biological pretreatment diminishes the conductivity, the higher is the possible concentration factor.

The explanation for the above mentioned phenomena could be that the osmotic pressure π and likewise the proportional factor b of inorganic salts are higher than π of organics or ammonia (see fig. 8-3). Volatile fatty acids, which contribute to the conductivity, are removed in the biological process. So the coincidence of adsorption, absorption and biodegradation diminishes the conductivity C_{LF} and the initial osmotic pressure Π_0. This effect decreases with decreasing leachate concentrations.

Due to the fact that biological pretreatment of raw leachate enhances the effectivity of the reverse osmosis, the membrane process should only be operated as the final step of a multi-stage treatment plant, if the following parameters are exceeded:

- conductivity > 10 mS/cm
- ammonia > 700 mg/l.

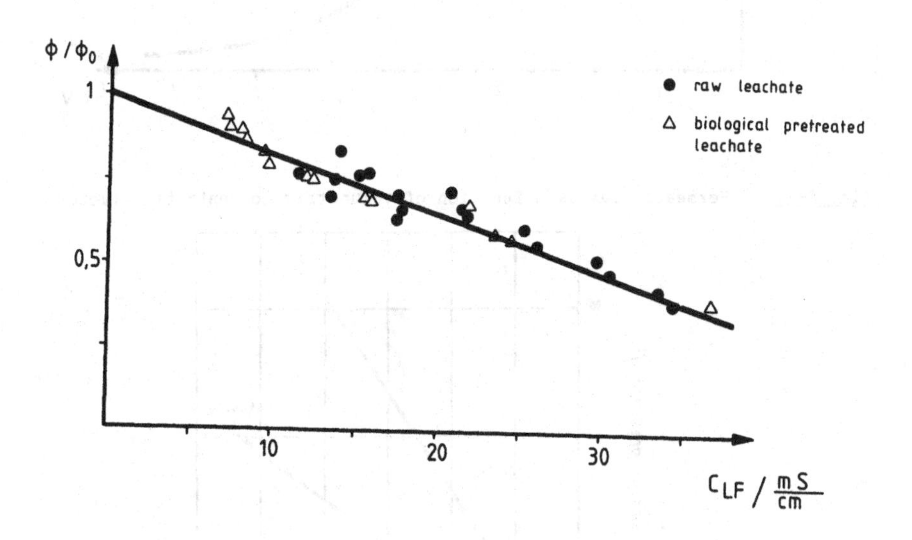

fig. 8-1: Flux decline as a function of feed conductivity

References

JANS, J. M., van der SCHROEFF, JAAP, A. (1987) "A treatment concept for leachate from sanitary landfills" Proceedings of the ISWA International Sanitary Landfill Symposium in Cagliari, Sardinia.

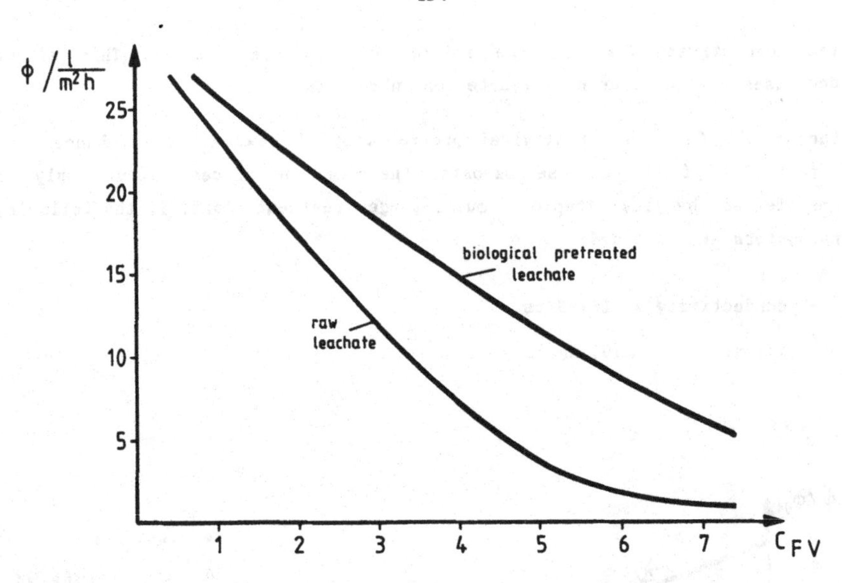

fig. 8-2: Permeate flux as a function of volumetric concentration factor

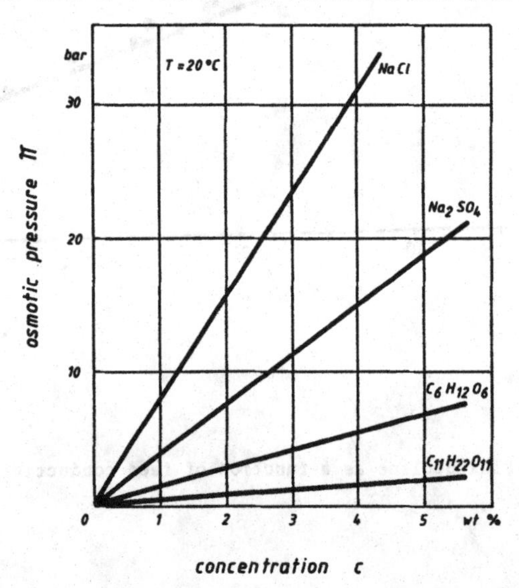

fig. 8-3: Osmotic pressure of some organic and inorganic solutions

SESSION E:

APPLICATIONS IN THE FOOD AND BEVERAGES INDUSTRY

MEMBRANES IN FOOD PROCESSING

Munir Cheryan

University of Illinois
Department of Food Science
Agricultural Bioprocess Laboratory
1302 W. Pennsylvania Avenue
Urbana, Illinois 61801, USA

SUMMARY

Membrane technology is well situated by virtue of its special features to play a major role in the food and beverage industry. Depending on the process chosen, one can either clarify (remove suspended matter) using microfiltration (MF), purify or fractionate macromolecules such as proteins in solution using ultrafiltration (UF), deplete salts using nanofiltration (NF) or electrodialysis (ED) and concentrate using reverse osmosis (RO). Dairy products and fruit juices have benefited the most from this technology in the past decade, but there is considerable activity in beverages (alcoholic, coffee, tea), animal products (gelatin, eggs, blood), grain processing (corn refining, soybean processing) and biotechnology (enzyme fractionation, membrane reactors). Among membrane processes, ultrafiltration probably is the largest application today in the food industry. Improvements in membrane chemistry and engineering design should see rapid growth in reverse osmosis and microfiltration applications in the next decade.

Effective Industrial Membrane Processes — Benefits and Opportunities, pp.157–180

INTRODUCTION

The application of membranes in the food and beverage industries increased dramatically in the 1980s. This is a remarkable achievement, considering that food manufacturers are traditionally quite conservative in their approach to innovation and utilization of "new, improved" technology. Indeed, membrane technology, from its birth in the early 1960s to the early 1980s, was viewed by the food industry with a mixture of interest and suspicion. Since the food industry operates on a smaller margin of profit than most other process industries, it was difficult to justify discarding what had always worked (e.g.,thermal evaporation) versus a new unit operation (e.g., reverse osmosis) that had not yet been proven to "work" in other industries, let alone the food industry.

Regulatory agencies, such as the U.S. Food and Drug Administration (FDA) and the U.S.Department of Agriculture, were unsure of how to treat this new technology, resulting in further reluctance by the food industry to adopt membrane separations. This explains why, for example, although cheesemaking by ultra-filtration (UF) had been developed and patented in France in 1969, it was only in the early 1980s that the U.S. dairy industry accepted it as a serious alternative to the traditional method. Today, there are probably few large-scale cheese operations that are not involved with membranes in some way, either for cheese-making, whey treatment, salt brine recovery or other processes.

The acceptance of membrane technology by the food industry today is due to several reasons: advances in the material science of membranes, which have resulted in more robust membranes better suited to sanitation standards and cleaning regimes used by the industry; improved module design; a better under-standing of the fouling phenomena; and increased awareness by food industry managers of the science and technology of membranes.

The 1990s will continue to see a major penetration of membrane technology in all its facets in the food and kindred industries. Recent estimates (Maubois 1989) place the installed area for UF in the dairy industry alone at over 150,000 square meters. In contrast, reverse osmosis (RO) applications are presently about 50,000 square meters, and microfiltration about 10,000 square meters. No data is available for the food and beverage industry as a whole, but it must be a good market, if one is to judge by the large number of companies that have entered the field just within the last decade all over the world.

Membrane technology as an art and a science is familiar enough that there is little need to cover the fundamentals in this paper; good reference texts are available (e.g., Belfort 1984, Cheryan 1986, Parekh 1989, Porter 1990, Rautenbach and Albrecht 1989, Sourirajan and Matsuura 1985). This overview will instead highlight some of the the technical factors that must be considered before one can successfully apply this technology in the food industry. Some case studies will also be presented of existing and potential applications of membranes that could help create new products or improve existing processes.

FACTORS AFFECTING MEMBRANE APPLICATIONS

Membrane material

There are over 130 different materials in the patent and scientific literature from which membranes have been made; many more are being added every year. The vast majority of the materials have been used for microfiltration membranes and much fewer have worked successfully for RO applications. Cellulose acetate (CA) and derivatives are still widely used, despite their real (and perceived) limitations.

Thin-film composite membranes containing a polyamide separating barrier on a polysulfone or polyethylene supporting layer, generally give better performance for RO applications with regard to temperature and pH stability and cleanability, but have almost zero chlorine resistance. These would be the material of choice for RO and NF applications, unless there was a specific fouling problem with them. Certain polyamides may be more prone to bacterial fouling than cellulosic membranes.

Polysulfone-type materials (polyether sulfone, polyphenylene sulfone, sulfonated polysulfone) are practically in a class by themselves for ultrafiltration. Vinylidine fluoride and acrylonitrile-based membranes are also used, though perhaps not as widely. However, the relative hydrophobicity of these materials has been blamed for their "fouling" tendencies, and thus one can also find these materials appropriately modified to make them less hydrophobic. It is interesting that as we understand more about the interactions between feed components and the membrane, cellulosic membranes are making a small comeback in some applications, perhaps because they are relatively more hydrophilic, less prone to nonspecific protein binding and usually somewhat cheaper.

MF membranes are available in a wide range of natural and synthetic polymers (polypropylene, polycarbonates, polysulfone, PVC, cellulose esters) and inorganic materials (alumina, zirconia/carbon composites, carbon/carbon composites, stainless steel and silica). The ceramic membranes in particular have considerably widened the range of potential applications, particularly in the biotechnology industry, where steam sterilizability and cleanability are of paramount importance.

Membrane Properties

Pore size and pore size distribution are obviously of critical importance in determining the performance of membranes. Microfiltration membranes have fairly well-established procedures for integrity testing and performance evaluation. However, this is not true for UF,NF and RO membranes which are usually rated quite arbitrarily by individual manufacturers. This is partially a reflection of the sometimes complex mechanisms involved in solute and solvent transport, especially with NF and RO membranes. It is not uncommon to find wide variations in the reporting of pore size distributions for the same membrane by different investigators.

The lack of standardization of classifying UF membranes can be overemphasized. It is not uncommon for like-rated UF membranes from different manufacturers to be exchanged for a particular application. The problem arises when fine separations are being attempted, i.e., those where the two proteins differ in molecular size by less than a factor of 10. It is important to remember that due to the shape of proteins in solution, a molecular weight difference of 10 times may mean a size difference of only 2-3 times.

Module Design

There are basically six different designs of membrane modules: tubular (with channel diameters greater than 3 mm), hollow fiber or capillaries (self-supporting tubes, usually 2mm or less internal diameters), plates, spiral-wound, pleated sheets and rotary modules. The latter four designs use flat sheets of membrane in various configurations.

In selecting a particular membrane module, the major criteria are: (1) the physical properties of the feed stream and retentate, especially viscosity, (2) particle size of suspended matter in the feed, (3) fouling potential of the feed stream, and (4) for dairy and food industries, sanitary design and cleanability. Many foods containing macromolecules display non-Newtonian behavior. Their viscosity will increase dramatically above certain concentrations, making pumping difficult and reducing mass transfer within the boundary layer. This will necessitate the use of modules that can withstand high pressure drops, eliminating most hollow fiber/capillary modules. On the other hand, these modules have extremely high packing densities (surface area:volume ratios) and comparatively low energy consumption. Feed streams containing large suspended particles would fare poorly in spiral-wound modules, owing to the spacers in their feed channels.

With rotary modules, the rotation of the filter cylinder sets up counter-rotating vortices ("Taylor vortices"), which can generate tremendous turbulence at the membrane surface, thus minimizing the effects of concentration polarization. With this type of a unit, the turbulence, cross-flow velocity and the applied transmembrane pressure can be independently controlled.

The food and beverage industries are fortunate to have several designs of equipment available which, between them, will probably be able to handle any fluid stream that can be pumped. Nevertheless, the trend in recent years seems to be away from plate designs and towards spiral modules, with tubular designs holding their own (e.g., with ceramic/inorganic membranes).

Fouling and cleaning

If there has been one single factor that has inhibited the large-scale use of membrane technology in the food industry, it is "fouling" of membranes. Fouling manifests itself as a decline in performance, usually a decline in flux under constant operating conditions and a possible change in the sieving properties of the membrane. This phenomenon is frequently confused in the literature with other flux-depressing phenomena, such as change in membrane properties with time due to deterioration, changes in feed solution properties and concentration polarization.

Minimizing membrane fouling must include consideration of the chemical nature of the membrane as well as physico-chemical properties of the feed stream. This is clearly shown by cheese whey, which is one of the most notorious foulants of membranes. Although it contains less than half the solids and one-third the protein of milk, cheese whey fouls membranes much more severely than does milk. Furthermore, hydrophilic membranes (e.g., cellulosics) generally foul less than hydrophobic membranes (e.g., polysulfone). However, one should not confuse "fouling" problems with "cleaning" problems. Several papers at this conference specifically address these problems.

APPLICATIONS

Table 1 is a list of applications in the food and kindred industries. This

Table I. Food industry applications of membrane technology

DAIRY

RO: Concentration of milk and whey prior to evaporation
Bulk transport
Specialty fluid milk products (2-3X/UHT)
NF: Partial demineralization and concentration of whey
UF: Fractionation of milk for cheese manufacture
Fractionation of whey for whey protein concentrates
Specialty fluid milk products
MF: Clarification of cheese whey
Defatting and reducing microbial load of milk
ED: Demineralization of milk and whey

FRUITS AND VEGETABLES

Juices -- Apple (UF,RO), apricot, citrus (MF/UF, RO, ED), cranberry, grape (UF,RO), kiwi, orange, peach (UF,RO), pear, pineapple (MF/UF,RO), tomato (RO).
Pigments -- anthocyanins, betanins (UF,RO)
Wastewater -- apple, pineapple, potato (UF,RO)

ANIMAL PRODUCTS

Gelatin -- concentration and de-ashing (UF)
Eggs -- concentration and reduction of glucose (UF,RO)
Animal by-products -- blood, wastewater treatment (UF)

BEVERAGES

MF/UF: Wine, beer, vinegar -- clarification
RO: Low-alcohol beer; coffee; tea

SUGAR REFINING

Beet/cane solutions, maple syrup, candy wastewaters -- clarification (MF/UF), desalting (ED), preconcentration (RO)

GRAIN PRODUCTS

Soybean processing: Protein concentrates and isolates (UF)
Protein hydrolyzates (CMR)
Oil degumming and refining (UF,NF)
Recovery of soy whey proteins (UF,RO)
Wastewater treatment

Corn refining: Steepwater concentration (RO)
Light-middlings treatment: water recycle (RO)
Saccharification of liquefied starch (CMR)
Purification of dextrose streams (MF/UF)
Fermentation of glucose to ethanol (CMR)
Downstream processing (MF,UF,NF,RO,ED, PV)
Wastewater treatment

CMR = continuous membrane reactor, ED = electrodialysis, MF = microfiltration, NF = nanofiltration, PV = pervaporation, RO = reverse osmosis, UF = ultrafiltration

includes reverse osmosis (for concentration,to complement or replace evaporation), nanofiltration (for desalting and deacidification), ultrafiltration (for fractionation, concentration and purification), microfiltration (for clarification instead of centrifuges, and sterilization instead of heat, and fractionation of macromolecules), electrodialysis (for demineralization instead of ion-exchange) and possibly even pervaporation for specialized applications, instead of extraction and/or distillation. Two industries will be highlighted in this paper: the dairy industry and the fruit juice industry .

The Dairy Industry

Figure 1 shows a general schematic of the possible applications of membranes in the dairy industry. Processing cheese whey, a by-product of cheese manufacture, was the first successful commercial application. "Nanofiltration" or "loose RO" has a specialized application in the dairy industry. Reverse osmosis usage is dependent on prevailing energy costs, since it is used primarily as a substitute for thermal evaporation. Not shown in Fig.1 are electrodialysis and pervaporation; the former has major commercial application (although not on the scale of the pressure-driven membrane processes), but the latter has yet to make its presence felt in the dairy industry.

Milk concentration by reverse osmosis

The use of reverse osmosis for concentration of milk has been studied since the early 1960s. The major attraction to the dairy industry (apart from the obvious savings in energy) was that, unlike thermal evaporation or freeze concentration, moisture removal was accomplished without a change of phase or having to use extremes of temperatures. The milk is exposed to minimum heat during concentration, which avoids protein denaturation, development of the "cooked" flavor and other heat damaging effects on the constituents of milk.

Figure 2 shows typical data obtained for the concentration of milk in a spiral wound thin-layer composite membrane. The asymptotic relationship between applied transmembrane pressure and flux is due to concentration polarization of proteins and fat, giving rise to increased hydrodynamic resistance to permeate flow and/or to higher osmotic pressure. The decrease in flux at higher pressures is probably due to a compaction of the polarized/fouling layer, giving rise to increased resistance, or to a much higher osmotic pressure (Cheryan et al 1990).

Fouling (i.e., a decrease in flux with time with under constant operating conditions) is generally much less with milk than with whey. The nature of the deposit is mostly protein, but inorganic salts such as calcium phosphate play an important role. This salt could act as a binding bridge between the membrane and the protein, leading to a higher hydraulic resistance of the protein layer.

The maximum solids concentration in the retentate is limited largely by the osmotic pressure, which is 600-700 kPa in normal milk. Most commercial modules have a limit of about 3-4 MPa, which means a maximum of 3-4X concentration before the flux drops too low to be economical.

Perhaps the greatest potential for RO in the dairy industry is in bulk milk transport, especially in those countries which have large distances between producing and consumption areas. Considering that milk is more than 85% water, pre-concentration of the milk prior to shipment to central dairies should result in considerable savings in transportation costs, as well as reducing chilling and

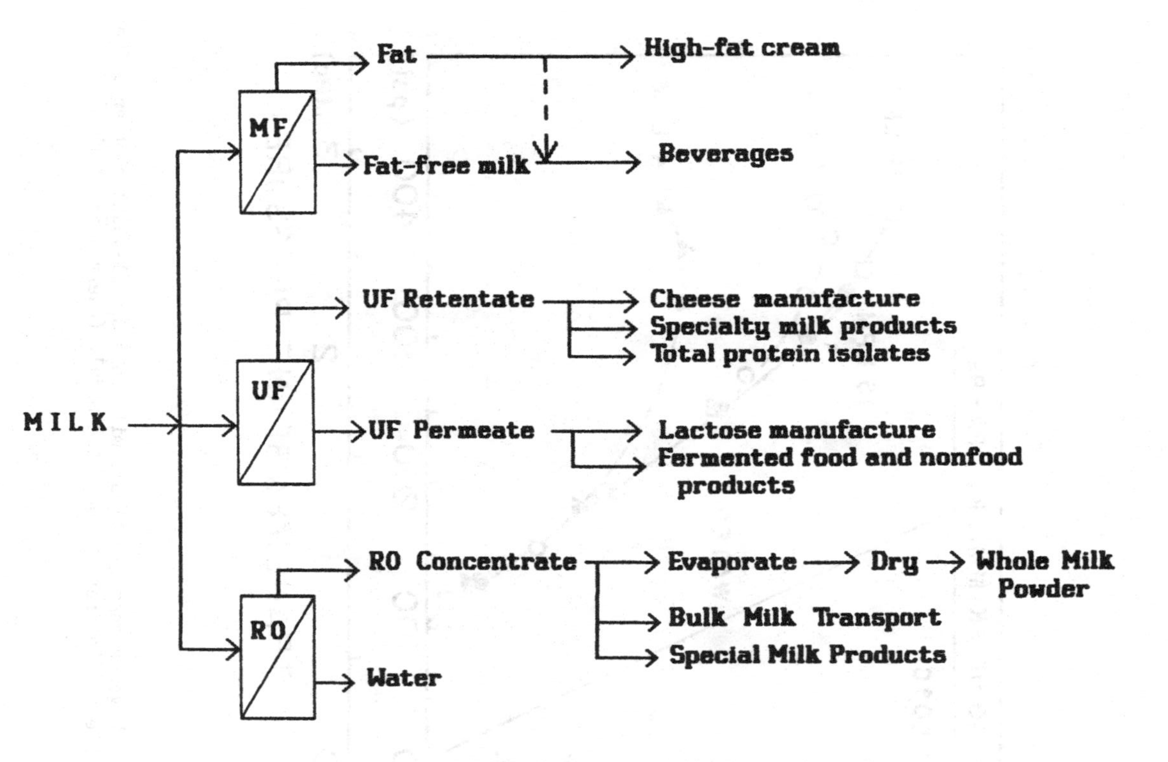

Fig.1. Fractionation of milk using membrane technology

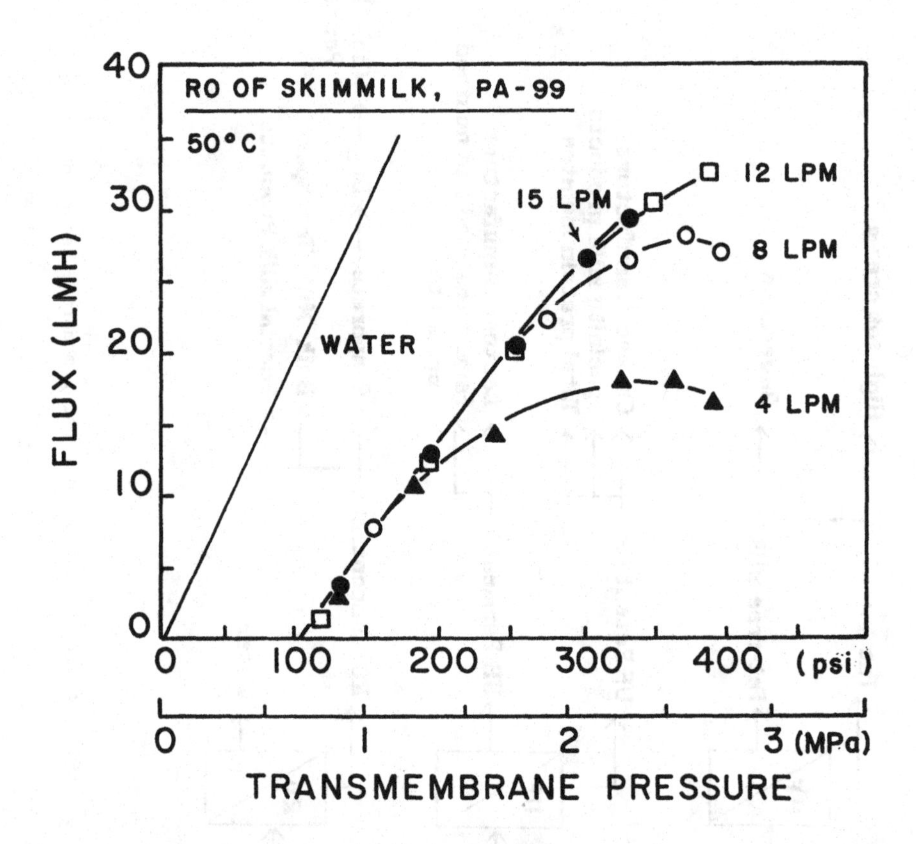

Fig.2. Reverse osomosis of skimmilk in a spiral-wound membrane. The variable is flow rate through the module (Cheryan et al 1990)

storage costs. RO-milk products, when reconstituted with good quality water, are indistinguishable from unconcentrated milks in flavor and other quality attributes. If accepted by the regulatory authorities in the countries concerned, this would result in several large RO plants in USA, Australia, Canada, India, China and possibly even the European common market, among others.

The removal of water from milk during the production of dried milk powder accounts for a significant portion of the product's final cost. Milk (either whole or skimmed) is usually concentrated to 45-50% total solids before spray drying. Because of the upper solids limitation mentioned above, RO cannot be used by itself as a complete substitute for conventional evaporation. RO, however, can be used as a pre-concentration step ahead of the evaporators to reduce the time and energy demands of evaporation, or as a means to increase capacity of existing evaporators. Table II compares the energy consumption of RO and thermal evaporation for a 2.5X concentration of milk.

Although the feasibility of concentrating milk by RO has been demonstrated for many years, it is surprising that there are few such installations in the world. There is ample evidence that, for equivalent capital costs, the operating cost of a RO plant is much less than a thermal evaporation plant (Cheryan et al 1990). The potential energy savings are enormous. In 1986, there were 680,000 metric tons of dried milk products and 864,000 tons of condensed and evaporated milk products produced in USA alone. This represents potentially about 5,500,000 tons of water that can be removed by incorporating a 2.5X RO process in the evaporation plants. Even if only 25% of the condensed, evaporated or dried milk products was processed with RO in this fashion, it represents a savings of at least 300 million MJ per year.

Milk fractionation by ultrafiltration

There is considerable potential in the manufacture of specialty milk-based beverages. Milk is the major source of calcium in the average diet and a significant source of high-quality protein and other nutrients. However, in these days of cholesterol- and fat-consciousness, the dairy industry is facing a challenge to provide healthy low-fat,low-cholesterol fluid milk products, without impairing the other desirable qualities of milk, such as the excellence of its protein and calcium and its "natural" image. Skimming off the fat will reduce calories and the cholesterol, but the resulting milk has poor texture and taste.

To compensate for the poor taste and to maximize the concentration of the desirable protein and calcium, we have developed a process for the manufacture of a low-fat, low-cholesterol, high-calcium fluid milk product using membranes (Figure 3). Since calcium in milk is mostly in the insoluble or bound form, or associated with impermeable casein micelles, concentrating skimmilk by UF will simultaneously concentrate the calcium and protein, while leaving the concentration of sodium, potassium and lactose unchanged, as shown in Table III. Fat can be added back to improve sensory quality if desired. The taste and texture of the UF-milk (called PRO-CAL, to reflect its higher protein and calcium contents) is superior to regular skimmilk, even skimmilk with added milk solids. The low-cholesterol product, with 1% added fat, can be used to make soft-serve ice cream and milk shakes. A similar UF-milk product called PhysiCal has been quite successful for several years in Australia.

"On-farm ultrafiltration of milk" has been studied in USA and France with a

Table II. Comparison of energy consumption and surface areas for RO-concentration of milk 2.5X at a feed rate of 1000 kg per hour (From: Cheryan et al 1987)

PROCESS	Area (m^2)	Energy (kcal/kg milk)
Thermal concentration	10.4	455
Open-pan boiling	10.4	455
Evaporator:		
Double-effect evaporator	25	209
Mechanical Vapor Recompression	32	136
Membrane Process		
Batch, single-pump	65	80
Batch, dual-pump	65	7
Continuous, one-stage	206	16
Continuous, three-stage	93	7

Table III. Composition of whole milk and PRO-CAL, a low-cholesterol, low-fat, high-calcium milk produced by ultrafiltration (Cheryan 1989). Values are per 100 gm serving

Component	Regular Whole milk	PRO-CAL ONE	PRO-CAL ZERO
Fat (gm)	3.5	0.87	0.1
Protein (gm)	3.5	6.1	6.1
Carbohydrate (gm)	4.8	4.8	4.9
Sodium (mg)	52	52	52
Potassium (mg)	161	161	161
Calcium (mg)	130	217	217
Cholesterol (mg)*	14	3.5	0.5
Calories	65	52	49
Calories from fat	48	15	4

* Calculated assuming 0.40% of the fat is cholesterol

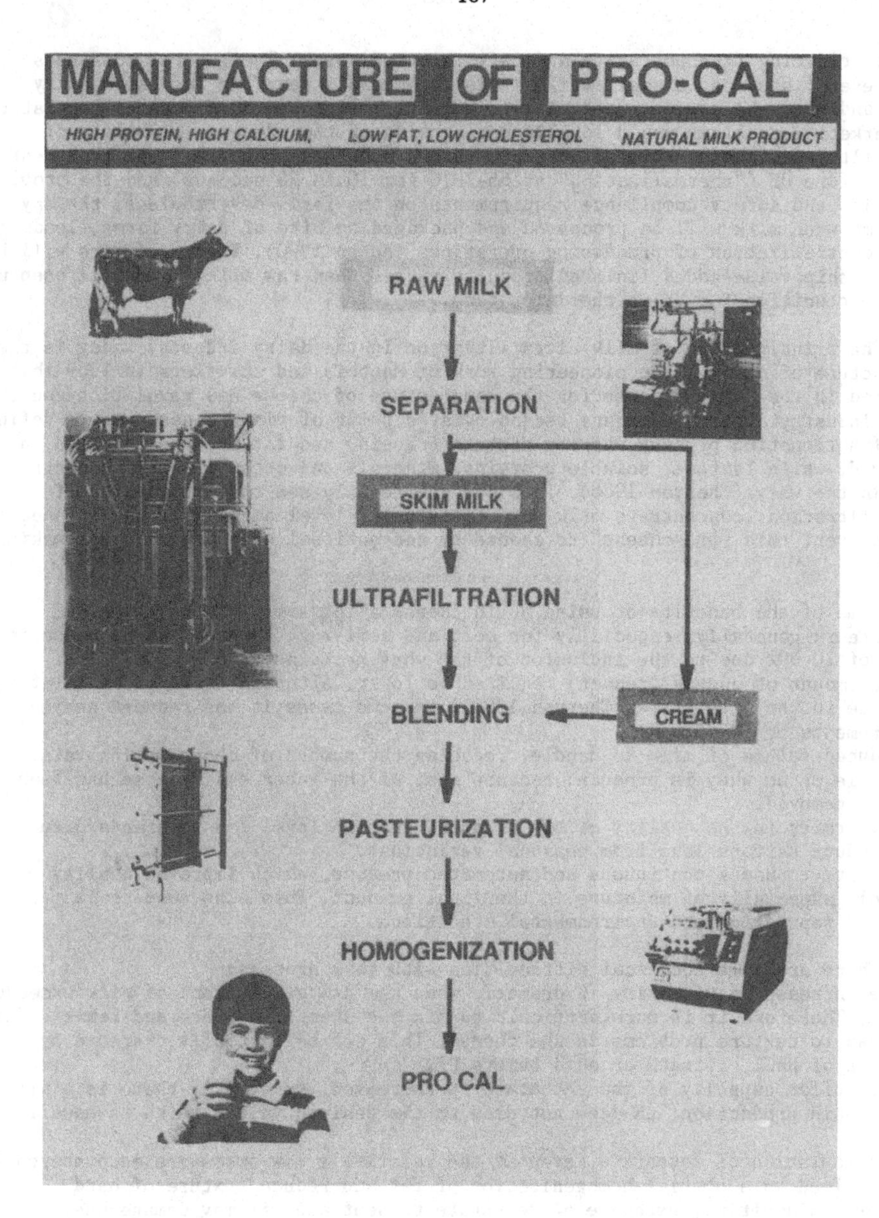

Fig.3. Manufacture of PRO-CAL (a low-fat, low-cholesterol, high-calcium milk) by ultrafiltration (Cheryan, 1989)

view to reducing transportation and refrigeration costs. The general consensus from several studies is that, while it is technically feasible for large dairy herds and when the concentrated milk is used for manufacture of cheese, regulatory and marketing problems would inhibit application of the process. Stability of ultrafiltered milk is satisfactory but there are doubts about the heat treatment given before UF ("thermalization" at 65-70°C for 10 to 20 seconds) and the problem of health and safety compliance requirements on the farm. Nevertheless, the day may come when milk will be processed and packaged on site at dairy farms, leading to de-centralization of processing operations (Honer 1990). Thus producers will be able to ship value-added finished products rather than raw milk. This will open up new opportunities for UF on the dairy farm.

The principal use of milk ultrafiltration in the dairy industry today is the manufacture of cheese. The pioneering work of Maubois and coworkers in 1969 that resulted in the "MMV" process for the manufacture of cheese has revolutionized the dairy industry. From a membrane technologist's point of view, cheese can be defined as a fractionation process whereby protein (casein) and fat are concentrated in the curd, while lactose, soluble proteins, minerals and other minor components are lost in the whey (Cheryan 1986). Thus one can readily see the application of ultrafiltration: concentrate milk to a protein/fat level normally found in cheese, then convert this "pre-cheese" to cheese by conventional or modified cheesemaking methods.

Some of the benefits of using UF in cheesemaking are:

-- There is generally, especially for soft and semi-soft cheeses, an increase in yield of 10-30% due to the inclusion of the whey proteins.

-- The amount of enzyme (rennet) required is lower, although there is conflicting evidence in the literature (Cheryan 1986); in some cases it has reduced enzyme requirements by 70-85%.

-- Reduced volume of milk to handle, reducing the number of cheesemaking vats.

-- Little or no whey is produced because most of the water and lactose has been already removed.

-- Uniformity in the quality of the product. Protein level can be standardized and the product suffers less from seasonal variations.

-- Ability to use a continuous and automated process, which improves quality control, especially of moisture in the final product. This also results in improved sanitation and environmental conditions.

There are some technical difficulties with this process:

-- The increase in viscosity is dramatic when the protein content of milk exceeds 12-14%. Therefore it is more difficult to mix the starter culture and rennet. This may lead to texture problems in the cheese. This can be partially overcome by addition of NaCl, citrate or acid before UF.

-- The buffer capacity of the retentate is increased, so even if there is a higher lactic acid production, pH does not drop to the desired value, which is usually pH 4.6.

-- Recirculation of retentate, even at the relatively low pressures encountered in UF, may lead to a partial homogenization of fat and reduce texture of hard cheeses. In addition, exposure of retentate to heat and air may damage whey proteins, which may affect water content and texture of the cheese.

-- Semi-hard and hard cheeses have a low moisture content, and consequently a high total solids content. A true "pre-cheese" cannot be made for these products. However, even a partial pre-concentration can have significant benefits.

Several cheeses have been made quite successfully in this manner (Cheryan 1986), among them feta, cottage, camembert, mozzarella, ricotta and quarg. However, the manufacture of hard cheeses (e.g., cheddar) by UF has long been considered the ultimate challenge for the cheesemaker and membrane technologist. Perhaps the only successful solution to date comes from CSIRO in Australia, after more than a decade of development. It is now marketed by APV under the "SiroCurd" name. By 1986, the first commercial cheddar cheese plant using the Sirocurd process was installed at the Murray Golburn dairy in Cobram (in Victoria, Australia) producing cheddar cheese that was reportedly indistinguishable from the conventional batch process. A larger plant designed to handle 1 million liters of whole milk per day was installed in 1990 at Land O' Lakes, Minnesota.

A schematic of the process is shown in Figure 4. Whole milk is standardized and pasteurized before being concentrated 5X in a nine-stage UF system utilizing polysulfone spiral modules (not all nine stages are used at start-up; a few are kept in reserve to be brought on line during the processing day to compensate for flux decline). The final three stages include diafiltration to reduce the lactose content of the retentate (this is necessary for proper texture and flavor development of the finished cheese).

The retentate (38-40% total solids) is split into two streams; about 10% is re-pasteurized and goes for fermentation with the starter culture; this is mixed with the rest of the retentate and enzyme (rennet) in a continuous coagulator. The coagulum is then cut and cooked in two syneresis drums where generation of curd and whey occurs, which is then pumped to a draining and matting conveyor and a standard cheddaring system before packaging. The permeate from the UF plant is sent to a RO plant to produce sterile water for cleaning and diafiltration. The major benefits are its continuous nature, higher yields and better control of moisture in the final cheese (Parrott, 1990).

Milk microfiltration

The development of inorganic/ceramic membranes has widened the applications of membranes for processing milk. Although a relatively small application today in terms of installed membrane area, this is expected to increase rapidly in this decade. It is interesting that, although several studies have shown that these inorganic membranes also suffer from typical fouling problems caused by interactions between the membrane, protein and/or minerals, it has not prevented them from being used for separations of milk components by a mechanism that is not fully understood.

The main applications of microfiltration in milk processing are fat separation, bacterial removal and concentration of caseinate. The bacteria and fat in milk are considerable larger in molecular size than all other components. A membrane with pores of 0.3-1.0 microns should be able to separate the fat and bacteria from the rest of the milk components, provided the membrane has a fairly uniform and narrow pore size distribution and the appropriate physico-chemical properties that minimizes fouling. However, attempts with polymeric membranes failed, because a "dynamic" or secondary membrane of the polarized particles quickly caused MF membranes to behave as UF membranes, i.e., the caseins and whey proteins were also being rejected. To minimize the formation of this secondary membrane, the units had to be operated at extremely high velocities (over 6 meters per second) and low transmembrane pressure (less than 40 kPa) to minimize compaction of the polarized layer.

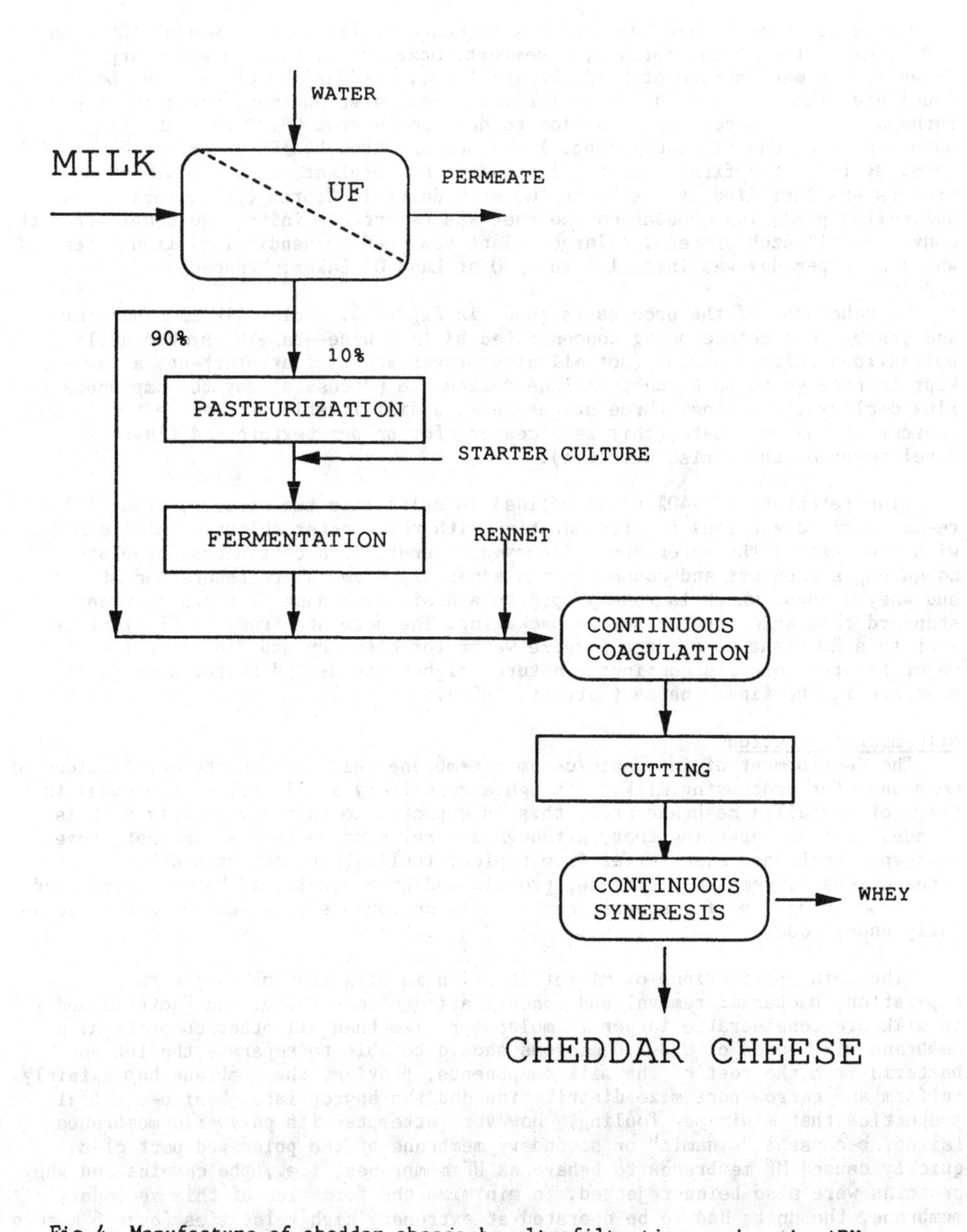

Fig.4. Manufacture of cheddar cheese by ultrafiltration, using the APV SiroCurd Process (Adapted from Parrott, 1990).

This appears to be a contradiction, since high velocities (especially in narrow-diameter tubes, e.g.,less than 3 mm) result in high pressure drops. A pressure profile along the length of a typical module is shown in Figure 5(a). Typically, permeate channels are operated at low (close to atmospheric) pressures. On the retentate side, high inlet pressure would cause fouling by compaction at the inlet of the module, while the low pressure at the outlet meant that some of the available membrane area towards the outlet was not being utilized effectively. The result is a precipitous drop in flux and a protein rejection of more than 90%.

The key to the eventual successful development of the "Bactocatch" process (a trade name for the process developed by Alfa-Laval) is the "constant pressure" or "high-flux, low-fouling (HFLF)" mode of operation (Malmberg and Holm 1988). As shown in Figure 6, this requires the simultaneous operation of a retentate pumping loop and a permeate pumping loop, to simulate a backwashing operation, but in a continuous manner rather than the periodic or intermittent manner practiced traditionally. With two parallel flows adjusted so that the pressure DROP was the same on the permeate and retentate sides of the module, the pressure profile would be more like that shown in Fig.5(b). The flux would remain high and protein rejection low, typically less than 3%.

This combination of tubular ceramic membranes (rated at about 1.4 micron pore diameter from Alcoa-Ceraver) and the double loop constant- pressure operation has produced dramatic results. Fluxes of 500-700 liters per square meter per hour can be achieved for a period of over 6 hours. If needed, the transmembrane pressure can be increased slightly by of 0.05-0.1 bar (0.75-1.5 psi) to compensate for fouling. Bacterial retention is 99% with the microbial load usually found in milk. On the other hand, there is no significant change in the concentration of other components, so the permeate is essentially bacteria-free skimmilk.

The Bactocatch process became commercial in 1988-89 with one commercial operation in Sweden. Its application is supposed to result in more stable pasteurized and refrigerated milk products. Perhaps it could be more useful in subtropical and tropical countries, where inadequate refrigeration and transportation facilities results in high microbial loads in the milk coming in to dairy plants. A Bactocatch system on the receiving dock to lower the microbial load can significantly improve the quality of the milk products in these countries.

Cheese Whey

Whey is the liquid fraction that is drained from the curd in cheese making. Every 100 kg of milk used in cheese manufacture results in 10-20 kg of cheese and 80-90 kg of whey. Even though individual components of cheese whey (proteins and lactose) have uses by themselves, whey itself was difficult to dispose of or utilize, due to its unfavorable lactose-to-protein ratio and high biological oxygen demand of 30,000-50,000 ppm. Ultrafiltration offered a way to improve the ratio in favor of the protein. Figure 7 shows a general schematic of possible applications of membranes in whey treatment.

Reverse osmosis of whey is done primarily to reduce energy consumption in whey drying plants. If used as a pre-concentrator, it is probably best to do a 2X concentration (to 12% solids) by RO, and the rest (12% to 45% solids) by evaporation. However, concentration to 18-27% total solids may be justified if bulk transport of whey is the primary goal. The limiting factors to high concentration by RO are osmotic pressure, viscosity and solubility of the lactose

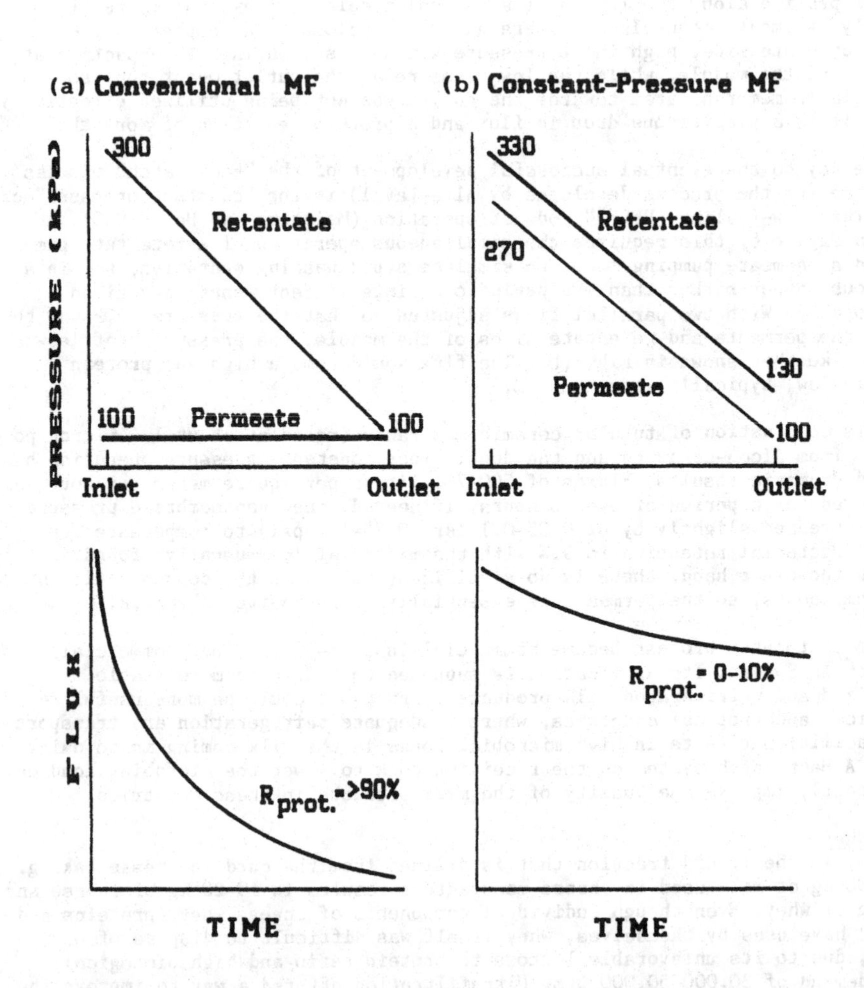

Fig.5. Typical pressure profiles in a ceramic tubular module during microfiltration of milk. (a). Conventional microfiltration (b) Constant-pressure microfiltration. R_{prot} is rejection of protein.

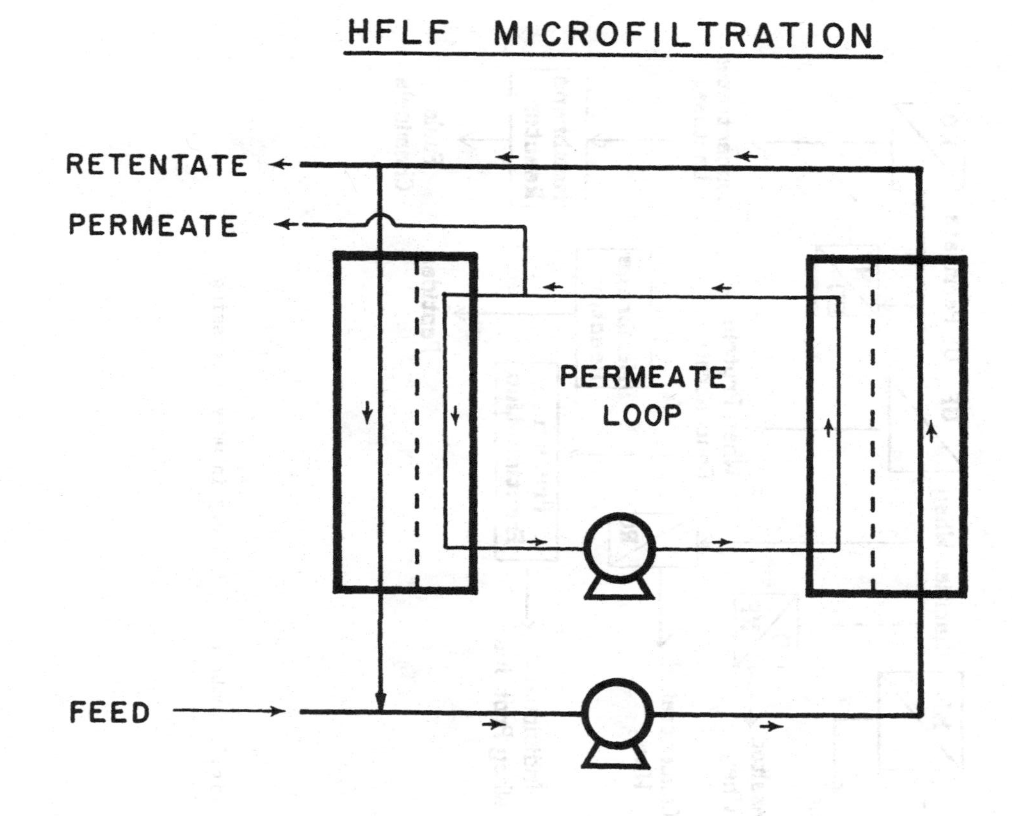

Fig.6. Schematic of constant-pressure microfiltration for high-flux, low-fouling (HFLF) operation.

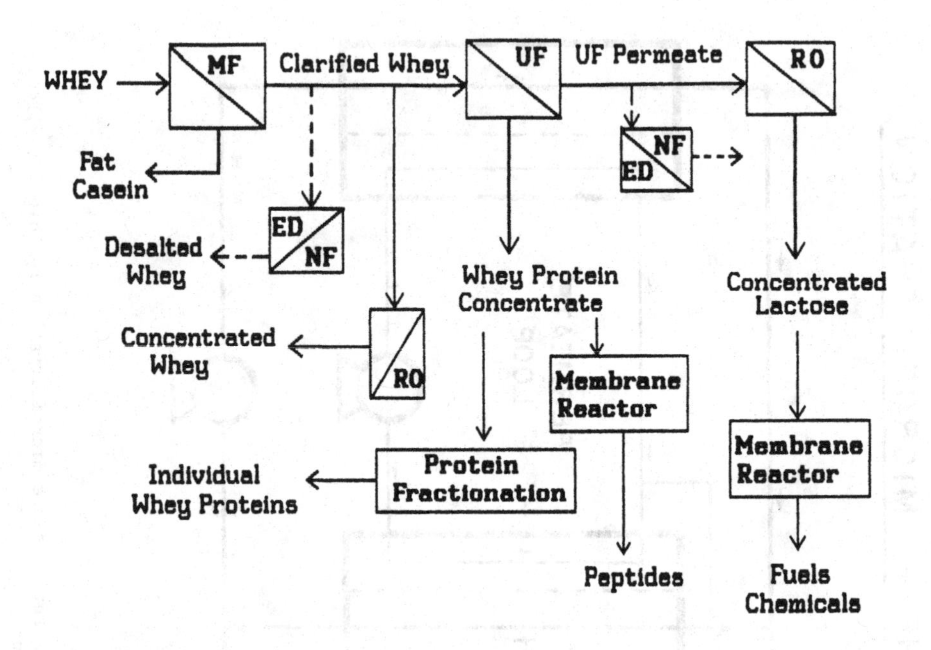

Fig.7. Membrane technology in whey processing

and calcium salts, which may precipitate out at high concentrations, especially if temperature and pH are not regulated.

Today, whey protein concentrates produced by ultrafiltration are well-established in the food and dairy industry. Owing to the relatively mild process conditions of temperature and pH, the functionality of the whey proteins remain good, giving rise to a wide range of applications. The initial protein content of 10-12% (dry basis) can be increased by ultrafiltration to result in 35%, 50% or 80% protein products, with a concomitant decrease in lactose and some salts. Further fractionation of the components into beta-lactoglobulin and alpha-lactalbumin fractions can be also be done.

Whey is a severe foulant of membranes. It appears to foul hydrophobic membranes (e.g., polysulfone) much more than hydrophilic membranes (e.g., cellulose acetate). Numerous studies (e.g., see Cheryan 1986) have shown that membrane fouling in whey ultrafiltration is due to proteins and salts. The appropriate pre-treatment is important for high fluxes. Heat, pH-adjustment, addition of calcium-sequestering agents, removal of minerals by ion-exchange or electrodialysis have been suggested.

The major application of MF is as a pretreatment for UF of whey. Whey usually contains small quantities of fat (in the form of small globules of 0.2-1 microns) and casein (as fine particulates of 5-100 microns). Centrifugal separation of whey does not completely remove the fat and casein fines. Thus when the whey is ultrafiltered, these components can have detrimental effects on the functional properties of the whey protein concentrates, especially the lipid-containing components. A process similar to the HFLF or "Bactocatch" described earlier, with tighter membranes of about 0.2 microns, can effectively remove substantial quantities of these undesirable components. Fat/protein ratios of 0.07-0.25 in whey can be reduced to 0.001-0.003 by microfiltration (van der Horst and Hanemaaijer 1990). In addition, some of the precipitated salts may be removed, and there is a considerable reduction in microbial load.

Fruit juices

There are three primary areas where membranes can be applied to the processing of fruit juices:

(1) Clarification, e.g., in the production of sparkling clear beverages using microfiltration or ultrafiltration

(2) Concentration, e.g., using reverse osmosis to produce fruit juice concentrates of >42°Brix,

(3) Deacidification, e.g.,electrodialysis to reduce the acidity in citrus juices.

The use of cross-flow microfiltration for clarification has been covered by Short and Skelton(1991) and electrodialysis for deacidification by Lopez-Leiva (1988). These topics will not be discussed in this paper. Reverse osmosis has not yet made a significant penetration into the fruit juice industry, even though the potential is vast. The U.S. harvest of oranges in 1985-86 season was 7.5 million metric tons, of which approximately 82% was processed and marketed as frozen orange juice concentrate. The major benefit of RO is that it avoids thermal damage of delicate aroma compounds.

Early work by Merson and Morgan (1968) with apple juice showed that cellulose acetate membranes showed good sugar retention, but aroma/flavor compounds permeated the membrane, which lowered the quality of the concentrate. Polyamide RO membranes retain more of the flavors than cellulose acetate. Fresh fruit juices of 10-15°Brix have osmotic pressures of 1.4-2MPa (215-300 psi), which limits concentration to 20-25°Brix with conventional RO. Pectins contributed to fouling and viscosity which also affect plant performance. Thus apple juice RO is usually preceded by depectinization by enzymes and clarification (usually by UF), which then makes it one of the easiest liquids to process by RO (Pepper 1990). Cleaning and restoration of capacity at the end of a process cycle is no problem. This suggests that spiral modules are probably the best design for this application.

With orange and other citrus juices, many of the flavor compounds reside in an oil phase present in the juice in the form of an emulsion. Since they are sparingly soluble in water, flavor retention should be much better when membrane-processed than with apple juice. Orange juice concentrates are today produced mostly by conventional multistage evaporation. Most of the citrus essences are stripped in the first effect, and in the later stages other compounds are destroyed or transformed into other undesirable compounds (e.g., furfurals).

The mid-1970s saw intense research activity in this area (Matsuura et al 1974, Watanabe et al 1978) primarily with cellulosic membranes. A high retention of sugars, acids, phenolics, nitrogen compounds and ash was obtained, but only 30-100% retention of volatile aroma compounds. Reverse osmosis with a polyamide composite membrane can concentrate juice without a significant loss of aroma, sugar or acids (Medina and Garcia 1988). However, conventional RO is again limited by osmotic pressure and viscosity considerations to less than 30° Brix. Therefore RO can be used as a pre-concentration step, with thermal evaporation completing the required concentration to 42° Brix. Adding RO ahead of the evaporators can increase evaporator capacity and reducing thermal treatment. The first commercial RO plant used PCI AFC-99 tubular membranes with a feed rate of 4200-9200 l/hour and a water removal rate of 2000 l/h. The feed was pasteurized orange juice with a dissolved solids content of 10-12° Brix, pH 3.2-3.7, and a pulp content of 2-7.5%. It was necessary to do a quick flush with alkali every 4-6 hours to remove the hesperidin that fouled the membranes (Pepper 1990). Flavor of the concentrate has been judged to be very good and comparable to single-strength fresh juice when rediluted.

Perhaps the most significant development in recent years is the production of highly concentrated (42-60°Brix) fruit juices using a combination of high and low-retention RO membranes. The osmotic pressure of the juices such as orange juice containing 10-12% soluble solids (10-12°Brix) is about 250-300 psi. It increases to 1500 psi at 42°Brix and 3000 psi at 60°Brix. The resulting high viscosity and osmotic pressures would result in very high energy consumption and require modules capable of pressures beyond present-day state of the art.

The "FreshNote" process was developed by SeparaSystems, a joint venture of du Pont and Food Machinery Corporation, to overcome these limitations (Cross 1988, Walker 1990). As shown in Figure 8, the basic concept is a two-stage process. Ultrafiltration is first used with the juice to separate out the pulp (the "bottom solids") from the serum which contains the sugars and flavor compounds. The bottom solids contains some soluble solids, all the insoluble solids, pectins, enzymes, orange oils and the microorganisms that would affect the stability of the concentrate. The UF retentate, about 1/10-1/20th the feed volume, is subjected to

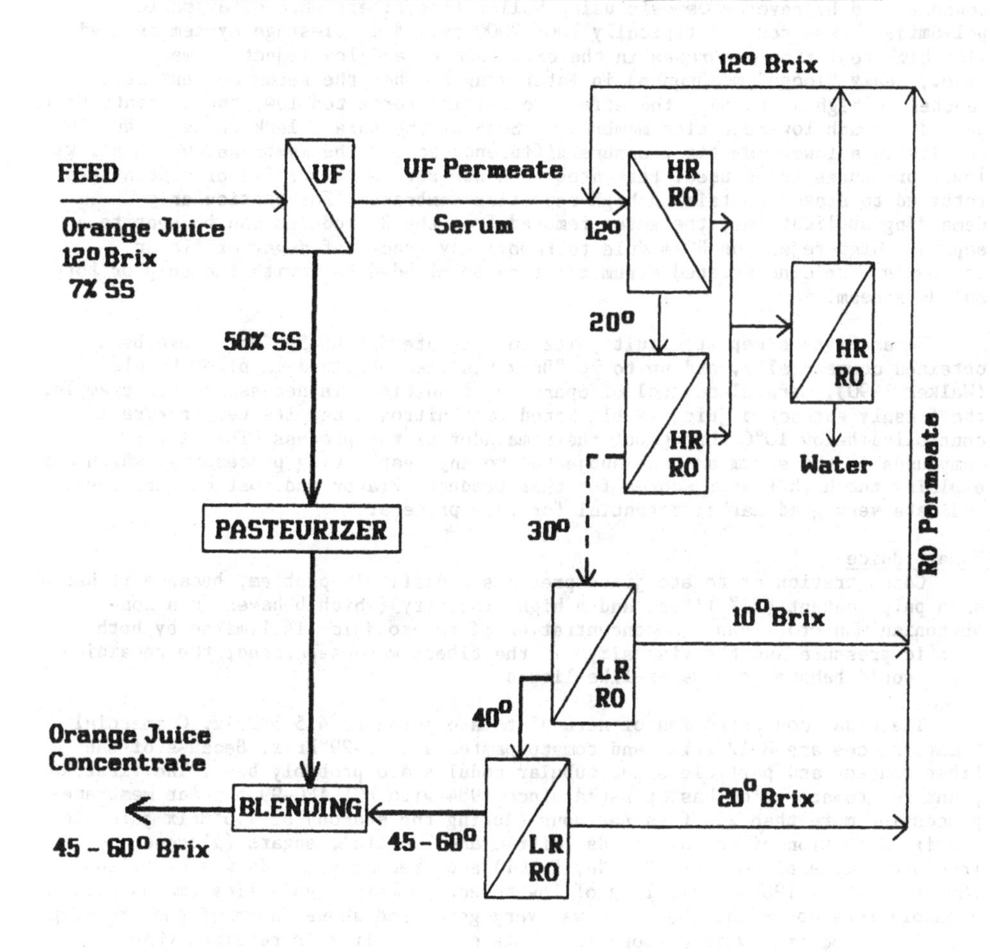

Fig.8. The "FreshNote" process for fruit juices, using a combination of high-rejection (HR) and low-rejection (LR) membranes. (Adapted from Cross 1988, Walker 1990).

a pasteurization treatment that destroys spoilage microorganisms and improves stability of the finished product when blended back with the RO-concentrated UF permeate.

The serum (UF permeate), which amounts to about 90-95% of the feed volume, is concentrated by reverse osmosis using hollow fine fibers made of aromatic polyamide. Pressures are typically 1000-2000 psi. A multi-stage system is used with high rejection membranes in the early stages and low rejection membranes (i.e., leaky "loose" membranes) in later stages. When the serum concentration reaches so high as to make the effective driving force too low, the concentrate is pumped through low-rejection membranes. Some of the sugars leak through, but this results in a lower osmotic pressure difference across the membrane, which allows lower pressures to be used. Permeates with non-zero sugar or flavor content are returned to stages containing high-rejection membranes. For particularly demanding applications, the water removed from the RO modules can be sent to a separate high-rejection RO module to remove any traces of sugar or flavor compounds. The concentrated serum can then be blended back with the pulp or bottom solids stream.

SeparaSystems reports fruit juice concentrates of 45-55 °Brix have been obtained commercially, and up to 70 °Brix has been obtained in pilot trials (Walker 1990). Careful control of operating conditions is necessary. For example, the freshly extracted juice is blanketed with nitrogen and its temperature is controlled below 10°C throughout the remainder of the process. The flavor compounds in the serum are not subjected to any heat during processing, which also explains the high flavor scores for this product. Flavor and cost comparisons indicate very good market potential for this process.

Tomato Juice

Concentration of tomato juice presents a difficult problem, because it has a high pulp content (25% fiber) and a high viscosity (which behaves in a non-Newtonian manner). Thus, RO-concentration of tomato juice is limited by both osmotic pressure and the viscosity. If the fibers were separated, the remaining serum would behave as a water-like liquid.

The usual concentration of natural tomato juice is 4.5-5°Brix. Commercial tomato sauces are 8-12°Brix, and tomato pastes are 28-29°Brix. Because of the fiber content and particle size, tubular modules are probably best. The first RO plant for tomato juice has operated since 1984 with PCI AFC-99 tubular membranes, processing more than 250 tons per hour (during the season) of 4.5°Brix juice to 8°Brix. Retention of organic acids (citric and L-malic), sugars (glucose, fructose), mineral ions (K, Mg, Na, P, Cl) and free amino acids was excellent (Gherardi et al 1986). Some loss of low molecular weight volatiles (methanol, ethanol) were observed. The color was very good, and showed none of the browning normally associated with evaporation. This color quality is retained when evaporated to 28-30° Brix paste because of the reduced time in the evaporator. This plant in Italy has operated for five seasons with the same set of membranes, despite aggressive cleaning (Pepper et al 1990).

The original work was stimulated by potential energy savings. Further development work has been done to reach 15-20°Brix, although direct reverse osmosis may not be the most economical way of reaching this concentration. Using a combination of HR (high rejection) and LR (low rejection) membranes as with the "FreshNote" process has not been successful so far because the reblended pulp and serum concentrate separate out on standing (Pepper et al 1990).

SUMMARY AND CONCLUSIONS

Membrane technology will continue to be applied on a large scale in the food and allied industries. As shown in Table 1, there are many areas in addition to the dairy and fruit juice industries that membranes have been used. Some industries, e.g., wet corn refining and oilseeds, have only recently begun to apply this technology in their operations. Considering the vast amounts of water used, these industries may experience the greatest growth of membrane applications in this decade.

REFERENCES

Belfort, G. (Editor). 1984. Synthetic Membrane Processes. Academic Press, New York.

Cheryan,M. 1986. Ultrafiltration Handbook, Technomic, Lancaster, PA.

Cheryan,M. 1989. PRO-CAL: a high-calcium, low-cholesterol milk produced by ultrafiltration. Unpublished report, University of Illinois, Urbana.

Cheryan,M., Sarma,S.C. and Pal,D. 1987. Energy considerations in the manufacture of khoa by reverse osmosis. Asian J.Dairy Res. 6:143-153.

Cheryan,M., Veeranjaneyulu,B.V. and Schlicher,L.R. 1990. Reverse osmosis of milk with thin-film composite membranes. J.Membrane Sci. 48:103-114.

Cross,S. 1988. Achieving 60 Brix with membrane technology. Presented at the 49th Annual Meeting, Institute of Food Technologists, New Orleans, June 19-22.

Gherardi,S., Bazzarini,R., Trifiro,A., Voi,A.L. and Palamas,D. 1986. Preconcentration of tomato juice by reverse osmosis. Int. Fruchtstaft-Union,Wiss.-Tech.Komm.,[Berlin]. 19:241-252.

Honer,C. 1990. Does ultrafiltration have a place on the farm? Dairy Field Today 173 (3): 52.

Lopez-Leiva,M.N. 1988. The use of electrodialysis in food processing. Part 2. Review of practical applications. 21: 177-182.

Malmberg,R. and Holm,S. 1988. Low bacteria skim milk by microfiltration. North European Food Dairy J. 1: 75-78.

Matsuura,T., Baxter,A.G., Sourirajan,S. 1974. Studies on reverse osmosis for concentration of fruit juices. J.Food Sci. 39: 704-711.

Maubois, J.L. 1989. Paper presented at North American Membrane Society annual meeting, Austin, TX. June 1989.

Maubois, J.L. 1989. Paper presented at North American Membrane Society annual meeting, Austin, TX. June 1989.

Medina,B.G. and Garcia,A. 1988. Concentration of orange juice by reverse osmosis. J.Food Process Engr. 10: 217-230.

Merlo,C.A., Rose W.W., Petersen,L.D., White,E.M. and Nicholson,J.A. 1986. Hyperfiltration of tomato juice. Pilot-scale high temperature testing. J.Food Sci. 51: 403-407.

Merson,R.L. and Morgan,A.I. 1968. Juice concentration by reverse osmosis. Food Technol. 22(5): 97-100.

Parekh, B.P. (Ed). 1988. Reverse Osmosis Technology, Marcel Dekker, Inc.NY.

Parrott,D.L. 1990. The application of spiral-wound ultrafiltration membranes in the APV Sirocurd process for the continuous production of cheddar cheese. Proceedings, The 1990 International Congress on Membranes and Membrane Processes, Vol.1, p.265-266.

Pepper,D., Orchard,A.C.J. and Merry,A.J. 1985. Concentration of tomato juice and other fruit juices by reverse osmosis. Desalination 53: 157-166.

Pepper,D. 1990. RO for improved products in the food and chemical industries and water treatment. Desalination 77: 55-71.

Porter,M.C. (Ed). 1990. Handbook of Industrial Membrane Technology, Noyes Publications, Park Ridge, NJ.

Rautenbach,R. and Albrecht,R. 1989. Membrane Processes. John Wiley, New York.

Short,J.L. and Skelton,R.1991. Cross-flow microfiltration in the food industry. Presented at International Conference on Effective Industrial Membrane Processes: Benefits and Opportunities, Edinburgh, Scotland. March 19-21.

Sourirajan, S. and Matsuura, T. 1985. Reverse Osmosis/Ultrafiltration Principles. National Research Council Canada, Ottawa, Canada.

Van der Horst,H.C. and Hanemaaijer,J.H. 1990. Cross-flow microfiltration in the food industry. State of the art. Desalination 77: 235-258.

Walker, J.B. 1990. Membrane process for the production of superior quality fruit juice concentrates. Proceedings, The 1990 International Congress on Membranes and Membrane Processes, Vol.1, p.283-285.

Watanabe A., Kimura,S. and Kimura,S. 1978. Flux restoration of reverse osmosis membranes by intermittent lateral surface flushing for orange juice processing. J. Food Sci. 43: 985-988.

CROSSFLOW MICROFILTRATION IN THE FOOD INDUSTRY

John L. Short Robert Skelton
Koch Membrane Systems

SUMMARY

The recent commercialization of cross flow microfiltration in the food industry includes such applications as; the clarification of highly colored fruit juices (grape, cranberry, cherry, etc.); the clarification of gelatin prior to concentration by ultrafiltration; and the filtration of liquid sweeteners for the beverage industry. Cross flow filtration shows particularly attractive economics in the displacement of diatomaceous earth filtration as had already been proven in the 80's, in the apple juice industry, worldwide. Higher and more consistent product clarity, lower operating costs, and ease of operation compared to rotary vacuum filtration, as well as elimination of the need to buy (frequently import) and subsequently dispose of used filtration aids (diatomaceous earth, etc.), have provided the incentive to switch to cross flow microfiltration.

NOMENCLATURE

Reverse Osmosis (R.O.)	A cross flow membrane filtration process which can filter dissolved salts from water (operates typically at 200-1000 psi, $1.4x10^6$kpa-$6.9x10^6$kpa).
Ultrafiltration (U.F.)	A cross flow membrane filtration process which can filter our macro-molecular solutes such as proteins from aqueous solution (operates typically at 20-150 psi, $1.4x10^5$kpa-$1.0x10^6$kpa).
Microfiltration (M.F.)	Filtration of insoluble material in the size range of 0.1 to 10 or greater from solution. May be "cross flow" or "dead end" (operates typically at 5-100 psi, $3.4x10^4$kpa-$6.9x10^5$kpa)

Effective Industrial Membrane Processes — Benefits and Opportunities, pp.181–189

NOMENCLATURE
continued

Retentate	**The liquid containing retained species which does not pass through the membrane during processing.**
Permeate	**The liquid containing dissolved species which passes through the membrane during processing.**
"Cross Flow" Filtration	**A cross flow or tangential flow filter has 3 streams; feed, retentate, and permeate. The flow of the feed/retentate is parallel or tangential to the filter surface.**
"Dead End" Filtration	**Conventional filtration with 2 streams; feed (flowing normal to the filter) and filtrate.**
Whey	**The residual fluid after cheese or casein production from milk.**
Whey Protein Concentrate (WPC)	**Whey proteins concentrated by ultrafiltration and subsequently dried. Protein to T.S. ranges from 35% to 75% and depending on intended application.**
Enzyme	**A natural catalyst which increases the rate of a specific chemical reaction without itself being used up.**
Pyrogens	**Breakdown product of bacterial cell walls which cause febrile reaction (fever) when injected into the blood stream.**

NOMENCLATURE
continued

"Fouling"	**Irreversible absorption of feed components on the membrane causing decrease in productivity. Only reversible by chemical cleaning.**
"Concentration Polarization"	**Build up of retained species as a layer at the membrane, controlled by operating conditions of pressure and flow.**
Brix	**% of sugar in solution, typically measured by refractometer.**
Asymmetric Membrane	**A membrane, which in cross section has smaller pores on one surface than the opposite surface, often a skinned structure having a tight surface skin which does the actual separation, supported by a porous substructure.**

INTRODUCTION

In the early Seventies, Ultrafiltration (U.F.) and Reverse Osmosis (R.O.), were being commercialized in food industry applications. One of the first developments in the application of these cross flow filtration operations in the food industry was carried out under a Federal Water Quality Administration grant to Crowley's Milk Company of Binghamton, New York. Abcor (as Koch Membrane Systems was previously known) acted as Process Engineering subcontractor for a pilot unit with a capacity of 500 pounds (227 kg) of whey per hour; followed by a full scale demonstration plant with a 10,000 pounds (4545 kg) per hour throughput, and finally a 250,000 pounds (113640 kg) per day commercial unit. (Horton et al 1970). The success of this work was such that the Ultrafiltration plant is still in operation and the Reverse Osmosis plant was only recently shut down after almost 20 years of operation.

The ability to fractionate whey into its major components protein and lactose, has developed into a major business in the intervening two decades. New membranes made of chemically and thermally stable engineering polymers (such as polyvinylidene fluoride, polyether sulfone, etc), module design improvements improving fluid dynamic efficiency and mass transfer (various "thin channel" designs including "plate and frame" and spiral wound) and market place demand for sophisticated functional proteins have caused this technology to mature rapidly. In the mid-seventies another application of cross flow membrane technology became commercial, namely the clarification of apple juice by means of ultrafiltration. An ultrafiltration membrane has the ability to both clarify (remove yeasts, molds, bacteria, colloids, and suspended solids) and "fine" (remove proteins, polyphenols, tannins, polysaccharides, and their complexes) to produce a stable sparkling clear juice (Heatherbell et al). The first commercial applications in juice processing of cross flow filtration technology took place in France and South Africa. Introduction to the apple juice industry in the USA happened in the early 80's and by the end of the decade, apple juice clarification by U.F. was commercial around the world.

CROSSFLOW MICROFILTRATION

As indicated above, many applications of ultrafiltration technology in the Food Industry are based on the ability of the membrane process to concentrate and purify valuable components in the feed (for example enzymes) increasing product yield (cheese making) or adding value to materials previously wasted due to their low concentration (the recovery of whey protein concentrates). Cross flow microfiltration has the ability to clarify streams containing high or low molecular weight solutes as an alternative to conventional filtration technologies using filter aids. Cross flow microfiltration may be used as a pretreatment for other membrane processes which achieve concentration of solutes such as Ultrafiltration or Reverse Osmosis.

GELATIN

A new application of cross flow membrane technology where a traditional unit operation (rotory vacuum filtration using diatomaceous earth) has been replaced, is in the production of gelatin. Gelatin is defined as the proteinaceous solution derived from the selective hydrolysis of collagen, the major constituent of the connective tissue in skin and bones.

Commercial uses for gelatin in the food, pharmaceutical and photographic industries depend on its ability to form gels, to change reversibly from gel to solution phase, its high viscosity in aqueous solution and its effectiveness as a protective colloid. The two types of gelatin, acid of alkaline pretreated, are subsequently neutralized and have excess fat skimmed off prior to extraction. After conversion of collagen to gelatin, purification takes place prior to concentration and drying.

The filtration of gelatin to remove dirt, coagulated protein, fats and other insoluble particles from the raw stock prior to concentration has typically been carried out by rotary vacuum filtration using diatomaceous earth at a rate of 20 kilograms/100 kilograms gelatin (dry basis). The development in the mid 80's of asymmetric, microfiltration membranes suitable for use in the food industry in both tubular and spiral configuration made the displacement of the conventional technology with cross flow filtration a reality.

The key features of these membranes include; fabrication from polyethersulfone, (an advanced engineering polymer with excellent chemical and thermal stability) cast on a polyolefin non-woven backing material for mechanical strength in flat sheet or tubular configuration, asymmetric structure to eliminate pore plugging, as well as controlled and even distribution of pores with a smooth surface designed specifically for the process/cleaning cycles required in commercial operation.

The commercially demonstrated benefits of cross flow microfiltration for the clarification of gelatin prior to concentration by ultrafiltration include the following:

__QUALITY__ - **Cross flow microfiltration provides a consistent quality product with no significant losses, no adjustment is required for feed variability.**

__UPKEEP__ - **Cross flow microfiltration provides an easier and cleaner method for clarification. It is an automated process with "clean-in-place" capability.**

__SAVINGS__ - **An economic analysis shows a payback of three years for a capacity of approximately 30,000 years kilograms/hour.**

The ability of the microfiltration membrane, when operated in cross flow, to retain suspended and insoluble material while allowing proteins of relatively high molecular weight to permeate freely with fluxes which translate to viable economics have been the key to this success.

SYRUPS AND SWEETENERS

In a similar fashion to the use of cross flow ultrafiltration to clarify apple juice, the depyrogenation of glucose (dextrose) solutions has been commercial for several years. Pyrogens are breakdown products of bacterial cells walls which cause a febrile reaction (fever) when injected into the bloodstream. The use of cross flow ultrafiltration for the removal of pyrogens is now accepted by the Japanese Pharmacopeia. Ultrafiltration membranes with a molecular weight cut-off of 5K to 10K are generally accepted for this application.

In the production of glucose/dextrose syrups from starch, filtration to remove suspended solids, etc. is required as the syrups are required to be clear and free from haze. As with the juice and gelatin industries the "conventional" technology was rotary vacuum filtration using diatomaceous earth as a filter aid. Whereas apple juice is generally 10 brix (% sugar), syrups are typically 30-40 brix and thus much higher viscosity. The loading and type of suspended solids is very different to juice which contains more fibrous material requiring "open channel" tubular membrane modules. It has now been shown commercially that spiral modules with feed channel spacers proven on high viscosity feed streams such as precultured milk in the dairy industry, work well for clarification of syrups. Multi stage "stages-in-series" plant design allows for automated continuous operation 20-22 hrs/day 2-4 hrs/day being reserved for "cleaning-in-place".

In summary the benefits of cross flow microfiltration are as follows;

Clarification quality - Cross flow membrane produces a superior quality solution of less than 2 NTU with essentially no suspended solids present.

Clarification consistency - The positive membrane barrier results in the same quality of filtration day in and day out. These are no upsets, break through etc. Thus, down stream processing can be enhanced (optimized) because it processes a uniform quality feed.

No diatomaceous earth (DE) - the primary economic incentive is the elimination of the need for DE for filtering the feed solutions. This significantly reduces the customer's operating costs and results in reasonable returns on investment on a substitution basis. For those cases in which plant expansion is being considered, an even more favorable situation for membrane processing would exist.

At many plant sites, D.E. disposal has become an expensive problem. No longer are plants permitted to dump in open unregulated areas or sell to farmers as a feed bulking substance. The consensus among the industry is that this situation will become more regulated with time, thus further increasing the disposal costs.

Decrease down time - When upsets or break through do occur, rotary vacuum filters may need to be recoated with D.E. media, thus resulting in extra labor and D.E. consumption.

Reduced wear - Because of the abrasiveness of the D.E. media, plants must annually replace pipe lines, valves and pumps through which the slurry is pumped.

Reduced Resin Fouling - Since the quality of membrane filtration is superior to the D.E. operation, down stream polishing filters will have longer life cycles and thus lower operating costs.

Table 1 illustrates the comparative economics for cross flow microfiltration versus conventional technology. As with the gelatin example a payback of approximately 3 years is typical, however not all the advantages can be directly translated into dollars and cents.

Table I
Dextrose Clarification

System Size 300 USGPM, 95 D.E.

	MF	ROTOVAC
System	$ 700,000	-------
Membranes	150,000	-------
Installation	150,000	-------
Total Installed Capital Cost:	$ 1,000,000	Existing

Annual Operating Costs (22 hrs/day, 350 days/year:

	MF	ROTOVAC
Labor	$ 22,000(5hr/day)	$55,000 (12hr/day)
Electrical	95,000(310 kw)	58,000 (190 kw)
Maintenance	5,000	25,000
Chemicals	15,000	2,000
D.E.	-0-	400,000
Membranes	75,000	-0-
	$ 212,000	$ 540,000

Annual Operating Savings	$ 328,000
Less Depreciation	$ (100,000)
Annuals Savings	$ 228,000
Less Taxes (36.4%)	83,000
Annual Total Savings:	$ 145,000

REFERENCES

Horton, B.S., Goldsmith, R.L., Hossain, S., and Zall, R.R., Membrane Separtions Processes for the Abatement of Pollution from Cottage Cheese Whey, Presented at the Cottage Cheese and Cultured Milk Products Symposium, University of Maryland, March 11, 1970.

Heatherbell, D.A., Short, J.L. and Struebi, P., Applejuice Clarification by Ultrafiltration, Confructa Vo. 22 (1977(No. 5/6, pp. 157-169.

REFERENCES

[illegible] R.D., [illegible], W.L., [illegible], S., [illegible], E.A. [illegible] Techniques for the Analysis [illegible] Proceedings [illegible] University of Maryland, March [illegible] 1970.

[illegible], Short, [illegible] and Stone, [illegible] Applications [illegible] Contracts Vol. 72 [illegible] pp. [illegible]-69.

USING MEMBRANES TO PURIFY WATER AND DE-ALCOHOLISE BEER

Dr. J. R. Cross (Elga Ltd, UK)

ABSTRACT

Reverse osmosis is a membrane process which can be used both for purifying water and removing alcohol from beer.

The sharp selectivity of the latest thin-film composite membranes enables them to retain the molecules responsible for the characteristic flavour, colour and aroma of beer, while allowing alcohol and water molecules to permeate through. "Diafiltration water" must be added continually to the beer during de-alcoholisation to maintain its normal composition.

This paper describes a case-history in which the use of reverse osmosis - both to de-alcoholise beer and purify the diafiltration water - resulted in substantial benefits for the brewery in terms of product quality, costs and plant operation.

INTRODUCTION

In common with most other membrane processes, reverse osmosis (RO) is a separation technique that operates without the application of heat, and without a change of phase or the addition of chemicals. It is unique among membrane processes, however, in being able to remove the complete spectrum of impurities found in water supplies (Figure 1).

FIGURE 1 APPLICATION RANGE OF MEMBRANE PROCESSES

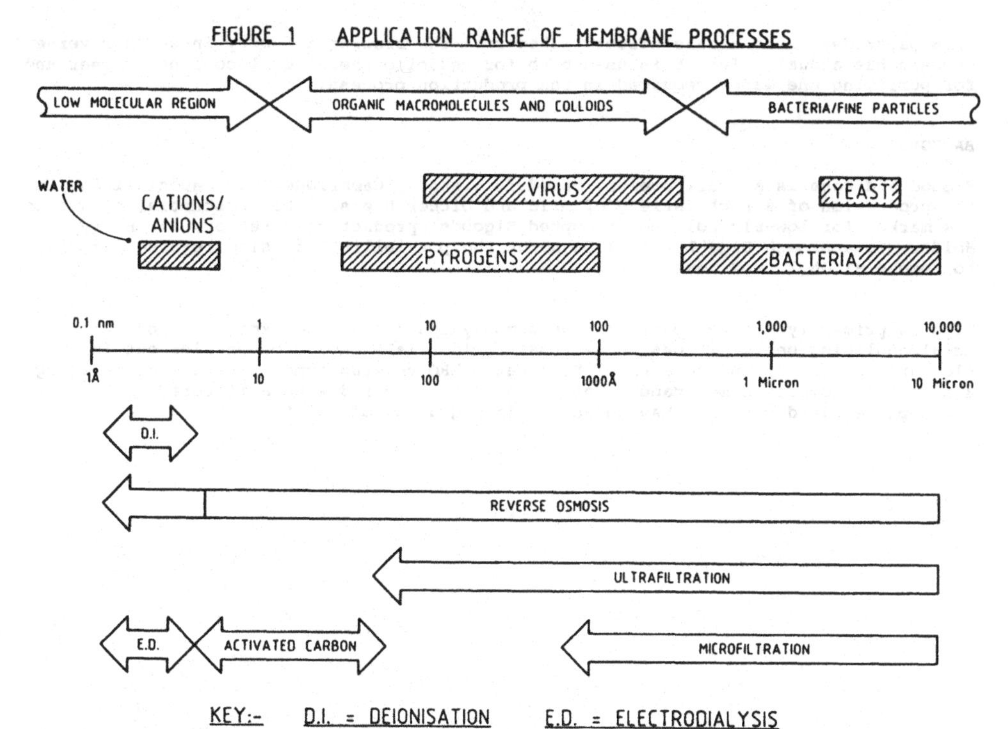

KEY:- D.I. = DEIONISATION E.D. = ELECTRODIALYSIS

Effective Industrial Membrane Processes — Benefits and Opportunities, pp.191–196

The latest RO membranes can take out more than 99% of the organic macromolecules and colloids from the feedwater, together with up to 98% of the inorganic ions. Under certain operating conditions, they can remove almost all the micro-organisms as well.

As a result of the low energy consumption and high performance of modern RO systems, reverse osmosis is well established as a cost-effective and environment-friendly water purification process. In some applications, RO is used as a stand-alone technique (with appropriate pre-treatment of the feedwater), whereas in others it forms part of a sequence of water treatment processes. In many cases, for instance, it is employed in tandem with deionisation to provide high-purity process water for industrial manufacturing.

Contemporary RO membranes not only have a better overall performance than their predecessors, but they also have a much sharper selectivity. Early forms of cellulosic RO membranes, for example, had a molecular weight cut-off of around 500 Daltons, whereas the current generation of thin-film composite membranes have a cut-off of approximately 100 Daltons.

The combination of high rejection rates and a finely-tuned selectivity has enabled thin-film composite RO membranes to be used for a novel separation duty which is quite different from water purification: the de-alcoholisation of beer.

Cellulosic membranes had proved to be unsuitable for this application because their high molecular weight threshold allowed constituents of the beer to pass through the membrane with the alcohol, resulting in a loss of flavour. But the tighter structure of composite membranes enables them to retain the molecules responsible for the characteristic taste, colour and aroma of beer, while allowing the alcohol molecules to permeate through.

This paper describes a case-history taken from the brewing industry in which reverse osmosis has a dual role: it is used both for reducing the alcohol content of beer and for purifying the water required in the production process.

BACKGROUND

Elgood and Sons is a small brewery based in Wisbech, Cambridgeshire, specialising in the production of a rich variety of mild and bitter beers. The substantial growth in the market for low-alcohol beer prompted Elgoods' production director, Sir Henry Holder, to approach Elga Ltd in 1984 about the possibility of using reverse osmosis to remove alcohol from beer.

Elga is primarily a water purification company and had had no experience of de-alcoholising beer. It was agreed that field trials should be carried out in Elgoods' laboratory, using a small Elgastat LabRO reverse osmosis unit incorporating a thin-film composite membrane. The RO unit would be fed with a filtered, flash-pasteurised pale ale having an original gravity of 1030°.

When reverse osmosis is used to remove alcohol from beer, the feedstock is pumped into a module containing the semi-permeable RO membrane. Alcohol and water are forced through the membrane, while the de-alcoholised beer is retained. During this process, the beer could become highly concentrated, so water must be added continually to maintain the normal composition. The dilution process is called 'diafiltration' and the make-up water is known as 'diafiltration water'.

Laboratory tests showed that each volume of beer required three volumes of water to reduce the alcohol content from 3.5% $^v/v$ to 0.6%. But when Elgoods tried adding mains water to the concentrate, the flavour of the beer was unacceptable. So the RO unit was used first to purify the diafiltration water before being employed for the de-alcoholisation duty. One advantage of using reverse osmosis permeate for diafiltration is that the low bacterial content of the water minimises the likelihood of spoilage.

The field trials proved to be successful, and in mid-1985 Elga was asked to supply a full-scale reverse osmosis plant capable of processing 10-barrel batches of low-alcohol beer. Elgoods decided to call the beer 'Highway' and to make it available both in bottles and on draught. The RO plant was commissioned in November, 1985 in good time for a pre-Christmas launch of the product.

PLANT AND PROCESS

The cornerstone of the system is an Elga Intercept RO 3 LF reverse osmosis plant. The modifications made to the RO unit - which is normally used for water treatment - were minimal: special, hygienically-designed RO membranes were fitted instead of the normal ones, and the valves and pipework in the beer lines were constructed from stainless steel. Using standard plant helps to keep capital costs to a minimum, while the modular structure of RO systems enables them to be readily scaled up to meet the demands of increased production.

Water must normally be pre-treated before passing through a reverse osmosis unit in order to protect the RO membranes from fouling. The pre-treatment package for the water purification cycle in the Elgoods system comprises a base-exchange softener followed by a filter.

The first stage in operating the system entails using the reverse osmosis plant to provide purified diafiltration water. Raw mains water passes through the base-exchange softener and the pre-filter before entering the RO unit. The purified water (permeate) - which typically has a conductivity below $20\mu Scm^{-1}$ - is then fed into the diafiltration tank (Figure 2A), while the concentrate is discharged to drain. At the same time, a normal brew of full-strength beer is prepared and pumped into the beer tank in readiness for de-alcoholisation.

In the second stage, (Figure 2B) the beer forms the feedstock for the RO unit, but in this cycle the concentrate is retained and fed back to the beer tank. The permeate - consisting mainly of alcohol and water - passes into the alcoholic permeate tank, and the losses are made up by continually pumping the diafiltration water into the beer tank. The recycling process is continued until the alcohol content of the beer reaches the required level.

The contents of the alcoholic permeate tank are discharged to waste. After an initial period of doubt, H.M. Customs and Excise now accept that reverse osmosis is an acceptable alternative to conventional technology for removing alcohol from beer. Duty paid on the beer is now refunded on a routine basis.

FIGURE 2 REVERSE OSMOSIS SYSTEM FOR PRODUCING LOW-ALCOHOL BEER
(A) USING R.O. TO PRODUCE DIAFILTRATION WATER

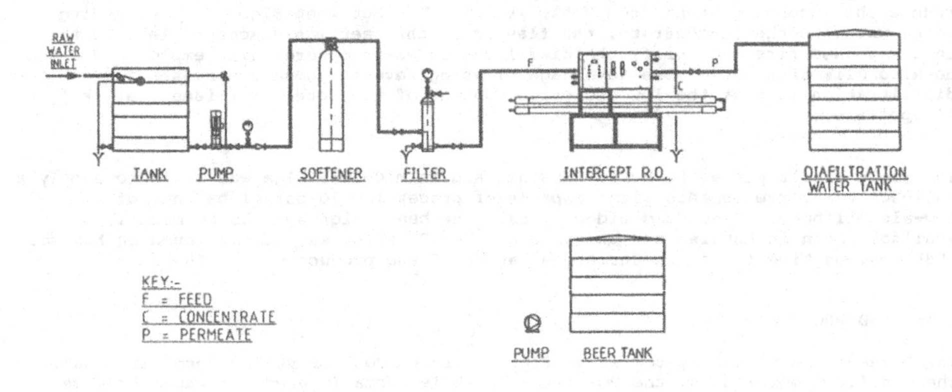

FIGURE 2 REVERSE OSMOSIS SYSTEM FOR PRODUCING LOW-ALCOHOL BEER
(B) USING R.O. TO DE-ALCOHOLISE BEER

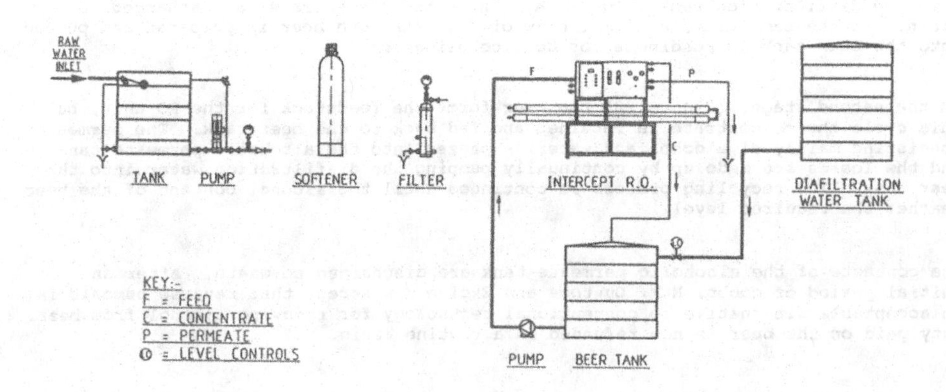

BENEFITS

The benefits obtained by using reverse osmosis to de-alcoholise beer fall into three categories: plant operation, product quality and costs.

One of the major advantages of RO plant is that it is reliable and very simple to operate. The only services required are electrical power and a water supply, and complete process control is provided by a positive displacement pump and a pressure control valve. By contrast, evaporators need careful monitoring of flowrates, steam supplies, vacuum pressures and temperatures if 'overcooking' of the beer is to be avoided. Moreover, evaporators are more difficult to install than RO units.

Reverse osmosis equipment requires little supervision - apart from end-of-run monitoring to check that the alcohol content of the final product has reached the required level. The membranes are cleaned with an alkaline detergent (Elgalite RF II) after 10-barrel batches of beer have been processed, and cleaning-in-place procedures are straightforward. There have been no problems with bacterial spoilage; the bottled product has a shelf-life of more than six months.

Unlike some de-alcoholisation techniques, reverse osmosis takes place at ambient temperatures, so the low-alcohol beer produced by RO is not tainted by 'cooked' or 'worty' flavours. Furthermore, as the starting material is full-strength beer produced by normal fermentation processes, the house flavour of the brewery is preserved in the final product.

With regard to economic benefits, reverse osmosis scores over conventional alcohol-removal technology both in requiring less capital expenditure and in having substantially lower running costs. The major operating costs of RO are membrane replacements (every two to four years) and the power consumed by the high-pressure pump.

Current estimates of energy costs are £17.00 per 100 barrels for a reverse osmosis plant - including the production of diafiltration water and cleaning in place - compared with £56.50 per 100 barrels for an evaporator. In 1989, Elgood and Sons won a PEP (Power for Efficiency and Productivity) Award sponsored by the Electricity Supply Industry for using a highly efficient electrically-powered production plant. The RO system was estimated to use 68% less energy per barrel than distillation.

COMMERCIAL OUTCOME

As a result of using reverse osmosis to produce low-alcohol beer, Sir Henry Holder received the 'Alfa Laval Food From Britain Award' for technical achievement in 1986. Soon afterwards, Elgoods won a national contract to supply bottled low-alcohol beer to a leading supermarket chain. the commercial success of Highway was followed by the launch of a new low-alcohol brown ale - Elgoods 'Brown LA' - in 1989.

Other breweries that have recently installed Elga reverse osmosis plant to produce low-alcohol beers include Harveys of Lewes and Youngs of Wandsworth; another RO plant is shortly to be commissioned at the Cornish Brewery Company Ltd in Redruth.

Harveys produce two varieties of de-alcoholised beer having their distinctive house flavour, one of which won a bronze medal in the Brewery Industry International Awards, 1990; Youngs' low-alcohol product has been favourably received by the Campaign for Real Ale.

CONCLUSION

Reverse osmosis has considerable attractions as a means of removing alcohol from beer which has been brewed by traditional methods. Not only is the process simple to operate, but running costs are low and the characteristic flavour of the beer is preserved. The relatively low throughputs of RO plant make reverse osmosis an ideal de-alcoholisation technique for smaller breweries wishing to produce low-alcohol versions of their best-selling beers.

ACKNOWLEDGEMENTS

The author would like to acknowledge the help received in the preparation of this paper from Sir Henry Holder, production director of Elgood and Sons Ltd, and P. T. Jordain, development manager, Elga Ltd.

The Elga Intercept RO 3 LF reverse osmosis plant at Elgoods' brewery.

SESSION F:

APPLICATIONS IN THE FOOD AND BEVERAGES INDUSTRY

ZIRCONIUM OXIDE BASED COMPOSITE TUBULAR MEMBRANES FOR ULTRAFILTRATION

W. DOYEN, R. LEYSEN, Nuclear Research Institute, Belgium
J. MOTTAR, G. WAES, Government Dairy Research Station, Belgium

SUMMARY

Since the breakthrough of ultrafiltration as an industrial process, a lot of membrane modification techniques, including surface treatments have been proposed in order to improve the UF-performance of these membranes. In this paper a new technique has been proposed, which is rather different from the already existing techniques: A polysulfone membrane is hereby "doped" with an inorganic oxide such as zirconium oxide.
It is shown in this paper that several properties of this "composite membrane", as compared to the 100 % polysulfone membrane are changed, e.g. pore size and pore size distribution, hydrophilicity and resistance towards fouling. These composite membranes have been successfully prepared in the tubular configuration by means of a casting-bob experimental set-up. The performances of these membranes for the filtration of skimmed milk and Gouda cheese-whey are being demonstrated on a semi-pilot scale.

Effective Industrial Membrane Processes — Benefits and Opportunities, pp.199–216

1. INTRODUCTION

During the last two decades, ultrafiltration has become an industrial process in a wide variety of fields of application: from the chemical industry (recovery of electropaint, latex), textile industry (indigo natural dye), paper industry, lubricating oils, the medical industry (end stage renal disease, blood processing) to the food industry (fruit juice clarification, whey processing, cheese making and water treatment).
The first type of membranes used for ultrafiltration purposes was made from a cellulosic type of polymer, which of course lacked chemical stability. Relatively soon new polymer types e.g. polysulfones, polyacrylonitriles were used to overcome this difficulty. However the major breakthrough for this type of polymeric membrane was the discovery of the asymmetric structure, yielding a good combination of process flux and cut-off values.
In the search to further improve the flux properties of the membranes, methods like grafting of hydrophilic groups onto the polymeric backbone (1,2), or more recently the making of thin film composite ultrafiltration membranes by interfacial polymerization of polyureas and polyurethanes onto polyethersulfone support membranes (3), amongst other techiques were used. These methods are aimed to reduce the deposit of foulant gel layers on top of the membranes during filtration. A drawback of grafted polysulfone membranes is that the grafted groups tend to disappear upon many successive cleaning procedures alternatively with alkaline and acid cleaning detergents. As a result the membranes become more hydrophobic and are more fouling sensitive.
The solution we have been putting forward in order to minimize fouling and thereby to increase the membrane's flux properties, was the incorporation of inorganic oxides like ZrO_2 into the polymeric matrix, leading to a composite membrane (4,5). Polysulfone was used as the binding polymer. Loading degrees up to 90 wt % of ZrO_2 can be reached.
Parallel to the development of the polymeric membrane types, but with some delay, ceramic type membranes were developed, starting with the zirconia on carbon, to the more recently developed alumina type of membranes.
The composite type of membrane we have been developing combines some of the interesting properties of both the polymeric and the ceramic type of membranes as for instance :
- they are still flexible just as the polymeric membranes.
- they have in some cases even sharper cut-off values than the polymeric and ceramic membrane types (4).

The major disadvantage of the composite membrane type is its chemical stability with respect to organics and its thermal stability which is limited to some 120°C. It is obvious that the origin of these properties lies in the use of polysulfone as the binding polymer and that when other more stable polymers are used, these problems are less occuring.
In this paper the major ultrafiltration properties will be discussed and the obtained results for the filtration of skimmed milk and Gouda cheese-whey of this novel membrane type will be compared with a commercial membrane type.

2. MEMBRANE PREPARATION - MATERIALS AND METHODS

The preparation of the casting dope differs somewhat from the one for preparing a 100 % polymeric membrane. First the binding polymer here polysulfone UDEL of type P 1800 NT 11 of Amoco Chemicals Europe, is dissolved in N-methyl-2 pyrrolidone as the solvent. To this solution, zirconium oxide powder, supplied by Magnesium Electron Ltd., with a grain diameter of some 10 microns, is added.

The zirconium oxide powder is dispersed, de-agglomerated and further milled in a ceramic ball mill to the submicron range. Subsequently the suspension is degassed under vacuum and filtered on a 90 micron screen.
Two suspensions were prepared with a different ZrO_2 to polysulfone weight ratio (respectively 80/20 and 90/10) but with the same polysulfone to solvent weight ratio of 18/82. The viscosity of these two suspensions equals respectively 2 and 35 Pa.s.
The membranes were prepared in the tubular membrane configuration. Two types of support tubes, of which one was already selected in a previous study (4), were used. The first one was a polycrystalline type of carbon (PC) tube with a respectively inner and outer diameter of 6 and 10 mm and a length of 1,250 mm. The second was a carbon fiber type of tube (CF) with a respectively inner and outer diameter of 6.2 and 8.5 mm and a length of 1,250 mm. Both type of support tubes were manufactured by the Le Carbone Lorraine company.
In order to prepare the tubular membrane configuration a casting-bob method was used (6). In this experimental set-up the support tube is placed onto a conveyer plate which can be moved downwards with a uniform linear velocity into a precipitation bath. The casting-bobs had an external diameter of 5.85 and 6.1 mm. The first casting-bob was used for the polycrystalline type of tube and the second for the carbon fiber type of support tube. They were fixed by a stainless steel spring on top of the set-up and were hanging some 5 cm below the tube and some 3 cm above the precipitation bath. Before starting the driving mechanism, 3 to 6 cm^3 of casting dope (here the suspension) was put into the downside orifice of the tube. Subsequently the driving mechanism is started with a speed of 0.033 $m.s^{-1}$. When passing the casting-bob a layer of composite membrane "in status nascendi" is left at the inside of the tube.
Furtheron the membrane is formed by immersion precipitation; by extraction of the solvent in the precipitation bath, with water as the non-solvent. Subsequently the membranes are posttreated by a treatment in a hot water bath to remove all remaining solvent.
In order to make it possible to evaluate the performance of this novel membrane type, modules were prepared containing 37 of such tubes, with a length of 1,088 m. The total geometric area exposed to filtration for such a module equals 0.73 m^2. Figure 1 gives a view of one of the end sides of such a module equipped with the membranes at the inside of the carbon fiber type of support tubes. The characterization tests were performed with these modules except for the determination of the cut-off values which were performed with a single membrane tube arrangement.

3. CHARACTERIZATION OF THE COMPOSITE TUBULAR MEMBRANES

3.1 Characteristics of the used Carbon Support Tubes

Prior to measuring the membrane coated carbon support tubes, the support tubes themselves were investigated by measuring their pure water flux and their retention for dextran molecules. The mean pore size of the polycrystalline carbon and the carbon fiber tubes equals respectively 0.1 and 2 microns. In figure 2 the pure water flux values as a function of transmembrane pressure measured at 40° C with microfiltered water are given.
In order to compare the characteristics of the different tubes we took as a reference condition a transmembrane pressure of 0.3 MPa and a temperature of 40°C. Under this reference condition the pure water fluxes of respectively the polycrystalline carbon and the carbon fiber type of tube equal some $0.14.10^{-3}$ and $4.45.10^{-3}$ $m^3.m^{-2}.s^{-1}$ or in more practical units 500 and 16,000 $l.h.^{-1}m.^{-2}$.

FIG.1: VIEW ON ONE OF THE END-SIDES OF THE MODULES.

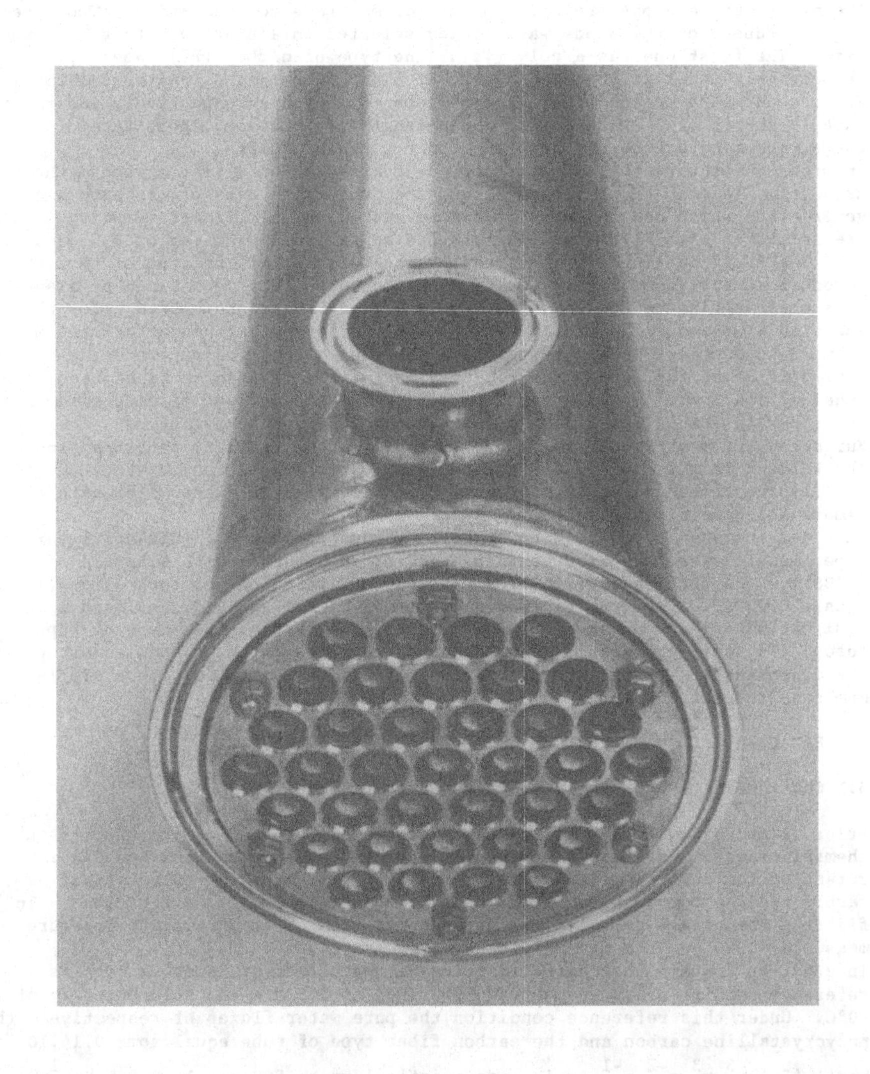

FIG.2: PURE WATER PERMEABILITY OF BOTH TYPES OF SUPPORT TUBES WITHOUT MEMBRANE.

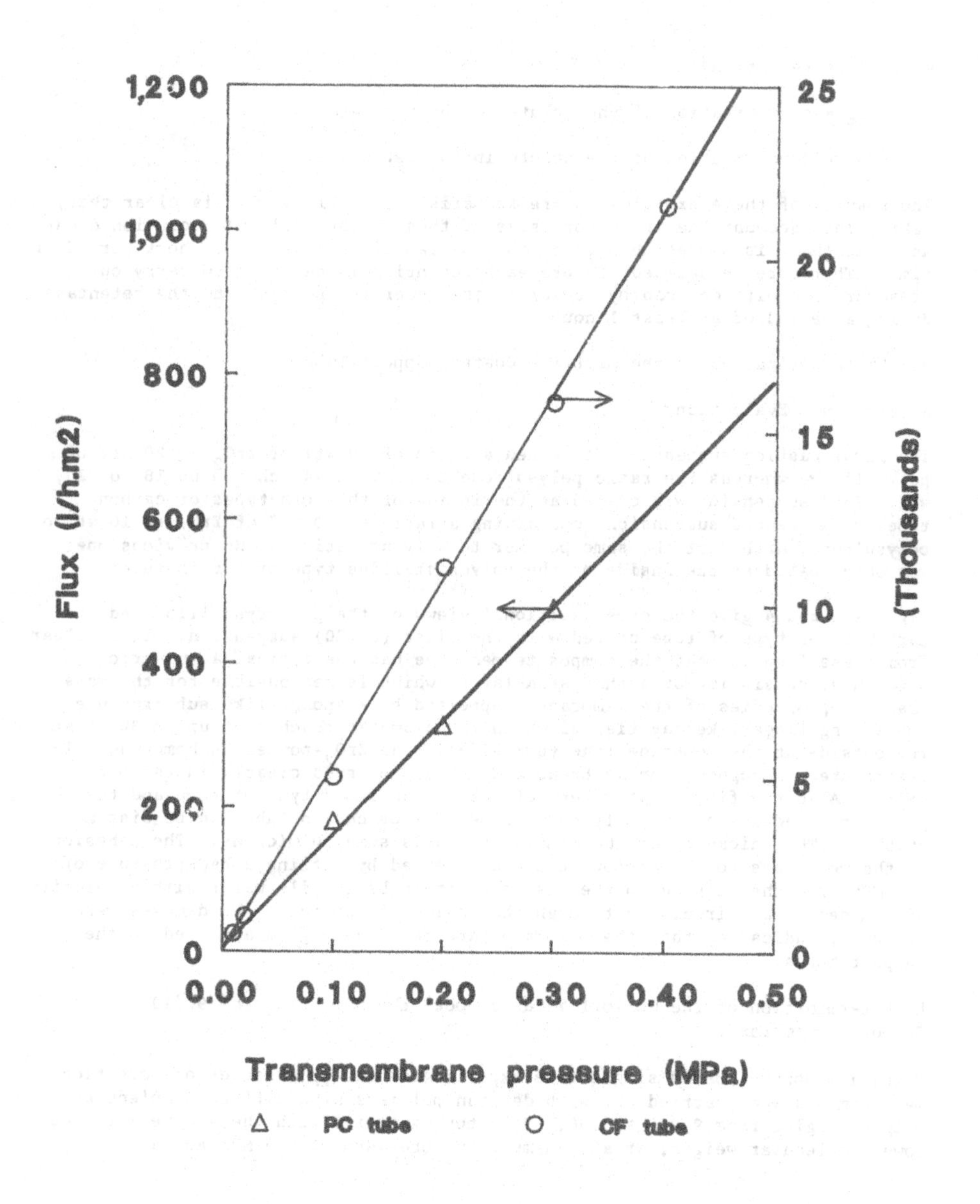

Retention measurements with dextran molecules were also undertaken to see whether and to what extend sorption phenomena occured. For this goal a 1,000 ppm dextran of molecular weight 70,800 Dalton was filtered across the tubes and retention was measured as a function of time.
Retention was defined by the classical formula [1],

$$R = 1 - \frac{C_P}{C_R} \qquad [1]$$

where R = rejection

C_P = concentration of the solute in the permeate

C_R = concentration of the solute in the retentate

The results of these experiments are summarised in table 1. It is clear that, taking into account the large pore size of these tubes (0.1 μm), sorption occurs and that there is a breakthrough of the dextran after a relative short period of time. Therefore we decided, before each retention measurement to carry out a presaturation with dextran by recycling the generated permeate to the retentate during a period of at least 1 hour.

3.2 Characterization of the Membrane Coated Support Tubes

a) Structure Evaluation.

The first casting suspension contained a ratio of 80 wt% of ZrO_2 to 20 wt% of polysulfone, whereas the ratio polysulfone to solvent was chosen on 18 to 82 wt%. This suspension was coated at the inside of the both types of carbon tubes. The second suspension, containing a ratio of 90 wt% of ZrO_2 to 10 wt% of polysulfone, with just the same polymer to solvent ratio as the previous one, was only coated at the inside of the polycrystalline type of carbon tubes.

Figures 3 and 4 give the cross-sectional views of the polycrystalline and the carbon fiber type of tube coated with the first (80/20) suspension. It is clear from these figures that the composite membrane has the typical asymmetric structure, consisting of a thin skin layer, which is responsible for the mass transfer properties of the membrane, supported by a sponge-like substructure containing fingerlike cavities of which the diameter reaches of up to 30 μm at the outside of the membrane (the tube side). The ZrO_2-powder is homogeneously distributed throughout the membrane and can be observed clearly in the pore-walls. Also the fibrous structure of the carbon fiber type of tube and the granular structure of the polycrystalline type of carbon tube can be distinguished. The thickness of the membranes equals some 50 microns. The adhesion of the membranes to the support tubes was tested by putting a backpressure of 0.7 MPa from the outside to the inside of the tubes, while water with a velocity of 4 $m.sec^{-1}$ was circulated through the inside of the tube. No damages were observed, indicating that the membranes are relatively good attached to the support tubes.

b) Determination of the Cut-off Value of both the 80/20 and the 90/10 Type of Composite Membrane.

Using a membrane coated single tube experimental set-up a series of retention measurements were carried out with dextran polymers with different molecular weight ranging from 9,000 to 168,000 Dalton, starting with the solute with the lowest molecular weight, at a transmembrane pressure of 0.3 MPa and a

TABLE 1: RETENTION OF THE EVALUATED CARBON TUBES FOR DEXTRAN 70,800 AS A FUNCTION OF TIME.

Time elapsed from start of the filtration experiment (minutes)	Retention (%)
3	51
15	4
30	0
60	0

Dextran concentration in the feed : 1,000 ppm
Temperature : 40 °C
Transmembrane pressure (TMP) : 0.2 MPa

FIG.3: SEM-PICTURE OF THE CROSS-SECTION OF THE COMPOSITE MEMBRANE COATED AT THE INSIDE OF A POLYCRYSTALLINE CARBON TUBE.

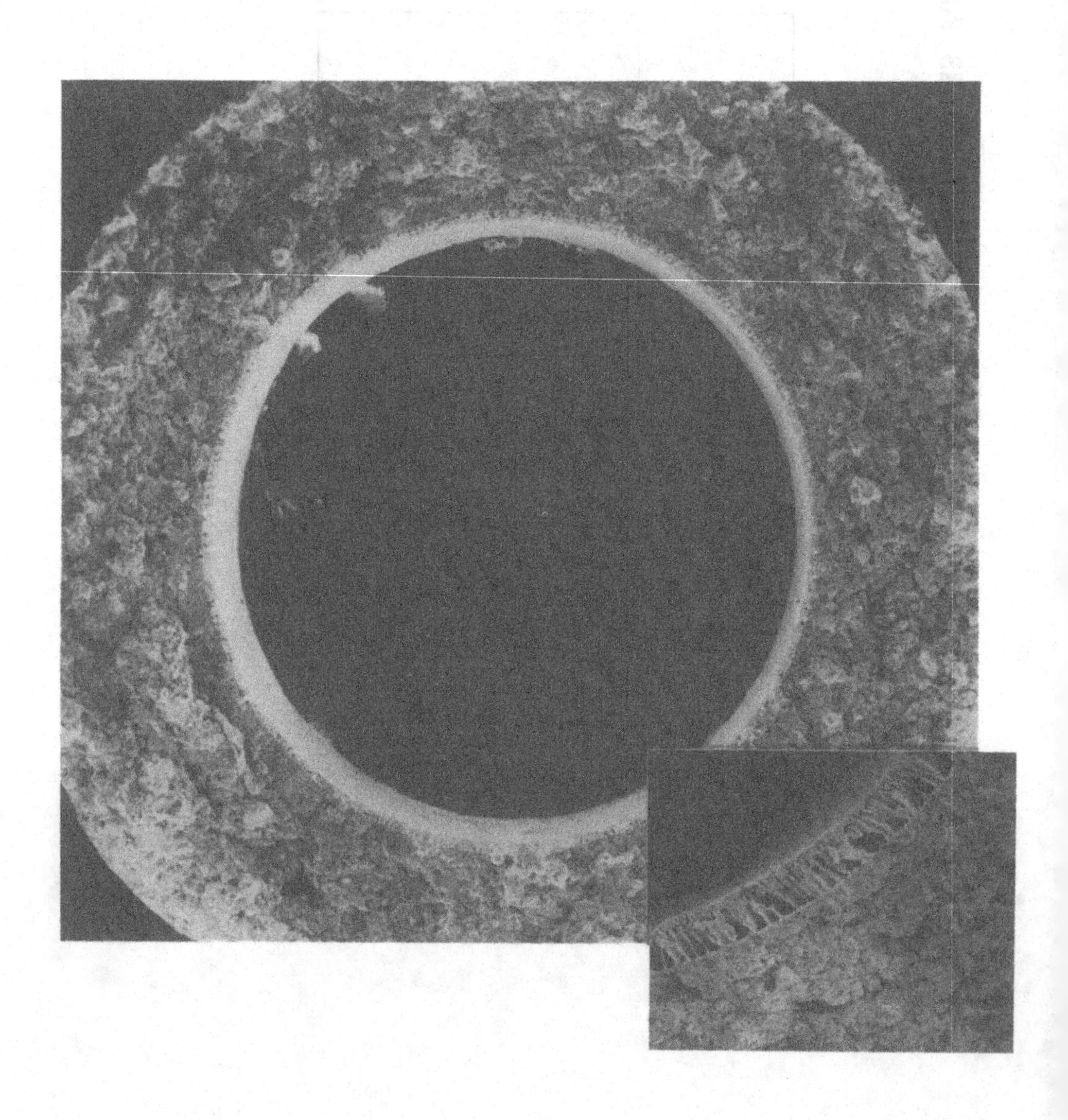

FIG.4: SEM-PICTURE OF THE CROSS-SECTION OF THE COMPOSITE MEMBRANE COATED AT THE INSIDE OF A CARBON FIBER TUBE

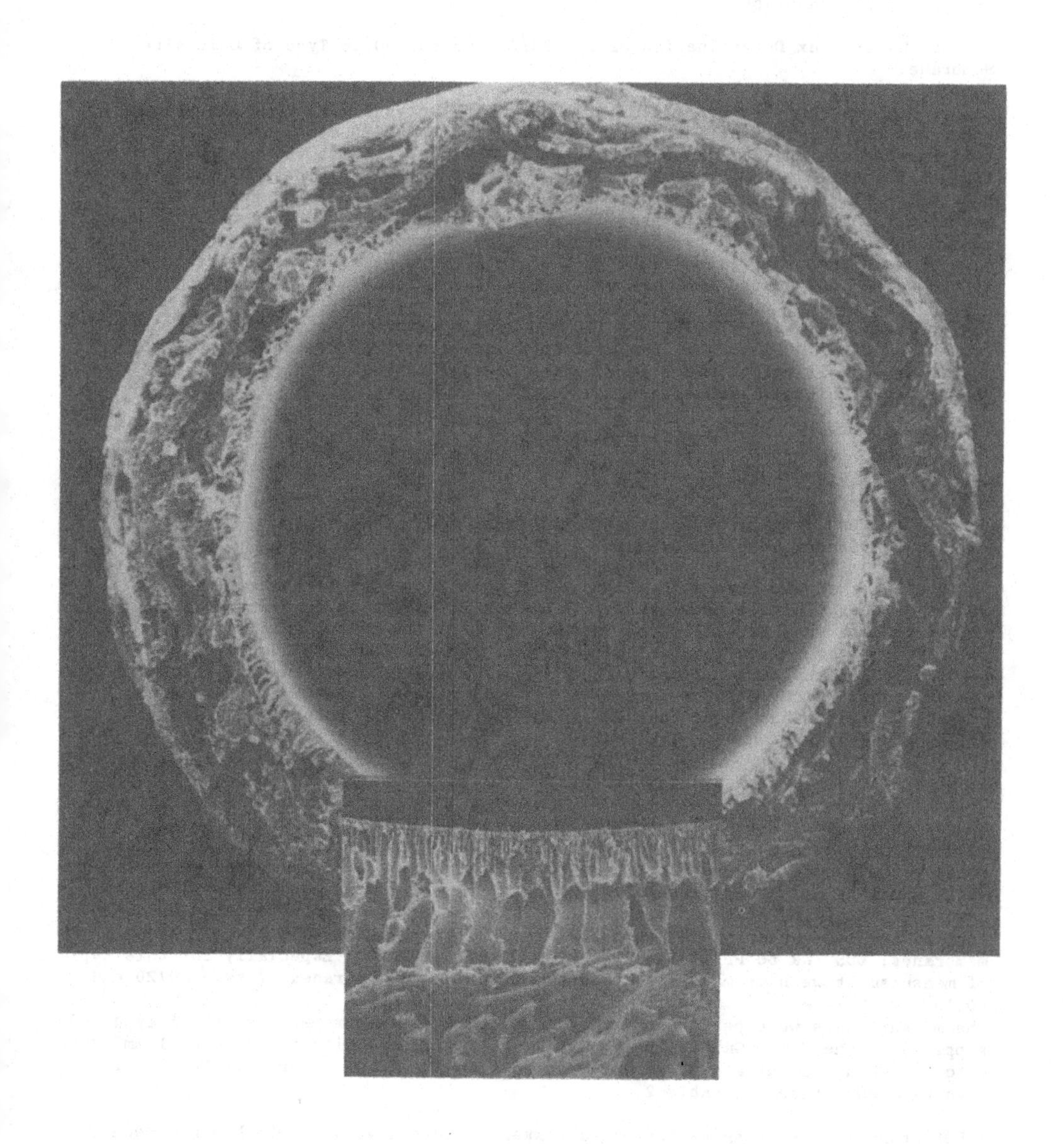

temperature of 40°C. The resulting retention curves are given in figure 5. The cut-off value which can be derived from these retention curves (90 % retention) for the 90/10 and the 80/20 type of membrane equals respectively 75,000 and 150,000. A similar cut-off value was found for the 80/20 type of membrane coated at the inside of the carbon fiber tube.
It has to be noticed from the slope of these retention curves that the pore size distribution of this membrane types is very narrow (3, 6, 7), e.g. one has a retention of only 12 % for dextran 40,700 and of already 92 % for dextran 70,800 with the 90/10 membrane.

c) Pure Water Flux Determination of the 80/20 and the 90/10 Type of Composite Membrane.

Figure 6 gives the pure water flux of both the polycrystalline type of carbon tube, and for the carbon fiber type of tube with the 80/20 membrane.
A first observation that can be made from this figure is that the pure water flux of the 80/20 membrane on the polycrystalline carbon tube nearly equals the pure water flux of uncoated support tube. The same was found for the 90/10 membrane. This result proves that the pure water flux of the two composite membranes is being masked by the polycrystalline carbon tube. This is not the case anymore when the carbon fiber type of tube is being used as support tube, as can easily be deduced from the water permeability data.

The observed pure water flux for a 150,000 Dalton membrane is relatively high in comparison with 100 % polysulfone membranes and with ceramic membrane types with a similar cut-off value. When looking at the flux (J) equation [2], a rearranged Hagen-Poiseulle-equation (8),

$$J = S.dp^2 \Delta P/32\mu l \qquad [2]$$

where S = surface porosity
dp = pore diameter
l = pore length, here the skin thickness
μ = viscosity of the solvent

it is clear that the reason that a membrane with a similar pore diameter (cut-off value) has a higher permeability, only can be sought in the fact that the skin-thickness must be smaller or the surface porosity must be higher. By comparing the cross-sectional views (SEM pictures) of both the composite membrane type and the 100% polymeric types hardly any difference in skin-thickness can be found, which is in the range of 0,5 to 1 micron. The reason for higher permeability has thus to be found in an increased surface porosity. Up to now, no experimental evidence is available.

d) Determination of the hydrophilic properties of the composite membrane types.

To get an idea to what extend the incorporation of ZrO_2 grains into a polysulfone membrane affects the hydrophilic character, water penetration experiments were executed, whereby the time to wet completely a dry membrane (contact angle equal to zero) was measured. It is evident that upon using this procedure, membranes should be compared with similar pore sizes or cut-off values. We have carried out measurements on the following commercially available flat sheet membranes; DDS* GR 60 PP, DDS* GR 40 PP, and Amicon PM10. Especially for this type of measurement we have made also flat sheet composite membranes of type 80/20 and 90/10.
The measurements were performed with a contact angle measurement system of type G40 supplied by the Krüss GmbH company, using water droplets with a volume of 1 mm^3 put onto the skin side of a dry (4h dried in an oven at 70° C) membrane. The results have been summarized in table 2.

* DDS stands for the Danske Sukker Fabrikker. The activity of DDS has now been taken over by DOW Danmark.

FIG.5: RETENTION CURVES OF BOTH THE 80/20 AND THE 90/10 TYPE OF MEMBRANE COATED AT THE INSIDE OF A POLYCRYSTALLINE TYPE OF SUPPORT TUBE.

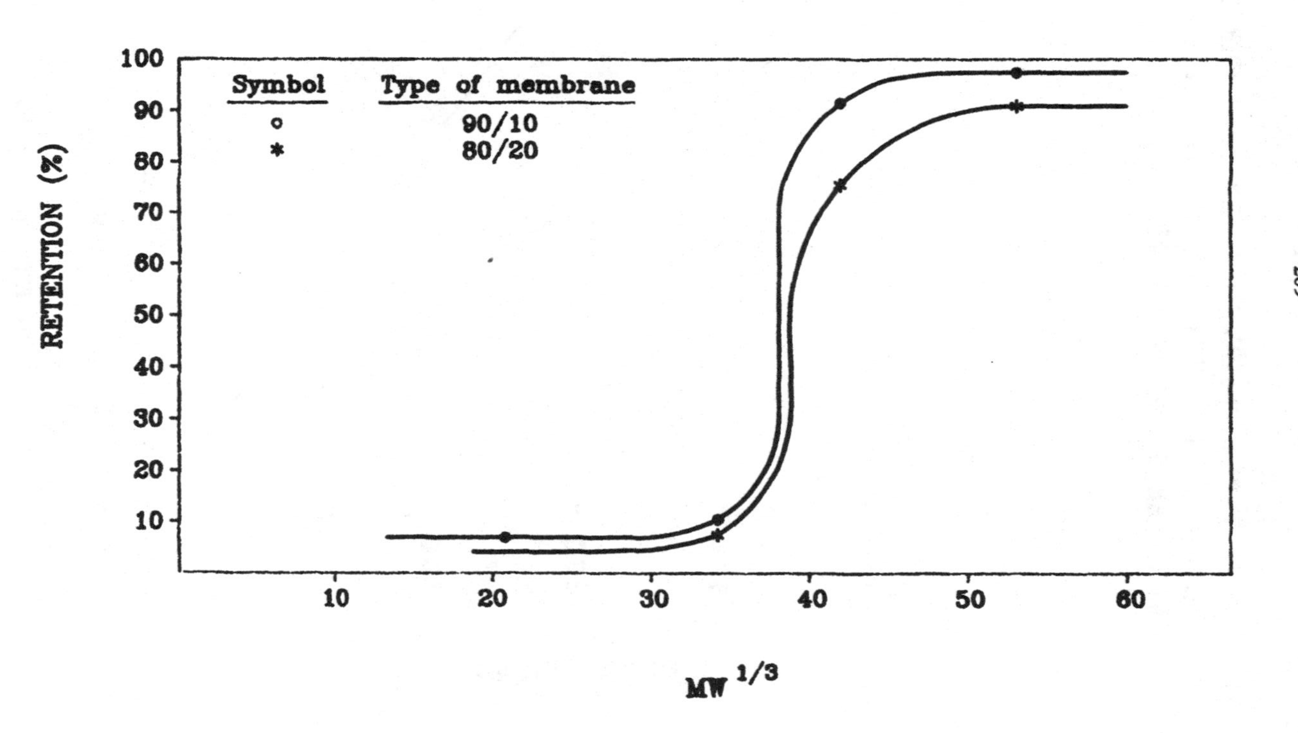

FIG.6: PURE WATER PERMEABILITY OF BOTH THE POLYCRYSTALLINE AND THE CARBON FIBER TYPE OF TUBE, COATED WITH THE 80/20 MEMBRANE TYPE.

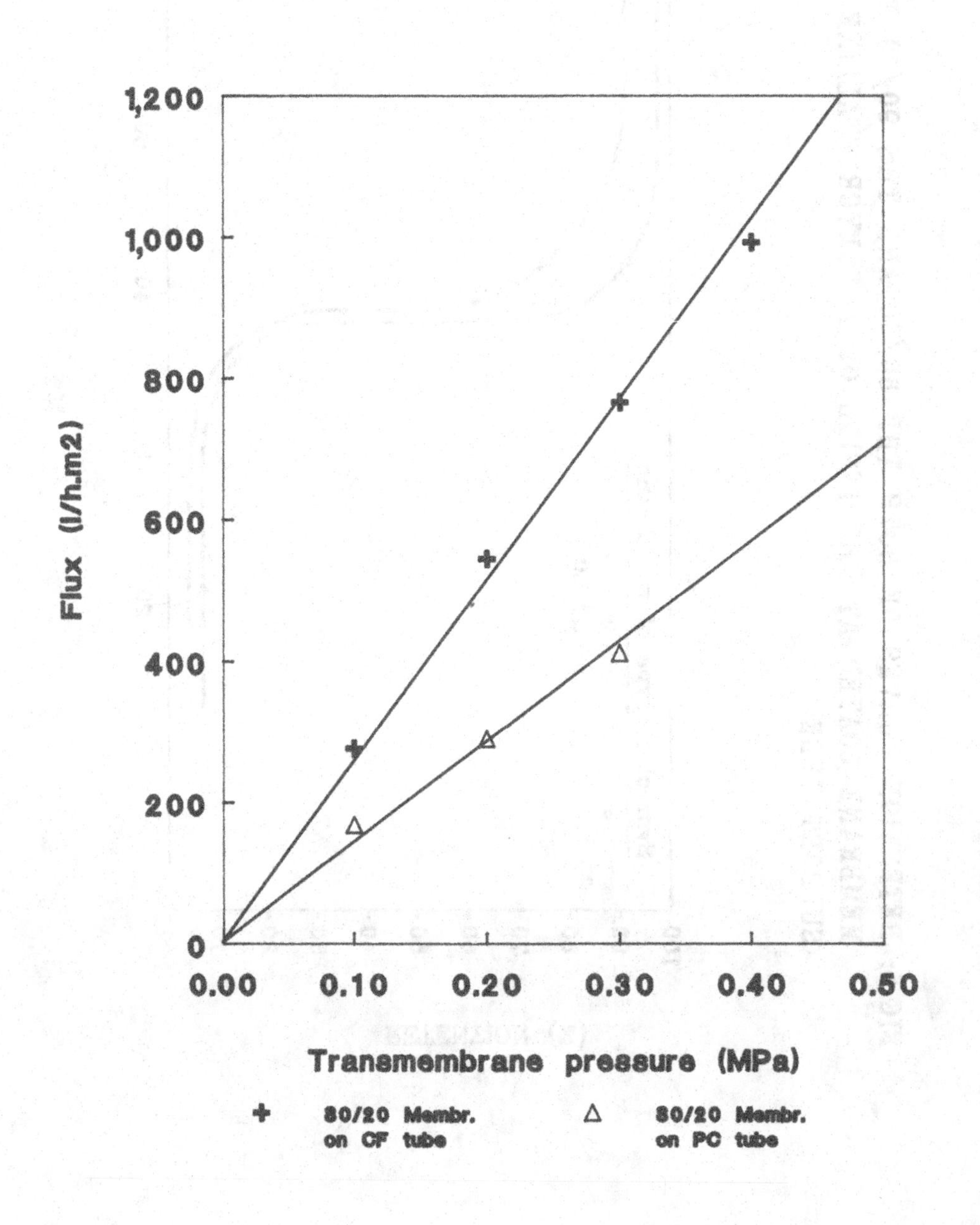

TABLE 2 : RESULTS OF THE WATER-DROPLET PENETRATION EXPERIMENTS.
(mean value of 3 different membrane samples)

Membrane type	Nature	Supplier or Origin	Given cut-off value (Dalton)	Penetration Time (Contact angle = 0°) (sec.)
DDSGR40PP	100 % poly-ethersulfone	DDS - DOW	100,000	57
DDSGR60PP	100 % poly-ethersulfone	DDS - DOW	25,000	300
PM10	100% poly-sulfone	Amicon	10,000	∞
90/10	90 wt% ZrO_2 10 wt% poly-sulfone	SCK	75,000*	16

From table 2 it is clear that wetting of the ZrO_2 "doped" membranes is the most easiest. This is a consequence of both the hydrophilic character of the composite membranes and possibly of the higher surface porosity.

4. PERFORMANCE TESTING OF THE COMPOSITE MEMBRANES

In order to evaluate the performance of this novel membrane type, filtration experiments were carried out with skimmed milk and Gouda cheese-whey. The experiments were carried out in a Tech-Sep semi-pilot installation containing 2 modules. These modules, described above in paragraph 2, have together a total geometric membrane surface area of 1.46 m^2.
For comparison, two reference modules of type Tech-Sep S37 were used, which contain the same number of tubes, but those tubes are somewhat longer (1.2 m). The total geometric membrane area for the reference modules together equals 1.64 m^2. The experiments were performed at 55°C with a linear velocity of 4 $m.sec^{-1}$ and a transmembrane pressure ranging from 0.3 to 0.6 MPa. After each experiment the membranes were cleaned with Ultrasil P3 type of cleaning detergents, supplied by the Henkel company alternatively an alkaline (type 13) and an acid detergent (type 75) and again an alkaline cleaning detergent, at a temperature of 60-80°C. To check the effectiveness of the cleaning, the pure water permeability was measured after each cleaning procedure, at 25°C and 0.4 MPa. The results given on the tables 3 and 4 are mean values for several experiments. The number of times each experiment was repeated is mentioned on each table.
Up to now only the polycrystalline carbon supported composite membranes were tested for their performance.
The protein retention was controlled by determining the various nitrogen fractions, using the technique of Aschaffenburg and Drewry (9). The nitrogen was dosed by means of the micro-Kjeldahl method.

4.1 Results obtained on Skimmed Milk

For these experiments the composite membranes were compared with the Tech-Sep membranes of type M1 (with a given cut-off value by the supplier of 60,000 -80,000). The results are summarised in table 3.
From the results it appears that the experimental membranes show comparable or even better filtration effectiveness (90/10 membranes) in comparison with the reference membranes; practical no caseïn losses were found and the whey-proteïn losses were very limited. Moreover, it was found that the mean flux value of the composite membranes type coated on the polycrystalline carbon support was 41 % higher for the same volumetric concentration factor of 6.
It has to be noticed that not only the initial direct fluxes (at low concentration factors) are higher for the composite membranes, but they are also higher at higher concentration factors. This is an indication that these membranes are less fouling sensitive.

4.2 Results obtained on Gouda Cheese-Whey

For these experiments, Tech-Sep membranes of type M4 (with a given cut-off value by the supplier of 20,000) were used as a reference. Prior to filtration, the whey was preheated for 30 min at 60 - 65°C to precipitate the $Ca_3(PO_4)_2$, which is the main foulant. The pH of the Gouda cheese-whey was 6.3. The results are summarised in table 4. It appears from the results that the whey-protein losses via the permeate, obtained with the 90/10 membranes were very limited. This was not the case with the 80/20 and the Tech-Sep membranes. Again, it was found that the mean fluxes were higher from 22% for the 90/10 membranes to 42% for the 80/20 membranes for the same concentration factor of 6.

TABLE 3: RESULTS OF THE ULTRAFILTRATION EXPERIMENTS ON SKIMMED MILK
(executed on batches of 300 l)

Membrane Origin	Membrane type	Nr. of exp.	Nitrogen losses in permeate			Volumetric concentr. factor	Direct fluxes (l/h.m2)	Mean fluxes (l/h.m2)	Pure waterflux after cleaning# (l/h.m2)
			Total Nitrogen (mg/100g)	Non-casein Nitrogen (mg/100g)	Non-protein Nitrogen (mg/100g)				
						1.15	57.0	57.0	After 1st exp.
Reference Membrane	M1(60.000 - 80.000)	2	44.7	43.5	37.4	2.01	49.0	55.3	280
						3.87	29.4	46.3	
						5.04	14.3	40.6	After 2nd exp.
						5.93	9.3	36.2	-
SCK Membranes						1.11	83.2	83.2	After 1st exp.
	80/20 membrane on polycryst. carbon tube	4	42.5	39.8	31.3	1.43	74.2	78.3	310
						2.00	60.2	70.8	
						3.33	43.3	62.7	
						5.00	27.5	55.2	After 4th exp.
						6.00	17.4	51.3	301
						1.11	82.0	82.0	After 1st exp.
	90/10 membrane on polycryst. carbon tube	13	39.0	38.3	31.6	1.43	70.6	75.8	279
						2.00	62.0	70.9	
						3.33	45.2	62.0	After 13th exp.
						5.00	32.0	55.5	285
						6.00	14.9	51.0	

Measured at T = 50 - 55 C, v = 4 m/sec. and TMP = 0.3 - 0.6 MPa

Measured at T = 20 C and TMP = 0.4 MPa

During one of the concentration experiments the concentration factor was rised to 20 and even up to 25 for the 80/20 membranes. Again, just as mentioned above, it was noticed that the direct fluxes at these high concentration factors decrease less in comparison with the reference membranes. The contrary was even noticed; at the beginning of the experiments (CF=1.11-1.21) the direct flux of the 80/20 membranes is some 46 % higher in comparison with the reference membranes, whereas at a concentration factor of 20 it has become even 134 % higher.

4.3. Conclusions of the experiments on Skimmed Milk and Gouda Cheese-Whey.

From the results obtained on skimmed milk and Gouda cheese-whey it has to be concluded that the "doping" of polymeric membranes with zirconium oxide grains leads to a composite membrane with interesting filtration characteristics. Higher flux values are found with the same (80/20 membrane) or even lower (90/10 membrane) amount of protein losses via the permeate. These experiments, were repeated several times, indicating that no performance degradation occured This was illustrated by the fact that first the pure waterfluxes of the membranes after each cleaning procedure did not change and secondly (this is not mentioned in the tables) that the protein losses via the permeate also did not increase during the several concentration experiments.

Out of these observations it can be concluded that little or no fouling is occurring. This can also concluded from the results of the cheese-whey experiments where, at high concentration factors, the direct flux of the 90/10 type of membrane is significantly less dropping back in comparison with the reference membrane.

5. CONCLUSIONS

In this paper a new composite membrane was introduced. The initial goal of its development being the improvement of the membrane flux properties. This was obtained by incorporating ZrO_2 grains into a polysulfone matrix. Two types of membranes were discussed with a composition of respectively 80 and 90 wt% of ZrO_2 into 20 and 10 wt% of polysulfone. The membranes were successfully coated by a casting-bob method at the inside of two types of carbon support tubes, respectively polycrystalline carbon and carbon fiber.
It was observed that the obtained membranes have relatively high pure waterflux, especially when taking into account their cut-off value. It appeared from the retention measurements that steep retention curves could be obtained, indicating a rather narrow pore size distribution. From water droplet penetration measurements it became clear that the composite membranes are wetted significantly faster than 100 % polysulfone membranes with a similar pore size. These results indicate that the composite membranes are hydrophilic in nature.

In the evaluation of this composite membrane type for skimmed milk and Gouda cheese-whey we see that their resistance to fouling is significantly higher than the reference membranes, resulting in mean fluxes being in the order of 50 $lh^{-1}m^{-2}$ for skimmed milk and 80 to 90 $lh^{-1}m^{-2}$ for Gouda cheese-whey for a concentration factor of 6. These results were obtained with the membranes coated at the inside of the polycrystalline carbon tubes. This kind of tubes limits the pure water fluxes of the composite membranes, as shown in the experimental section. We therefore envisage starting experiments using the fiber carbon support tubes which we know do not limit the pure water flux.

TABLE 4 : RESULTS OF THE ULTRAFILTRATION EXPERIMENTS ON GOUDA CHEESE WHEY
(executed on batches of 500 l)

Membrane Origin	Membrane type	Nr. of exp.	Nitrogen losses in permeate			Volumetric concentr. factor	Direct fluxes (l/h.m2)	Mean fluxes (l/h.m2)	Pure waterflux after cleaning' (l/h.m2)
			Total Nitrogen (mg/100g)	Non-casein Nitrogen (mg/100g)	Non-protein Nitrogen (mg/100g)				
Reference Membrane	M4(20.000)	2	47.7	46.1	38.8	1.21 1.52 2.59 3.15 4.65 6.04 (20.00*)	70.3 65.8 62.4 62.3 62.3 62.4 (17.3*)	70.3 67.0 65.7 65.3 64.9 64.7 (60.8*)	After 1st exp. 285 After 2nd exp. 269
SCK Membranes	80/20 membrane on polycryst. carbon tube	4	49.8	46.8	35.8	1.11 1.43 2.50 3.33 5.00 6.00 (20.00*) (25.00*)	102.4 99.5 89.0 82.9 80.5 74.9 (40.4*) (33.8*)	102.4 100.7 96.6 94.4 92.5 91.9 (88.9*) (87.4*)	After 1st exp. 301 After 4th exp. 296
	90.10 membrane on polycryst. carbon tube	13	35.1	35.0	30.2	1.11 1.43 2.00 3.33 5.00 6.00	109.6 110.0 78.7 62.5 58.1 41.4	109.6 109.8 97.9 86.4 82.3 79.2	After 1st exp. 279 After 13th exp. 288

Measured at T = 50 - 55 C, v = 4 m/sec. and TMP = 0.3 - 0.6 MPa

' Measured at T = 20 C and TMP = 0.4 MPa

* values of only one single experiment

ACKNOWLEDGEMENTS

The authors wish to acknowledge the skilful contributions made by Ings. A. Bassier and P. Traest of the R.S.Z., Melle and by B. Molenberghs, J. Stroobants and Mrs. A. Kelchtermans of the SCK in the preparation and characterisation of the membranes. Special thanks to Lic. S. Taghavi for the structural characterisation of the membranes by means of the Scanning Electron Microscope and to Ing. Ph. Vermeiren and Mr. J-P. Moreels for the hydrophilicity measurements. The authors would also like to thank the Flemish Government for the financial support.

REFERENCES

(1) C. Friedrich et al., "Asymmetric reverse osmosis and ultrafiltration membranes prepared from sulfonated polysulfone", Desalination, 36, 39-62, (1981).

(2) Kevin E. Kinzer, "Phase inversion sulfonated polysulfone membranes", Journal of Membrane Science, 22,1-29, (1985).

(3) K.B. Hvid et al., "Preparation and characterisation of a new ultrafiltration membrane", Journal of Membrane Science, 53, 189-202 (1990).

(4) W. Doyen et al, "New composite tubular membranes for ultrafiltration", to be published in the Journal of Membrane Science.

(5) W. Doyen et al, "Method for preparing a composite semi-permeable membrane", European Patent nr. 0 241 995 B1 (1987).

(6) Mark C. Porter, Handbook of Industrial Membrane Technology, Noyes Publications, Park Ridge, New Jersey, USA (1990).

(7) H.K. Lonsdale, "The growth of membrane technology", Journal of Membrane Science, 10, 121 (1982).

(8) A.G. Fane et al., "The relationship between membranes surface pore characteristics and flux for ultrafiltration membranes", Journal of Membrane Science, 9, 245-262 (1981).

(9) R. Aschaffenburg and J. Drewry, "New procedure for the routine determination of the various non-casein proteins of milk", 15th Int. Dairy Congress, 3, (5), 1631, (1959).

HISTORY AND DEVELOPMENT OF PACKAGE DESALINATION PLANTS FOR DEFENCE AND OFFSHORE APPLICATIONS

Author: D J Spurr
Water Products Department
Biwater Treatment Limited

Co-sponsored by The Institution of Chemical Engineers and The European Membrane Society.

Organised by British Hydranautics Research Group.

Effective Industrial Membrane Processes — Benefits and Opportunities, pp.217–226

INTRODUCTION

This paper reviews the development of RO technology for the production of potable water by desalination of seawater.

The history of membrane manufacturing methods and the materials used are described.

A number of package plants are taken as examples, and lessons learned are discussed.

MEMBRANE MATERIAL

Major strides in R & D, particularly in manufacture, formulation and chemistry of polymers have resulted in significant enhancements to membrane permeabilities and selectivities, resulting in smaller membrane area requirements and hence smaller permeator sizes.

In the late 1950s cellulose acetate membranes were found capable of rejecting electrolytes from aqueous solutions and that their semi-permeabiltiy far exceeded that of any other films tested. The large membrane thickness, however, meant that reasonable flux valves would not be achieved and therefore their commercial use in RO was impractical.

These early membranes are sometimes referred to as 'symmetric' membranes because the membrane structure is uniform throughout its depth. However, almost all commercial membrane systems use 'asymmetric' membranes that are either of flat sheet or hollow fibre geometry.

Asymmetric membranes consist of a thin dense skin which offers minimal resistance to flow but maintains the necessary solute retention characteristics. This thin skin is supported by a porous substructure offering insignificant resistance to flow but providing integral support for the thin skin.

If the thin semi permeable skin and the porous support structure are composed of the same material the membrane is known as integrally skinned. If they are made of different materials they are known as composite membranes.

In the early 1960's, Loeb and Sourirajan prepared a cellulose acetate membrane which was a major breakthrough in RO membranes. This was the first integrally skinned asymmetryric membrane. In consisted of a 0.2 to 0.5 micron thick dense layer, supported by a 50 to 100 micron thick porous substructure. It was manufactured by the phase inversion method.

In spite of intensive research over a fifteen year period no superior membranes were developed. Cellulose acetate, as a membrane material, combines three essential requirements for a practical RO membrane:

* high permeability to water

* low permeability to salts

* excellent film-forming properties

The phase inversion method of production means that for an integrally skinned membrane the dense support skin has to be fairly thick (approximately 1000A). Unfortunately, the thicker the skin, the greater the resistance to mass transport and, therefore, the lower the permeability. Hence permeability and selectivity are inextricably linked in an inverse relationship; an improvement in one giving rise to a deterioration in the other.

In an attempt to overcome this, the composite membrane was developed. An active layer is deposited on a porous support structure. The active layer provides a barrier to flow and performs the separation whilst the porous substructure maintains physical strength.

A thinner active layer could now be applied giving increased higher permeabilities without a reduction in selectivity.

In 1977 the first composite membrane was introduced. It was asymmetric in that it consisted of an active polyamide skin with a thickness of 025 to 050 microns deposited on a porous polysulphone support membrane.

Other composite membranes have been developed, for example Toray uses a polyether active skin on a polysulphone base. There are recent reports of ICI having developed an active skin 0.2 microns thick based on poly ether ether sulphone on a polyester backing. This new membrane reportedly has a number of advantage: it has improved resistance to Chlorine, greater thermal stability and is able to withstand pressures up to 100 atmospheres.

In general however, RO is a mature technology and, in the major application area of sea/brackish water desalination, development has reached a plateau.

MEMBRANE STRUCTURE

Commercial membrane systems are constructed in relatively few configurations, each with its own advantages and disadvantages.

* Hollow fibre

* Spiral wound

* Plate and frame

Hollow fibre units consists of bundles of small hollow fibres with diameters of the order of 0.1 mm and bores of about .04 to .045 mm. Several thousand of these fibres are formed into a U shape, the closed end of which is sealed whilst the open ends are formed into a resin pot. These are enclosed in a pressure vessel in an arrangement that is similar in appearance, if not scale, to a shell and tube heat exchanger. The pressure vessels are typically 4 to 8 inches in diameter and 5 feet long.

Feed water is fed under pressure to the outside of the hollow fibre: water then passes through the tube wall and is collected in the centre of the tube as permeate (product water).

The main advantage offered by this configuration is that large membrane surface areas can be installed in small volumes, typically 12000 ft^2/ft^3.

The large membrane areas allow larger volumetric flows, which in turn create larger pressure drops, causing a drop in membrane performance.

Spiral wound membranes are fabricated from flat sheets of membrane material, separated into alternate feed and permeate flow channels by an impermeamable layer and a gauge mesh. This sandwich is then wound round a perforated tube for permeate collection.

Spiral wound membranes are generally considered more robust than hollow fibre.

They are less susceptible to irrevocable blockage than the hollow fine fibre, owing to the more turbulent flow.

Plate and frame membranes are sheets of membrane alternately spaced by feed and permeate flow channels. The arrangement requires a large volume for a given surface area. Its principle advantage is that for highly fouling fluids it allows effective cleaning and easier membrane replacement.

Different membrane materials have different physical and chemical characteristics.

Cellulose acetate membranes are subject to hydrolysis by water at pH values outside the range 4 - 6. Membrane hydrolysis results in greater salt passage. It is usually necessary, therefore, to adjust the pH of feed water to within the range 4.5 to 5.5.

Polyamide membranes by contrast, are very resistant to hydrolysis and can tolerate water pH values of 4 - 11 normally and 2 - 12 for short periods.

Cellulose acetate is also a good nutrient base for microorganisms and is therefore liable to attack and destruction by such species.

Polyamide membranes are again, by contrast, resistant to biological attack but they are not resistant to traces of Chlorine in concentrations greater than 0.1 mg/l, whereas cellulose acetate membranes can tolerate up to 1 mg/l of Chlorine normally and a considerably higher concentration on a shock dose basis.

It is therefore normal to control biological growth on cellulose acette membranes by maintaining a chlorine residual concentration in the feed water.

Finally, cellulose acetate membranes have a tendency to compaction, so it is normal practice to allow space at design stage for additional membranes to restore output.

ALTERNATIVES TO RO

The main alternative desalination method is mechanical evaporation. RO is at a disadvantage with respect to weight and space, but offers the advantages of higher rejection rates hence lower seawater flow requirements and no external heat source.

Designs of evaporative type fresh watermakers area available which do not require an external heat source. These recover the heat otherwise lost in the seawater cooling stream. they do not show an advantage of space and weight savings over RO, they are less reliable and require more power.

The advantages of lower demand for seawater flow and independence from an external heat source are generally thought to outweigh the disadvantages of space and weight. Numberous comparisons have been made that indicate power comsumption of RO is 10% to 50% of alternative methods of desalination.

Generally RO is quoted as requiring 9.5 kw to produce 1 m^3 of potable water from sea water without energy recovery. Large scale multi-stage flash distilation typical requires 16 kw/m^3, and the comparable figure for vapour compression distillation is 24 kw/m^3. There are many factors to take into account and these figures are inevitably a simplification, but they cannot hide rthe fact that RO is considerably less energy intensive than the alternatives. Hence RO units are currently being supplied for in the North Sea and ships.

COSTS

The capital costs of an Ro system are significally influenced by a number of factors, including extent of instrumentation and control, use of single or double stage system and the capacity of the unit.

Operating costs also depend on the nature of the plant and the chemical costs in particular are significantly influenced by the degree of pretreatment.

One of the major costs is membrane replacement. For design purposes, a three year membrane life is generally assumed although experience shows lives in excess of 5 years are possible. The reason why membrane life predictions are made cautiously is that operation and maintenance of the plant has by far the most significant effect on membrane life.

BASIC CONFIGURATION

The simplest possible RO desalination plant consists of a cartridge filter, a high pressure pump, RO module and a reject stream with a control valve.

Enhancement can include a conductivity monitor on the permeate line linked to an automatic dump valve to prevent out of specification water entering supply, and anti scalent or acid dosing of the feed to permit higher recovery rates.

MOD PLANTS

At the end of the 1970's, Ames Crosta Babcock, now part of Biwater Treatment had a good track record of supplying RO plants for a number of purposes. Between 1971 and 1981 we had designed and built over 60 Package RO plants for a diverse range of applications including, in addition to potable water, insulator washing water, plating waste recovery, process water for food industry. A development programme on behalf of the Ministry of Defence had commenced. When the Falklands emergency began, a number of merchant ships were called into military service. The requirement to transport large numbers of troops to the south Atlantic meant that many of the ships involved had inadequate fresh water storage and production systems.

The requirements of the units were that they should be simple to operate, and robust. the short time scale meant that materials other than the ideal had to be used. The lessons learned were:

* Two units are preferable to one. Incase of failure only fifty per cent of output is lost, and one unit can usually be cannibalised to keep the other operational.

* Pre-treatment of feed water is of great importance in obtaining maximum RO membrane life. As a short term measure, disposable cartridge filters can be used, but in the long term sand filtration is more economical.

* Positive displacement reciprocating pumps are a source of vibration. This can be reduced by paying attention to the suction conditions to prevent cavitation, the use of pulsation dampers on the discharge, and supporting the unit on flexible shock absorbant mounts.

A development programme for standard MOD RO plants was already under way when the Falklands started, but the lessons learned were incorporated before finalisation of the standard designs.

Design requirements for MOD include maintenance of output quality and volume for a wide range in feed water TDS (from 30,000 to 45,000) and temperature (from 0°C) to 25°C), resistance to high shock loadings, very low air or vibration born noise levels.

TYPE 23

The original design allowed for a two stage plant. In the light of Falklands experience this was modified to two independent first stage streams, with a single second stage stream to provide water for specialist requirements. Eight of these plants have been built, two more are presently under construction and due for delivery in August of this year.

SSK SUBMARINE

The plant is a two stage design incorporating spiral would polyamide membranes. Because of the interdependancy of the two stages and space constraints the plant is relatively complex.

The two stages of this plant were required because membrane technology at the time of design could not manage water to potable standards from a single stage system. Membrane technology has now moved on, but two stages offer the continuing advantage of self compensation of product flow for a wide range of feed water temperature encountered between the surface of the sea and at levels beneath the surface.

MCMV

These plants are very simple, and consist of little more than the basic requirements previously outlined. The main specialist part comes in the selection of materials to provide a very low magnetic signature.

OFFSHORE PLANTS

On manned offshore platforms, provision of potable water is increasingly by desalination of seawater with the alternative of bunkering being used only as a fall back.

Desalination in the North Sea typically requires reduction of TDS mg/l from a concentration in the range 35,000 to 40,000 to less than 500.

This is well within the capability of a single pass reverse osmosis membrane.

Offshore oil platforms have a number of special requirements in addition to the very high quality demanded by the MOD.

Safety was brought into focus by the Piper Alpha disaster. Although we had been used to including explosion proof motors, flame retardent insulations and non toxic plastics etc, ad tightening up of design monitoring has been discernible.

Working for Offshore applications is at once a pain and a pleasure. It is an irritation at the bid stage to have to cross every T and dot every I, discussing almost every single nut, bolt and washer for a competitive bid that remains potential and somewhat speculative until the order is signed. At the same time the engineering discipline is regorous and can be enjoyable.

Manpower is especially expensive offshore and automation is therefore a high priority. Whilst nothing is impossible , a sensible line has to be drawn. When an automatic plant goes wrong, it is considerably more demanding to put right than a manual plant. the situation is further complicated since an operator who is familiar only with monitoring a number of 'status normal' indicators has not learned to know and love his watermaker.

Some operational requirements are readily automated: filter backwashing for instance. the pretreatment plant can be further enchanced by duty/standby provision so that it is not necessary to interrupt the supply of pre-treated feed to the watermaker whilst backwashing is in progress.

Periodically, membranes require washing in order to maintain performance. It is possible to arrange for this need to be calculated automatically by an on-board computer linked to instruments sensing hours run, feed pressure, permeate conductivity and temperature, and for the routine to be accomplished automatically with no need for manual initiation.

It is preferred however, for the operator to keep a daily log book and monitor the performance of this plant, initiating any required procedures manually.

The comments above would probably not apply for very large plants, but to date the largest offshore package we have produced consisted of two plants each with a maximum output of 75 m^3/d giving a total output of 150 m^3/d. One unit at a time is taken out of service for routine maintenance.

CONCLUSION

Seawater desalination by reverse osmosis is a mature technology, design is largely a routine matter with no new ground being broken.

Future improvements will come from new membrane materials and manufacturing methods, permitting higher recovery rates for a given pressure. It is probable that maximum feed pressures will rise above the present ubiquitous 1000psi.

Biwater Treatment, to whom I wish to say thank you for allowing me the time and resources to prepare this paper, are now seeking once again widen its RO activities to the levels achieved in the 1970's and early 1980's.

SESSION G:

BIOTECHNOLOGY APPLICATIONS

THE RELEVANCE OF FILTRATION IN LIVING CAPILLARIES TO INDUSTRIAL MEMBRANE PROCESSES

M.K. Turner
Department of Chemical & Biochemical Engineering,
University College London, Torrington Place, London WC1E 7JE

Why should a meeting about the benefits and opportunities of "Effective Industrial Membrane Processes" present a brief paper on "Ultrafiltration in Living Capillaries" ? The selective processes of the natural membranes may encompass passive and facilitated diffusion as well as active transport, and they may function consistently over many years. They may also be self-repairing. Nevertheless it would strain credibility to propose that there was any immediate comparison which could be drawn on these grounds with the synthetic membranes which are the subject of this conference.

What does bear comparison is the nature of their interaction with macromolecules. A synthetic ultrafiltration membrane is known to reject the protein albumin only after an initial period of permeability. The albumin which adsorbs to the membrane quickly reduces its own flux to a much greater extent than it affects the hydraulic permeability. In essence the membrane becomes more selective (Ingham et al, 1980; Howell et al, 1981). As Professor Michel's paper shows, the natural capillary membranes behave in a similar way. Their selectivity arises because they adsorb protein from the solution which bathes them, and the controlled "fouling" which occurs is crucial to their performance.

The comparable studies with synthetic membranes treat fouling and selectivity as separate phenomena. Too often they ignore the biological and chemical properties of the macromolecules which are adsorbed, even though the process is more complex than a gel-layer theory could describe (Reed & Sheldon, these Proceedings). Hence the need for an occasional look at the biological studies of capillary membranes.

REFERENCES

Howell, J.A., Velicangil, O., Le, M.S. & Zeppelin, A.L.H. (1980) Ann. N.Y. Acad. *Sci*. 369, 355.

Ingham, K.C., Busby, T.F., Sahlestrom, Y. & Castino, F. (1980) In *Ultrafiltration Membranes and Applications*" (Cooper, A.R. ed.) *Polymer Science and Technology* 13, 141. Plenum Press, New York.

Effective Industrial Membrane Processes — Benefits and Opportunities, pp.229–239

ULTRAFILTRATION IN LIVING CAPILLARIES

An Example of Simple Membrane Processing in a Biological System (Capillary Filtration)

C C Michel
(Department of Physiology & Biophysics
St Mary's Hospital Medical School, London)

SUMMARY

This paper reviews the process of ultrafiltration through the walls of capillary blood vessels. Capillaries are 5μm to 20μm in diameter and their walls are made of a single layer of flattened endothelial cells which act as a barrier to plasma proteins but allow water and low molecular weight hydrophilic solutes to exchange between the capillary blood and the surrounding tissues. Net fluid movements are driven by differences in hydrostatic pressure and by differences in osmotic pressure which arise from differences in concentration of the plasma proteins. The interaction between the surface coat of the endothelial cells and plasma proteins is important for the maintenance of normal hydraulic permeability and molecular selectivity of the capillary ultrafilter. Capillaries of the glomeruli of the kidney are specialised for ultrafiltration. They have high hydraulic permeabilities but the driving force for ultrafiltration is relatively low allowing filtration equilibrium to be approached or even achieved during a single transit. The relatively low pressures may contribute to the long term efficiency of the glomerular capillaries as ultrafilters and the part played by ultrafiltration in overall functioning of the kidneys.

INTRODUCTION

In contrast with most processes associated with biological membranes, this paper considers a relatively simple form of membrane separation which occurs at the inner surfaces of blood vessels. In the higher vertebrates (fish, amphibia, reptiles, birds and mammals), blood vessels are lined with a single layer of flattened cells, the endothelium. This layer separates the circulating blood, which is a suspension of red cells, white cells and platelets, from the fluid which bathes the cells of the surrounding tissues. In the larger blood vessels, the surrounding tissues are components of the vessel wall but in the capillaries, the endothelium is the wall itself and it is here in the capillaries that nutrients are delivered to the tissues and waste products are removed. In all capillaries except those of the brain, the exchange of low molecular weight solutes through the capillary walls is by passive diffusion. Water can also diffuse rapidly between the capillary blood and the tissue fluids and since the hydrostatic pressure inside the capillaries is greater than that in the tissues, the question arises as to how the fluid components of the blood are retained within the circulation. The answer to this question was provided almost a century ago by Ernest Starling (1896).

The fluid component of the blood, the plasma, has dissolved within it a relatively high concentration of plasma proteins. These are macromolecules and most of them have radii greater than 3.5 nm. Starling noted that the concentration of plasma proteins in the tissue fluids was considerably less than that of the plasma. He estimated the osmotic pressure which would arise from this difference in protein concentration and concluded that it would offset the greater hydrostatic pressure within the capillaries. Thus, fluid is held in the circulation by a balance of the differences in hydrostatic pressure and protein osmotic pressure across the capillary walls. Since differences in plasma protein concentration across the capillary walls can be sustained under low rates of steady filtration, the capillary walls act as ultrafilters.

A consequence of Starling's hypothesis, which has been generally confirmed, is that tissue fluid volume varies in a predictable way with changes in capillary pressure and plasma protein concentration. Under normal conditions, the circulation is regulated so that net fluid movements are kept to a minimum but in some tissues, net fluid movements between blood and tissues are an integral part of their function. This is most clearly seen in the mammalian kidney where relatively large volume of fluid are filtered from one set of capillaries and then later reabsorbed into a second set. Before discussing the process of ultrafiltration in the kidney, let us examine the ultrastructure and ultrafiltration properties of capillary walls in general.

CAPILLARY ULTRASTRUCTURE AND CAPILLARY ULTRAFILTRATION

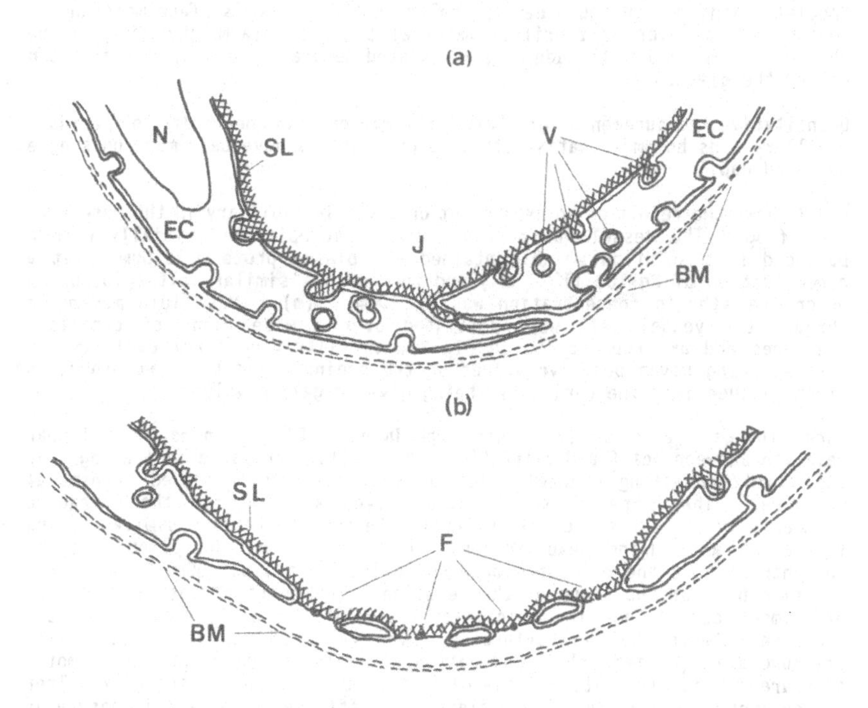

Figure 1 Diagrammatic representations of capillary walls of (a) continuous endothelium, and (b) fenestrated endothelium. EC, endothelial cell; SL, Surface layer; J, junction; V, vesicle; BM, basement membrane; F, fenestration.

When viewed under the electron microscope, two distinct types of capillary wall can be distinguished (Fig 1 a & b). In voluntary and involuntary muscle, skin, connective tissue and nerves, the endothelial cells form a continuous layer and while they may be flattened and thinned, the cell membranes are always distinct and separated by a layer of cytoplasm. The cells are joined together by specialised junctions (which are permeable to low molecular weight tracers) and enclosed by a continuous basement membrane. Endothelia of this type are said to be continuous.

The endothelia of capillaries in the kidney, in secretory glands and in the inner layers of gastrointestinal tract are made up of attenuated cells which in some regions appear to be penetrated by circular holes or fenestrations. The cells are joined together by specialised junctions to form a continuous layer and they are enclosed by a continuous basement membrane. This type of endothelium is said to be fenestrated. The fenestrations themselves may appear to be partly closed by a thin electron dense diaphragm but appear permeable to low molecular weight tracers.

Within both types of endothelial cells there are vesicles which can be labelled by electron dense tracers but whose role in transport is unclear. Special staining techniques have revealed that the inner surface membrane is coated with a layer of fibrillar material to which plasma proteins may be absorbed. Believed to be made of glycosylated membrane protein, this is often called the glycocalyx.

Quantitative measurements of fluid filtration can be made on perfused capillary beds but more satisfactory determinations have been made on single perfused capillaries.

Fig 2 shows the result of an experiment on a single capillary in the mesentery of a frog. The vessel was perfused via a micropipette initially with a balanced salt solution which contained the plasma protein, albumin, at a concentration of 8g 100 ml^{-1} (○) and then with a similar salt solution in which the albumin concentration was 2g $100ml^{-1}$ (●). Net fluid movements through the vessel wall were estimated over a wide range of capillary pressures and are represented in Fig 2 with filtration (from capillary to tissue) being given positive values on the ordinate and fluid reabsorption (from tissues into the capillary) being given negative values.

Three things are clear from this experiment. First, there is a linear relation between net fluid filtration and capillary pressure indicating that the fluid conducting channels are not distorted by increased transmural pressure. The slope of the relation gives a value for the hydraulic permeability, L_p, of the capillary wall. Second, L_p is very similar during filtration and during reabsorption. Third, changing the plasma albumin concentration in the perfusate between 2g $100ml^{-1}$ and 8g $100ml^{-1}$ shifts the position but not the slope of the relation. The shift of position in this experiment corresponds to approximately 70% of the difference in osmotic pressure between the two perfusates. The value of the intercept on the pressure axis (as zero fluid filtration) is related to the protein osmotic pressure of the perfusate and the reflection coefficient of the ultrafilter to the perfusate protein. The reflection coefficient is a useful measure of the molecular sieving properties of the ultrafilter and may be estimated from the square root of the ratio the intercept pressure to the perfusate osmotic pressure.

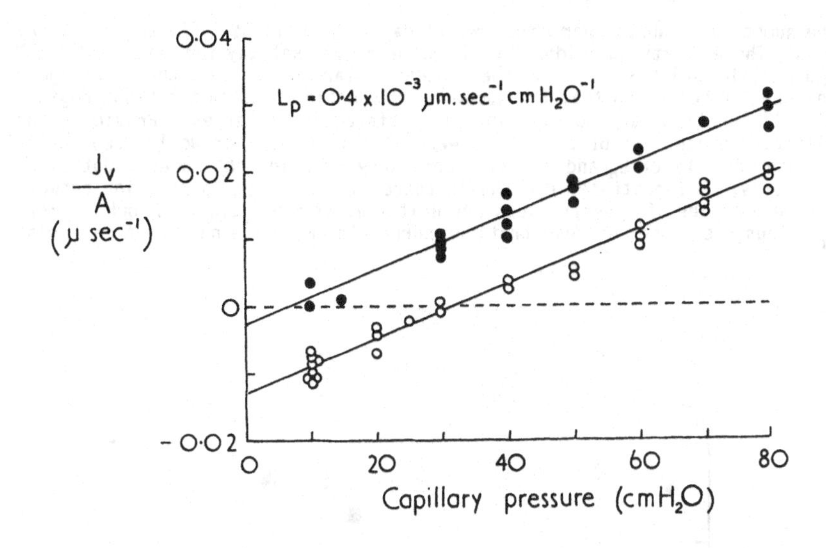

Figure 2 Relation between net fluid movements per unit area of capillary wall (J_v/A) and capillary pressure in a single frog mesenteric capillary perfused with Ringer solutions containing serum albumin at concentrations of 2g 100 ml^{-1}(●) and 8g 100 ml (○). From Michel 1981).

When small vessels are damaged, they lose their molecular sieving properties and hydraulic permeability, L_p, rises while σ_m, the reflection coefficient to a chosen macromolecule, falls. In both continuous and fenestrated endothelium, this increased permeability with tissue injury is associated with the development of relatively wide gaps between the endothelial cells.

In normal (undamaged) tissues, estimates of σ to plasma albumin are similar in many different capillaries though values for L_p may vary over two to three orders of magnitudes. In Figure 3, the mean value of σ to albumin for capillaries in a particular organ or tissue have been plotted against the mean value for L_p for the same capillaries. The relative constancy of σ_{Alb} in the face of wide variations in L_p suggest that the filtering properties of individual channels are similar in different capillary walls but the number of channels per unit area can vary very considerably indeed. It is also worth noting that the four highest values of L_p shown in Fig 3 represent fenestrated capillaries while other values reflect the permeability of capillaries with continuous endothelium. It is not difficult to account for the variations in the number of channels per unit area of capillary wall in different capillary beds. There is strong evidence that the principal pathway for water and small hydrophilic solutes lies in the intercellular clefts and there are good reasons for believing that the number of discontinuities in the tight regions of cell junction vary considerably in different capillaries. Providing the filtering properties of the clefts were also similar, this would account for the variability of L_p and relative constancy of σ in continuous endothelia. Similarly, in fenestrated endothelia there is a positive correlation between the mean number of fenestrations per unit area of capillary wall and the mean L_p. Thus the number of fenestrations correlates with the number of channels.

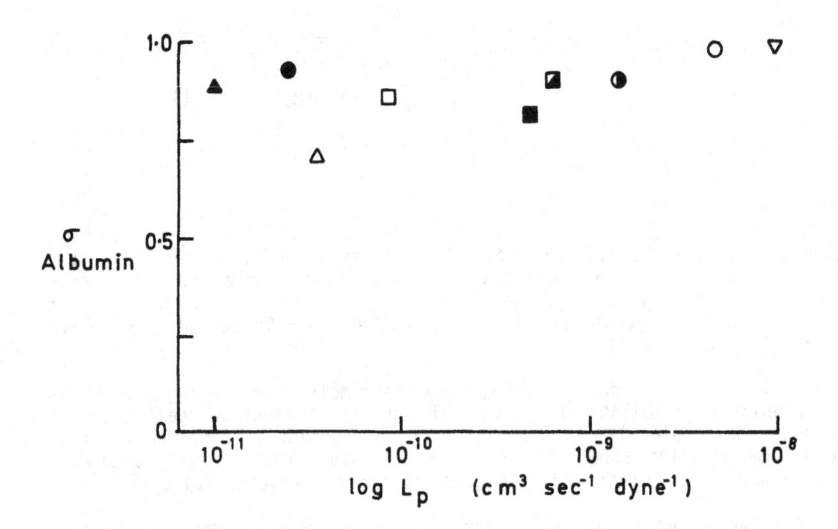

Figure 3 Values of σ to serum albumin for capillary walls in different capillary beds plotted against the corresponding values of L_p for the same vessels. (From Michel 1988)

But what is the nature of the filtering channel and can it have a similar structure in continuous and fenestrated endothelia? There are good reasons for believing that an important component of the filtering channels is located in the glycocalyx on the luminal surface of the endothelium. Such a component would cover both the entrance to the fenestrations and the clefts and so influence the permeability of both types of endothelia. It has been noted earlier that plasma proteins bind to the glycocalyx and if vessels are perfused with solutions containing no plasma proteins, L_p is found to increase and σ to non-protein macromolecules (such as dextrans, or Ficoll) falls. This increased permeability is not accompanied by conspicuous changes in the ultrastructure of the clefts or other parts of the endothelium but can be reversed by reperfusing with plasma or even solutions containing very low concentrations (>0.1g $100ml^{-1}$) of serum albumin. Serum albumin carries a net negative charge but it does possess positively charged lysine arginine side chains and if the arginine groups are chemically modified, the effects of albumin on permeability are lost (Michel et al, 1985). Furthermore, albumin has been shown to bind both reversibly and non-reversibly to sites on the endothelial surface. Removal of the reversibly bound albumin is associated with increased permeability (Schneeberger & Hamelin, 1984). Mild enzyme treatment (with pronase) which disrupts the luminal glycocalyx but appears otherwise to leave the endothelium intact, also increases permeability (Adamson, 1990).

Taken together, these observations suggest that serum albumin and other plasma proteins bind reversibly to negatively charged groups on the surface of the endothelium and arrange molecules of the glycocalyx into an ultrafilter. It has been suggested that a most effective filter would be formed if the albumin molecules converted a random arrangement of fibrous molecules into a regular lattice at the endothelial surface. Such a structure would greatly influence the L_p and σ to macromolecules of both fenestrated and continuous endothelium.

Support for this idea has been gained by finding that in the absence of other proteins, a highly cationised form of ferritin would mimic the effect of plasma proteins on permeability. The great advantage of cationised ferritin is that individual molecules can be identified in electron micrographs and it was possible to carry out an ultrastructural post-mortem after carrying out measurements of changes in permeability on single vessels. In this way, it was shown that cationised ferritin formed a layer at the luminal endothelial surface and changes in permeability were associated with change in the number of ferritin molecules in this layer rather than changes in luminal ferritin concentration (Turner et al, 1983). The effects of cationised ferritin on permeability could be contrasted with the effects native ferritin which does not bind to the endothelial surface (Fig 4). Measurements of the increase in σ to the neutral macromolecule, Ficoll 70, which accompanied perfusion with cationised ferritin also favoured the concept that the cationised ferritin was ordering a relatively sparse and random arrangement of fibrous molecules into a more evenly spaced lattice (Michel & Phillips 1985).

Thus there are strong grounds for believing that the ultrafilter at capillary walls is partly built out of the molecules to which it acts as a barrier. It is suggested that the ultrafilter is a regular structure for there appears to be advantages in a filtering membrane being isoporous. In normal healthy tissues, variations in the hydraulic permeability result from variations in the number of channels per unit area of capillary wall and this appears to be achieved by variation in the openness of the intercellular clefts of continuous endothelia and variations in the number of fenestrae per cell in fenestrated endothelia.

GLOMERULAR FILTRATION

The capillaries most specialised for ultrafiltration are those of the glomerulus of the kidney. By way of background, it should be said that each kidney is built up of a million urine producing units called nephrons. Each nephron can be considered as a blind ended tube of epithelial cells. The blind end is the glomerular end and projecting into it is a tuft of capillaries from which fluid is filtered and passes into the tubule. As fluid flows down the nephron, its composition is modified and most of its volume is reabsorbed by the action of the epithelial cells. The reabsorbed fluid then passes back into the blood of peritubular capillaries which lie in series with the glomerular capillaries. On average, 125ml of filtrate are formed every minute by the glomerular capillaries of the two kidneys of a healthy adult and 124ml of this filtrate is usually reabsorbed by the tubular epithelial cells. Each human adult kidney weighs between 130 to 160g but the glomeruli account for less than 1.0g. They receive nearly all the blood flowing into the kidney, filtering approximately one fifth of the fluid of the plasma into the nephrons. In spite of their tiny aggregate mass, the glomeruli capillaries have an aggregate surface area of just less than one square metre per kidney with the porous regions probably occupying just over 20% of this.

A closer look at the glomeruli reveals specialisations of the endothelium, the basement membrane beneath it and the associated epithelial cells. The capillaries of each glomerulus arise from a single afferent vessel and join up to form a single efferent vessel. Their endothelium is characterised by large fenestrations which occupy over 20% of their total surface. The endothelium sits upon a relatively thick and highly specialised basement membrane which consists of specialised collagen molecules and negatively charged glycosaminoglycans. Beneath the basement membrane are modified epithelial cells which are narrow and elongated and wrap around the capillaries like the closed fingers of a fist. The narrow spaces between them have been called filtration slits and while they are in the path between capillary and lumen of the tubule, molecular sieving occurs at fenestrations and upper surface of the basement membrane.

Glomerular capillaries have the highest L_p of all microvessels. They also have very high values of σ to plasma proteins such as serum albumin. As in other capillaries the glomerular barrier is greatly influenced by low concentrations of serum albumin and the L_p of isolated glomeruli has been shown to be greatly increased (without obvious changes in ultrastructure) when albumin is removed from capillary perfusate. (Fried et al 1986). Because it is possible to introduce a micropipette into the space surrounding the glomerular capillaries and so sample the filtrate as it is formed, there are detailed data on the molecular sieving characteristics of glomeruli. Thus, both the size and the charge of a macromolecule determine how easily it can be filtered. The glomerulus is a more effective barrier to negatively charged macromolecules than to neutral macromolecules of the same molecular size. The importance of molecular charge has proved to be of considerable significance in understanding changes which occur in certain diseases of the kidney.

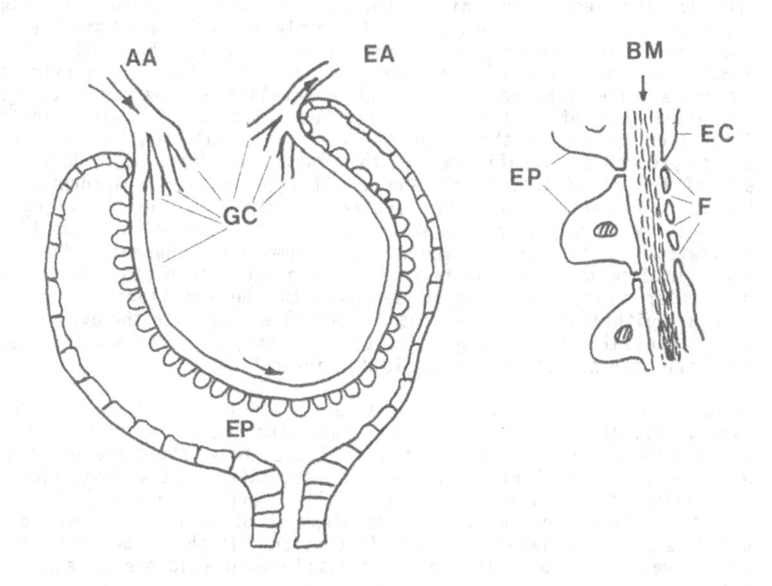

Figure 4 Diagrammatic representations of (a) glomerular capillary and its relation to the tubular epithelial cells, (b) ultrastructure of glomerular membrane. AA, afferent arteriole; EA, efferent arteriole; GC, glomerular capillaries; EP, epithelial cells; EC, endothelial cells; F, fenestrations; BM, basement membrane.

Lesions in the glomeruli result in the appearance of plasma proteins in the urine and the plasma proteins carry net negative charge. It has been shown that in some forms of glomerular disease, the appearance of proteins in the urine is associated with the loss of charge selectivity and not size selectivity. Thus the sieving of neutral molecules (such as dextrans) is unchanged while the passage of negatively charged molecules is increased. This suggests that negatively charged groups on the glomerular filter play a most important part in the filtration process. Attention has focused on the glomerular basement membrane and evidence from enzyme degradation experiments has implicated the glycosaminoglycan molecules, such as heparin sulphate, as key components of the structure of the molecular filter.

It has been noted that both L_p and the aggregate surface area of the glomerular capillaries are high. Surprisingly, however, the driving force for filtration is not large and considerably less than was once believed. Direct measurements of glomerular capillary pressure have shown that it lies in the vicinity of 45-60 mm Hg and falls only 1-2 mm Hg along its length (Brenner et al 1976). Since the osmotic pressure of the plasma proteins (which opposes filtration) is 20-25 mm Hg in the plasma entering the kidney and the hydrostatic pressure on the downstream side of the glomerular filter is 10 mm

Hg, the effective pressure driving filtration is only 10-25mm Hg at the beginning of the glomerular capillaries. The mean driving force, however, is considerably less than this. During its passage along the glomerular capillary one fifth to one quarter of the plasma is filtered and the proteins are concentrated raising their osmotic pressure to 45 mm Hg. Thus, the osmotic pressure of the plasma may rise to levels which bring filtration to a halt, a state referred to as filtration equilibrium. It may be thought that the attainment of filtration equilibrium represents a rather inefficient filtration system and this might be true if the only function of the kidney was to produce an ultrafiltrate of the plasma. In fact, the kidney reabsorbs 99% of the fluid that it filters and the process of reabsorption by the tubular epithelial cells is 'active' requiring metabolic energy and a proportional supply of oxygen which is carried to the cells by the blood. If filtration equilibrium is reached in the glomerulus, changes in filtrate rate occur in direct proportion to changes in glomerular blood flow. Since the glomerular capillaries are in series with the capillaries supplying the tubular epithelial cells, increased blood flow increases the oxygen supply to the epithelial cells in proportion to the filtration rate and thus match the metabolic demands of increased fluid reabsorption.

It has been speculated that the red cells, which carry O_2 in the blood, may have a subsidiary role in the glomerulus. The red cells are highly flexible disc-like cells and in their unstressed state, they have diameters comparable with those of capillaries. It has been suggested that as they flow through the glomerular capillaries, the red cells sweep along the surfaces of the endothelial cells preventing the development of unstirred layers of plasma proteins above the regions of high filtration. It should be said, that while an interesting hypothesis, there is little hard evidence to support it at present.

Indirect calculations suggest that in healthy human adults, glomerular filtration rate varies by no more than a few per cent over a 24-hour period (Robinson 1988). Thus, the glomerular capillaries of the two kidneys filter approximately 65,000 litres of fluid per annum. The filtering surfaces are probably cleaned by macrophages and while some glomeruli are lost compensation occurs by growth and there is little impairment of function until the fifth or sixth decade.

The combination of high L_p, large filtration surface area and relatively low filtration pressures may all contribute to the long term performance of glomeruli but of paramount importance is cellular regulation by phagocytes and new growth. At present, we have little understanding of how this is achieved.

REFERENCES

Adamson, R.H. (1990) 'Permeability of frog mesenteric capillaries after partial pronase digestion of the endothelial glycocalyx', J.Physiol. 428, 1-13.

Brenner, B.M., Deen, W.M. and Robertson, C.R. (1976) 'Determinants of glomerular filtration rate', Annual Review of Physiology, 38, 9-19.

Fried, T.A., McCoy, R.N., Osgood, R.W. and Stein, J.H. (1986) 'Effect of albumin on glomerular ultrafiltration coefficient in isolated perfused dog glomerulus', Am.J.Physiol., 250, F901-F906.

Kanwar, Y.S. and Farquhar, M. (1979) 'Presence of heparin sulphate in the glomerular basement membrane', Proceedings of the National Academy of Sciences Washington, **76**, 1301-1307.

Lassen, N.A. (1970) 'Contribution to discussion', In Crone. C. and Lassen, N.A., eds. Capillary Permeability, p549, Munksgaard, Copenhagen.

Michel, C.C. (1981) 'The flow of water through the capillary wall'. In: Water Transport Across Epithelia. Edited by H.H. Ussing, N.Bindster, N.A.Lassen and O.Sten-Knudsen. Copenhagen, Munksgaard, p268-279.

Michel, C.C. (1984) 'Fluid movements through capillary walls'. In: Renkin, E.M. and Michel, C.C. eds. American Handbook of Physiology, Section 2 Vol IV Microcirculation. American Physiological Society Washington DC pp 375-409.

Michel, C.C. (1988) 'Capillary permeability and how it may change'. J.Physiol. **404**, 1-29.

Michel, C.C. and Phillips, M.E. (1985) 'The effects of bovine serum albumin and a form of cationised ferritin upon the molecular selectivity of the walls of single frog capillaries', Microvascular Research, **29**, 190-203.

Michel, C.C. and Phillips, M.E. (1987) 'Steady state fluid filtration at different capillary pressures in perfused frog mesenteric capillaries', J.Physiol, **388**, 421-435.

Michel, C.C., Phillips, M.E. and Turner, M.R. (1985) 'The effects of native and modified bovine serum albumin on the permeability of frog mesenteric capillaries', J.Physiol, **360**, 333-346.

Renkin, E.M. and Gilmore, J.D. (1973) 'Glomerular filtration'. In Handbook of Physiology, Renal Physiology, eds: Orloff, J. & Berliner, R.W.) pp 185-248, American Physiological Society, Washington DC.

Robinson, J.R. (1988) 'Reflections on renal function (2nd edition)' Blackwells Scientific Publications, Oxford.

Schneeberger, E.E. and Hamelin, M. (1984) 'Interaction of circulating proteins with pulmonary endothelial glycocalyx and its effect on endothelial permeability', American Journal of Physiology, **247**, H206-H217.

Starling, E.H. (1896) 'On the absorption of fluids from connective tissue spaces', J.Physiol., **19**, 312-326.

Turner, M.R., Clough, G. and Michel, C.C. (1983) 'The effects of cationised ferritin and native ferritin upon the filtration coefficient of single frog capillaries. Evidence that proteins in the endothelial cell coat influence permeability', Microvasc.Res., **25**, 205-222.

PROCESS OPTIMIZATION OF AN ENZYME MEMBRANE REACTOR WITH SOLUBLE ENZYMES UP TO INDUSTRIAL SCALE

P Czermak (Akzo Research Laboratories, F R Germany)
W J Bauer (Centre de Recherche Nestle, Switzerland)

ABSTRACT

With the example of the enzymatically catalyzed hydrolysis of lactose the process optimization of steam-sterilizable dialysis membrane reactors is carried out up to industrial scale.
The expected conversion of lactose in the membrane reactor is dominated by the mass transfer resistance and the real flow in the reactor. Therefore a model for real reactors is developed to describe the transport reaction behaviour of the membrane reactors which considers the non-linear kinetics of the native enzyme, the real mixing conditions in the reactor and the mass transfer over the membrane. The calculation is carried out with the help of a coupled numerical solution.
By experimental investigations the enzyme kinetics, the mass transfer in the membrane, the hydrodynamics and the conversion are measured. The model permits the calculation of important process parameters.
For the hydrolysis of lactose by ß-galactosidase from Kluyveromyces marxianus the theoretical calculations with the developed model show good agreement with experimental results.

NOTATION

A	m^2	membrane surface area
c	$mol \cdot m^{-3}$	concentration
d_i	m	inside diameter of hollow fibre
D	$m^2 \cdot s^{-1}$	diffusion coefficient
D_m	$m^2 \cdot s^{-1}$	effective diffusion coefficient in the swollen membrane

Effective Industrial Membrane Processes — Benefits and Opportunities, pp.241–253

E_a	$J \cdot mol^{-1}$	activation energy
k_o	$m \cdot s^{-1}$	mass transfer coefficient
K_M	$mol \cdot m^{-3}$	Michaelis constant
K_I	$mol \cdot m^{-3}$	inhibition constant
L	m	length of hollow fibre
K,M,N		number of tanks
n		number of hollow fibres
r	$mol \cdot s^{-1}$	reaction rate
t	s	time
$\dot{V}$	$m^3 \cdot s^{-1}$	volumetric flow rate
V	m^3	volume
V_{max}	$mol \cdot s^{-1} \cdot m^{-3}$	maximum reaction rate
X		conversion
$Y_{p/s}$		molar yield of product on substrate
σ^2		variance of residence time related to dimensionless time
θ		dimensionless time
Re		Reynolds number

INTRODUCTION AND DESCRIPTION OF THE PROBLEM

Most conventional continuous enzyme reactors (e.g. fixed bed, fluidized bed reactor) cannot be sterilized without destruction of the enzymes and have to be operated at temperature and pH conditions on which the microbial growth is decreased (Prenosil et al 1987). It is possible to solve this problem with the use of steam-sterilizable dialysis membrane reactors, where the enzyme solution is kept separate from the substrate by the membrane (Czermak et al 1988a).

Steam-sterilizable dialysis membrane reactors have been developed in which enzymes can be used for enzymatically catalyzed conversions of low-molecular substrates (Czermak et al 1988b). For the enzymatically catalyzed conversion of lactose to glucose and galactose the dialysis membrane reactor has been scaled up to industrial scale (Czermak 1990; Patent 1988).

The principle of the reactor is based on the physical separation of substrate and enzyme by a semi-permeable membrane. Substrate as well as product can pass through this barrier by diffusion while the enzyme is retained by the membrane. The substrate solution flows on the tube side of the hollow fibre membrane whereas the enzyme solution flows on the shell side. The solution on the shell side is circulated over an external volume. Under these conditions, a continuous flow of substrate can be converted and the product drawn off continuously (figure 1).

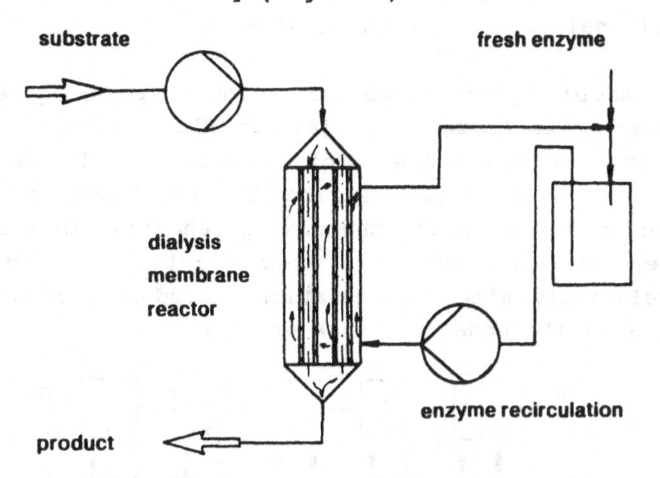

Figure 1: Simplified flow sheet of the enzymatically catalyzed process using the dialysis membrane reactor

For the hydrolysis of lactose it is expected that the efficiency of the reactor depends mainly on the mass transfer through the membrane and on the kind of flow and mixing on the tube side as well as the shell side of the reactor and in the recirculation system.

For the description of the conditions in the dialysis membrane reactor a mathematical model has to include the non linear enzyme kinetics of the hydrolysis of lactose, the mass transfer and the real mixing behaviour in the reactor (Czermak et al 1990a,b).
Thus, for precise calculations and the optimization of the reactors with a model the following experimental investigations are nescessary:

- determination of the residence time distributions,
- determination of the mass transfer through the membrane,
- determination of the enzyme kinetics,
- lactose hydrolysis in the dialysis membrane reactor.

MATHEMATICAL MODEL

The mathematical model which has been developed asume different degrees of mixing in the dialysis membrane reactor and the recirculation circuit. In general the degree of mixing in a real reactor will lie between the two extremes of plug flow and ideal mixing. Different kinds of models are known to represent the mixing behaviour of real flows (Levenspiel 1979).

For the enzymatically catalyzed conversion of lactose the dialysis membrane reactor is operated in laminar flow regimes. Because of that fact, in view of a scale-up of the reactor and with the knowledge of the residence time distributions the tanks in series model has been chosen to model the mixing behaviour in the dialysis membrane reactor. In addition to this the tanks in series model leads to relatively simple calculation procedures. Figure 2 shows the principle of the model (Czermak et al 1990a).

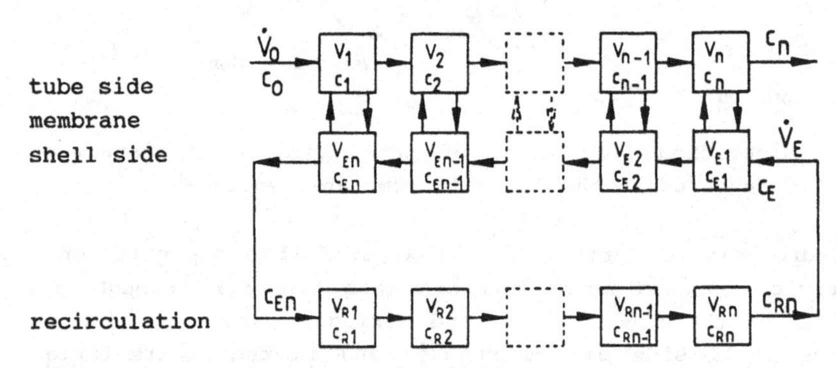

Figure 2: Tanks in series model to describe the real flow in the dialysis membrane reactor

The reactor is divided into three different compartments, the tube side and the shell side of the reactor and the recirculation circuit of the enzyme solution. Each compartment is divided again in a finite number of ideal tanks. With the assumptions of a stationary process, isothermal conditions and no ultrafiltration, for example for the first tank on the tube side on the shell side and in the recirculation volume the substrate mass balance gives:

tube side: $\dot{V}_0 \cdot (c_o - c_1) - k_o \cdot A \cdot (c_1 - c_{En})/n = 0$

shell side: $\dot{V}_E \cdot (c_E - c_{E1}) + k_o \cdot A \cdot (c_n - c_{E1})/n - r(c_{E1}) = 0$

recirculation: $\dot{V}_E \cdot (c_{En} - c_{R1}) - r(c_{R1}) = 0$

For the product mass balance gives the same equation except the rate of product formation is given by:

$$r(p) = Y_{p/s} \cdot r(c)$$

where $Y_{p/s}$ is the molar yield of product on substrate.

The calculations were realized with a coupled numerical solution. The system of non linear equations of the shell side and the recirculation was solved numerically with the Newton procedure (Czermak 1990).
So, for precise calculations it is necessary to know the mass transfer coefficient, the number of tanks - hydrodynamics - in all compartments of the reaction system and the kinetics of the biochemical reaction.

RESIDENCE TIME DISTRIBUTION

Levenspiel (1979) has given the relationship between the number of tanks and the variance of the residence times by the following equation:

$$\sigma^2 = \sigma^2_{exit} - \sigma^2_{entrance} = 1/N$$

The measurement of the residence time distributions was carried out with the pulse response technique. Concentrations of the tracer, blue dextran (MW 2000000 dalton), were measured at the entrance and the exit of each reactor compartment with an online spectrophotometer (Czermak 1990).

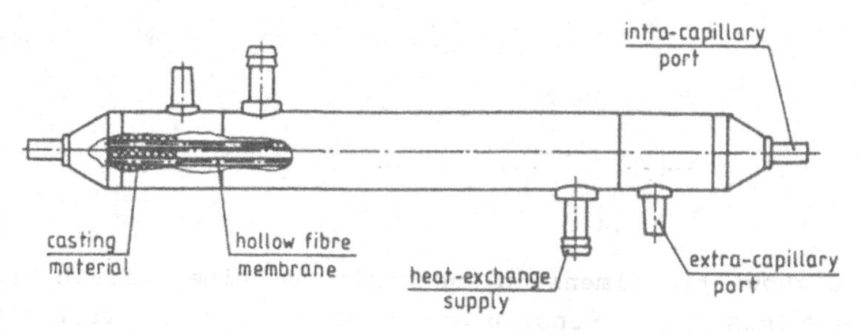

Figure 3: Sketch of the laboratory reactor (A = 0,12 m^2)

Reactors from laboratory scale (0,12 m^2 membrane area) over pilot scale (3,4 m^2) to industrial scale (9,4 m^2) were investigated. Figure 3 shows a sketch of the laboratory reactor, figure 4 of the reactor for the industrial scale (Czermak 1990; Patent 1988). Figure 5 gives an impression of the membrane reactor for the industrial scale.

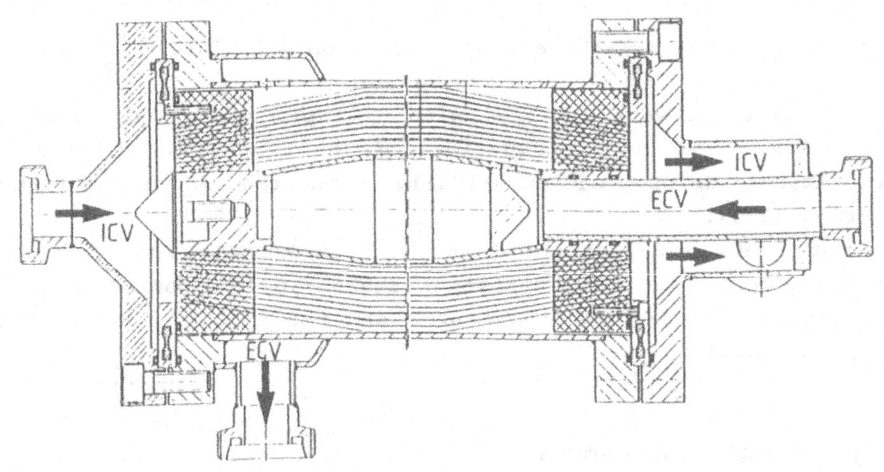

Figure 4: Sketch of the membrane reactor for the industrial scale (A = 9,4 m^2)

Figure 5: Photograph of the membrane reactor for the industrial scale

Figure 6 shows the dimensionless residence time distributions $E(\theta)$ plotted versus the dimensionless times θ for the tube and shell side reactor exit of the laboratory reactor.

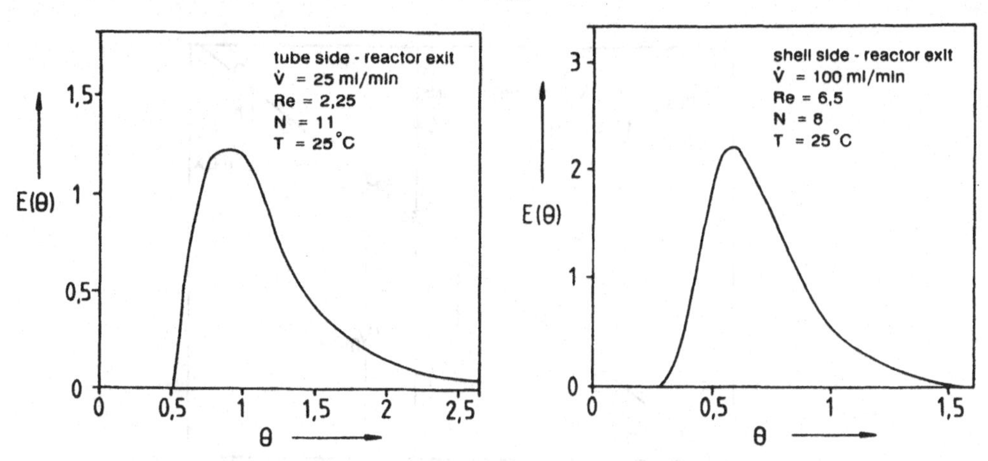

Figure 6: Residence time distribution of tube and shell side of the dialysis membrane reactor ($A = 0,12\ m^2$)

ENZYME KINETICS

The hydrolysis of lactose in the dialysis membrane reactor is carried out with technical grade ß-galactosidase from Kluyveromyces marxianus var. marxianus (Hydrolact L50; Boehringer, Ingelheim) and var. lactis (Maxilact LX5000; Gist Brocades, Delft). It is possible to describe the kinetics of the hydrolysis of lactose with the non-linear Michaelis-Menten kinetics with product inhibition (Prenosil et al 1987):

$$r_s = \frac{dc_s}{dt} = \frac{v_{max} \cdot c_s}{K_m \cdot (1 + c_p/K_i) + c_s}$$

Figure 7 and figure 8 show results for the kinetics of the used enzymes. Table 1 gives an overview over the experimental results of the enzyme kinetics (Czermak 1990).

Table 1: Enzyme kinetics for the ß-galactosidase (50 mM potassium-phosphate buffer; 1 mmol/l $MgCl_2$; pH = 7; T =25°C)

	K_m (mM)	K_i (mM)	V_{max} (mM/min)	E_a (KJ/mol)
Maxilact LX 5000	19,95	50	0,89	45,4
Hydrolact L50	14,8	12	0,34	39,6

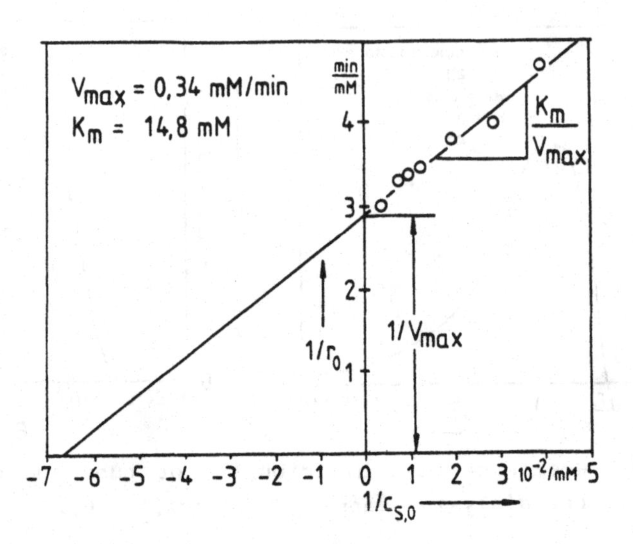

Figure 7: Lineweaver-Burk plot of the experimental results for Hydrolact L50 (T = 25°C; pH = 7,0; 50 mM potassium-phosphate buffer; 1 mmol/l $MgCl_2$)

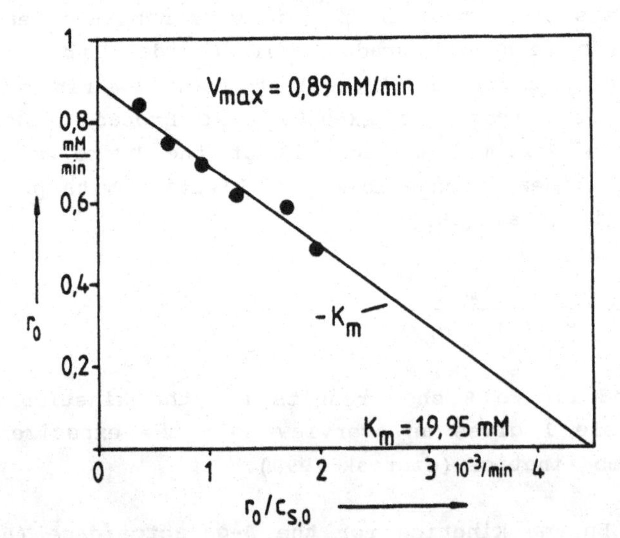

Figure 8: Eadie Hofstee plot of the experimental results for Maxilact LX5000 (T = 25°C; pH = 7,0; 50 mM potassium-phosphate buffer; 1 mmol/l $MgCl_2$)

MASS TRANSFER

The effective diffusion coefficient in the cellulosic hollow fibre membrane we used (Cuprophan[R], Akzo/Enka, FRG) has been evaluated by using an modified model from the literature (Qi et al 1985; Czermak 1990).
Figure 9 shows the simplified experimental set-up for the determination of the mass transfer. For this purpose, the substrate solution is pumped through the hollow fibre tube side in a closed circulation system and the concentration decrease of the solute is measured as a function of the time.

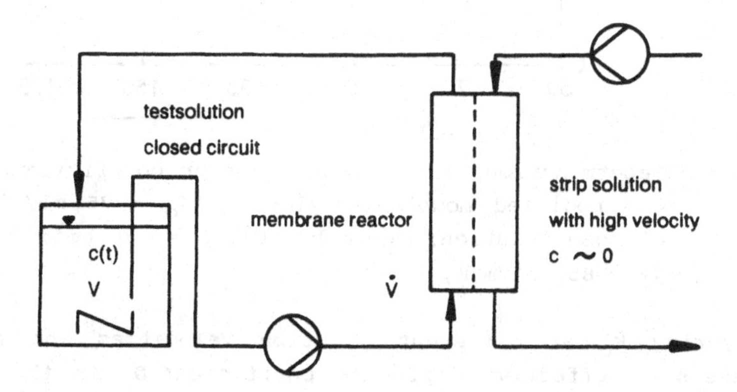

Figure 9: Simplified flow sheet of the experimental set-up for k_o determination

With the assumptions:
- isothermal conditions, no ultrafiltration
- concentration in the strip solution c = 0
- concentration difference between reactor entrance and end is very small,

and the balances for the hollow fibres and the whole reservoir it is possible to determine an overall mass transfer coefficient k_o.

$$k_o = - \frac{\dot{V}_o}{\pi \cdot d_i \cdot n \cdot L} \cdot \ln \left[1 - \frac{V}{V_o \cdot t} \cdot \ln \left[\frac{c(t=0)}{c(t)} \right] \right]$$

Figure 10 shows the results of an experiment with glucose as test substance (Czermak 1990).

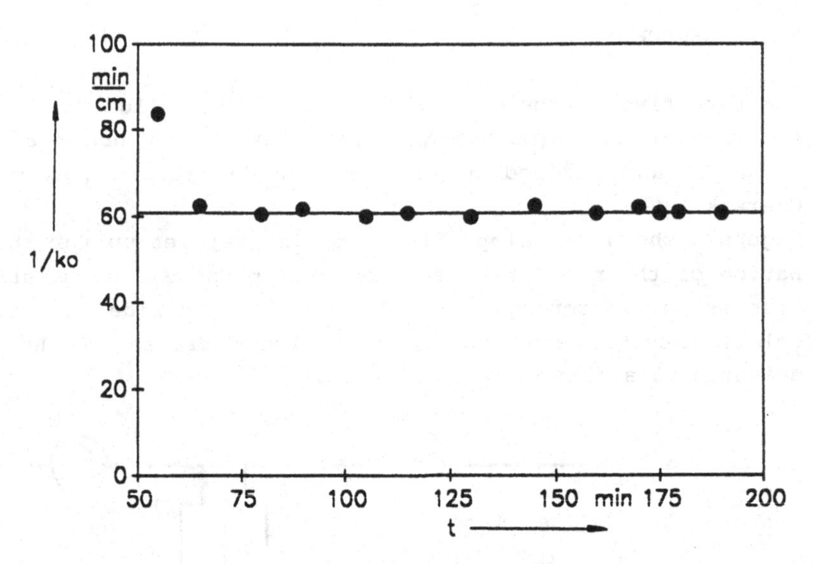

Figure 10: Determination of the mass transfer coefficient k_o with the modified model for glucose ($\dot{V}_o$ = 75 ml/min; 30mM glucose solution; Cuprophan G1, A = 280 cm^2; T = 25°C; $\dot{V}_E$ = 850 ml/min)

By measuring k_o at different solution velocities and membrane thicknesses an effective diffusion coefficient D_m in the swollen cellulosic hollow fibre membrane can be determined (Czermak 1990).

The following effective diffusion coefficients have been determined related to the swollen membrane (Cuprophan^R, Akzo/Enka; valid for 25°C):

lactose: $D_m = 8{,}1 \cdot 10^{-11}\ m^2/s$

glucose/galactose: $D_m = 12 \cdot 10^{-11}\ m^2/s$

HYDROLYSIS OF LACTOSE

Figure 11 shows results of calculations with the tanks in series model and experimental results for a hydrolysis of lactose in a 0,15 M lactose solution (50 mM potassium phosphate buffer, 1 mmol/l $MgCl_2$) on a laboratory scale. The conversion X is plotted versus the substrate flow rate V relative to a 1 m^2 membrane area. For the calculations the experimental results for the enzyme kinetics (K_m, K_i, V_{max}), for the hydrodynamics (number of tanks - N, M, K) and for the mass transfer (D_m) shown above were used. The figure shows a good correlation between experimental results and calculation values. An excess amount of enzyme (250 IU enzyme per ml enzyme circuit volume) was filled into the enzyme recirculation circuit and the reactor shell side. The large excess of enzyme means that the reactor was operated in a diffusion-limited way, i.e. the limiting factor with respect to the conversion achievable is the diffusion rate of the lactose through the membrane.

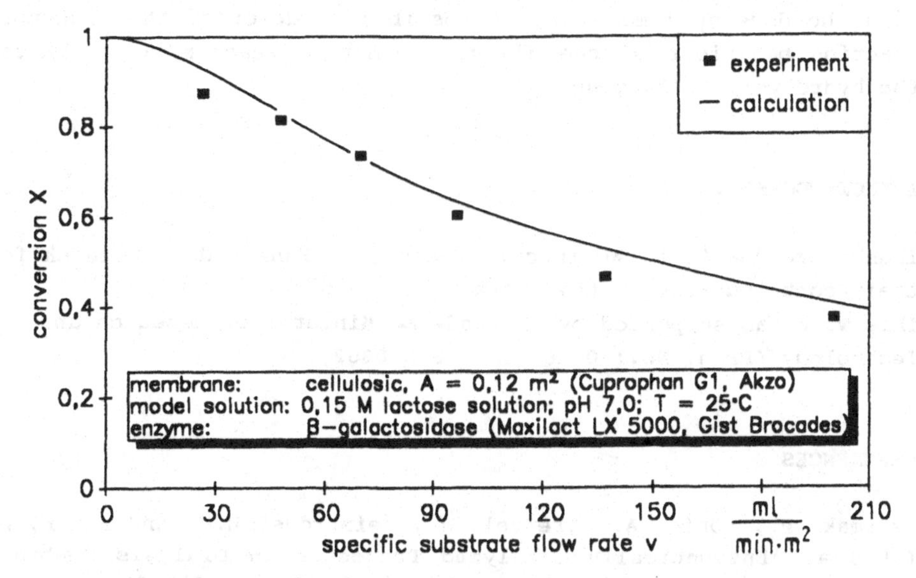

Figure 11: Comparison between experimental and calculated values for a hydrolysis of lactose on a laboratory scale (N=11; M=5; K=15; $\dot{V}_E$ = 230 ml/min; V_{max}= 5 mol/(m^3· s)

CONCLUSION

In view of the simple handling of the dialysis-membrane reactor and the possibility of a sterile reaction control it is possible to use the reactor for a wider variety of applications e.g. shown by Kirstein et al (1980) and Pronk et al (1988) and up to the immobilization of whole cells (Schönherr et al 1988).

Additionally, progress in the development of synthetic dialysis membranes will extend the application range of the dialysis-membrane reactor since it will then also be possible to use cellulase-containing aqueous fungus enzymes which has so far not been possible when using cellulosic membranes. In the same way,

the development of new modules particularly with respect to improved hydrodynamics will make it possible to increase the efficiency of the process even further.
In future, there will be an additional interesting field for the dialysis-membrane reactor, i.e. the sector of organic synthesis where some application examples already exist.

With the developed model it is possible to describe the transport reaction behaviour of the dialysis membrane reactor not only for the hydrolysis of lactose.

ACKNOWLEDGEMENT

Thanks are due to R. Wollbeck, D. Eser, D. Bahr and R. Wozasek for their contributions to this work.
This work was supported by the Federal Minister of Research and Technology (Proj. No.: 0318771A0, FIW 0862).

REFERENCES

Czermak, P., König, A., Tretzel, J., Reimerdes, E.H. and Bauer, W. (1988 a) 'Enzymatically Catalyzed Processes in Dialysis Membrane Reactors' Forum Mikrobiologie, vol. 11, no. 9, pp. 368-73

Czermak, P., Eberhard, G., König, A., Tretzel, J., Reimerdes, E.H. and Bauer, W. (1988 b) 'Dialysis Membrane Reactors for Enzymatic Conversions in Biotechnical Processes: Functional Principles and Examples for Application', DECHEMA Biotechnology Conferences, vol. 2, (Behrens, D., Ed.), VCH Verlagsgesellschaft, Weinheim, pp. 133-145

Czermak, P. and Bauer, W. (1990 a) 'Mass Transfer and Mathematical Modelling of Enzymatically Catalyzed Conversion in a Dialysis Membrane Reactor', Engineering and Food, vol. 3, (Spiess, W.E.L., Schubert, H., Ed.), Elsevier Applied Science, London - New York, pp.468-477

Czermak, P., Bahr, D. and Bauer, W. (1990 b) 'Verfahrenstechnische Optimierung der kontinuierlichen enzymatischen Laktosehydrolyse im Dialyse-Membranreaktor' Chem. Ing. Tech., vol. 62, no. 8, pp. 678-679 and MS 1885/90

Czermak, P. (1990) Entwicklung und verfahrenstechnische Optimierung eines Enzym-Membranreaktors für die Hydrolyse von Laktose, Fortschritt-Bericht VDI, Reihe 14, no. 64, VDI Verlag, Düsseldorf

Kirstein, D., Beßerdich, H., Kahrig, E. (1980) 'Hollow Fibre Enzyme Reactor Part II: Glucose Oxidase Reactor for the Production of Gluconic Acid from Glucose', Chem. Techn., vol. 32, pp. 466-468

Levenspiel, O. (1979) The Chemical Reactor Omnibook, OSU Book Stores Inc., Corvallis

Patent (1988) DE-3831786.9, EP-360-133-A

Prenosil, J.E., Dunn, I.J., Heinzle, E. (1987) 'Biocatalyst Reaction Engineering', Biotechnology, vol. 7a, (Rehm, H.J., Reed, G., Ed.), VCH Verlagsgesellschaft, Weinheim, pp. 489-545

Prenosil, J.E.; Stuker, E.; Bourne, J.R. (1987) 'Formation of oligosaccharides during enzymatic lactose hydrolysis: part I: state of art', Biotechnol. Bioeng., vol. 30, pp. 1019-1025 and 'part II: experimental', Biotechnol. Bioeng., vol. 30, pp. 1026-1031

Pronk, W., Kerkhof, P.J.A.M., van Helden, C., van't Riet, K. (1988) 'The Hydrolysis of Triglycerides by Immobilized Lipase in a Hydrophilic Membrane Reactor', Biotechnol. Bioeng., vol. 32, pp. 512-518

Qi, Z., Cussler, E.L. (1985) 'Hollow fiber gas membranes', AIChE J., vol. 31, pp. 1548-1553

Schönherr, O.T., Van Gelder, P.T.J.A. (1988) 'Culture of Animal Cells in Hollow-fibre Dialysis Systems', Animal Cell Biotechnology, vol. 3, (Spier, R., Griffiths, J.B., Ed.) Academic Press, London, pp. 337-353

Separation Techniques with Porous Glass Membranes: A Comparison of Dextrane and Protein Separations

P Langer, S Breitenbach, R Schnabel
(Schott BioTech, 6500 Mainz, Germany)

ABSTRACT

Porous glass hollow fibre membranes can be used for the separation of molecular mixtures as well as in microfiltration. In both cases, the degree of retention of a certain particle is mainly determined by the pore size of the membrane; the particular chemical surface modification of the membrane will have an additional influence. In this work, mixtures of dextranes have been used to evaluate the separation capacity of porous glass membranes. Interesting effects can be observed when higher molecular weight dextranes are used. A possible explanation is the flexibility of such large molecules and their deformation in a fluid flow field. Modelling transport through a membrane as a motion of a sphere (the diameter of which is the Stokes' diameter) through a cylindrical pore, the retention of dextranes and of proteins can be described by the same theory.

INTRODUCTION

Porous Glass Membranes

Porous glasses are not to be confused with sintered glass: they are produced from a homogeneous glass by effecting a phase separation (through thermal treatment) and leaching out one of the phases. By this technique, nearly cylindrical pores of well defined and uniform diameters are produced. This process originated from an earlier process which was aimed at the production of non-porous silica glass (Vycor) from leached borosilicate glass of pores generally about 4 nm in diameter [1]. The difference in the process for the manufacturing of Bioran porous glass membranes lies in the leaching step which is conducted in a way as to yield a high specific pore volume and also generally larger pore diameters as re-

Effective Industrial Membrane Processes — Benefits and Opportunities, pp.255–266

quired for the separation of biomolecules. Furthermore, by choosing an appropriate glass composition, pores up to 10 μm are feasible. The glass composition used for the membranes discussed below yields pores between approximately 10 and 300 nm in diameter and has the advantage that pore size distributions are extremely sharp with a standard deviation of less than 5 % from the mean value [2]. The porous glass matrix contains more than 96 % SiO_2, other elements present being B, Al and Na [3].

Porous glass beads have been used for many years in selected applications in chromatography. Porous glass may be derivatised by similar procedures as silica (see Unger [4], Chang et al. [5]) so that all common stationary phases can be prepared. In addition to this granular material, porous glass membranes have been developed [2] to meet increased demands from biotechnology.

These membranes (Bioran) are manufactured as hollow fibre membranes, bearing the advantages of well defined hydrodynamic conditions, large membrane surface per module volume and good mechanical strength. The fibres have an internal diameter of 0.3 and an outer diamter of 0.4 mm. They are assembled in modules of 0.05 and 0.2 m^2 membrane area, upon request also in units of 1.0 m^2. The potting material used is silicon rubber which is non-toxic for all applications in biotechnology. The membrane modules can be steam sterilized repeatedly at 120°C.

Due to the fact that glass membranes are symmetric isotropic membranes, they have a large internal surface area. This permits the binding of enzymes or affinity ligands at a high density (for a number of references on this topic, see the book by Messing [6]). For this purpose, Bioran membranes are provided with chemical surface modifications, examples of which are amino or epoxy groups covalently linked to the glass surface. Other surface modifications are hydrophobic or diol groups. The latter have been proven to be very effective in reducing non-specific protein adsorption to a very low level, thus reducing fouling and improving the long term performance of these membranes in ultrafiltration as well as in microfiltration [7].

Ultrafiltration with glass membranes

We have previously reported a detailed characterization of glass membranes by fractionation of protein mixtures [8]. We have subsequently developed a model to describe the passage of protein molecules through a porous hollow fibre membrane [9]. This model was based only on the relative sizes of the molecule and the pore diameter (measured by mercury porosimetry), and it was demonstrated that it described the experimental data quite accurately without requiring any adjustable parameters. It was therefore concluded that a protein molecule permeating a membrane behaved like a solid sphere penetrating a tunnel of cylindrical cross section: the rate of permeation will be affected by collisions with the pore walls, but specific interactions (attraction or repulsion) with the walls are absent (the membranes used in the experiments had the low-adsorption surface modification).

To further strengthen this hypothesis, we have now used polydisperse dextrane solutions, covering a broad range in molecular weight between 5 000 and 2 000 000 D. By using gel permeation chromatography analysis, continuous retention curves may be obtained for the entire range of molecular weight of interest.

MATERIALS AND METHODS

Standard Bioran membrane modules as described above were used in an ultrafiltration setup in which the retentate and the permeate were recirculated to the feed reservoir in order to maintain constant concentrations. The two main parameters that were usually held constant during experiments were the filtration pressure P and the volume reduction VR (stage cut), i. e. the ratio between permeate and feed flux as defined in Fig. 1. The parameters that had to be adjusted by the computer controlling the experiments were the feed and retentate fluxes delivered by two peristaltic pumps.

The dextrane mixture used in all experiments consisted of the following components (obtained from Polymer Standard Service, Pfeiffer & Langen, Pharmacia):

Molecular Weight (Peak)	Relative Amount (%)
560,000	2.5
350,000	15.5
150,000	31.0
52,000	31.0
31,500	12.5
20,500	5.0
11,000	2.5

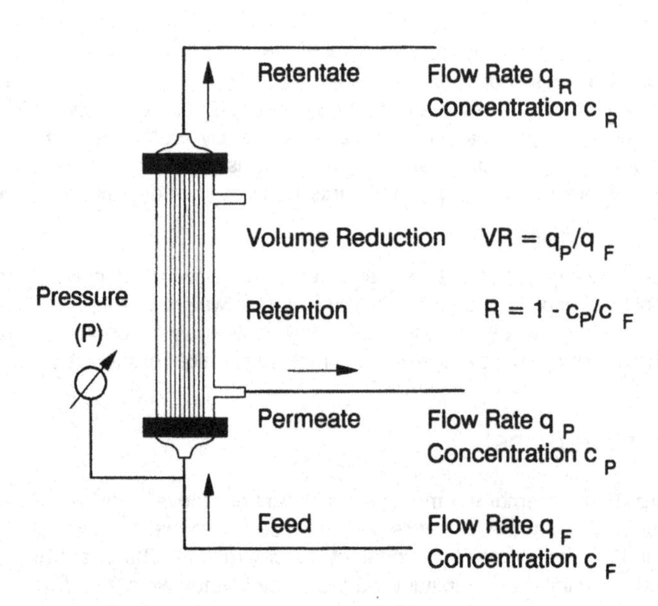

Figure 1 Definition of main parameters in the filtration setup for dextrane experiments (we use the terminology suggested by the European Society of Membrane Science and Technology [10]).

The mean molecular weight of the mixture was 180 000 D, the polydispersity index approx. 4. Thus, a molecular weight range from 5 000 to 2 000 000 D could be covered. The dextrane solution was prepared in 0.3 % NaCl and 0.02 % NaN_3, total dextrane concentrations used were between 0.5 and 6 %.

Samples from feed and permeate were taken over a period of 24 h and analyzed by gel permeation chromatography (Spectra Physics / Waters; column: PSS Hema Bio 1000 8x300 + 8x600 mm; eluent: 0.3 % NaCl / 0.02 % NaN_3 in water; flow rate: 0.7 ml/min; refractive index detector). A deconvolution of the permeate and feed chromatograms yielded retention data over the entire range of molecular weight. Chromatograms were calibrated by dextranes of a narrow molecular weight distribution.

RESULTS AND DISCUSSION

Virtually identical results were produced from samples taken at any time within the 24 h period of the experiment. This indicates that the membrane is not fouled during filtration. Typical results of retention curves obtained are shown in Fig. 2.

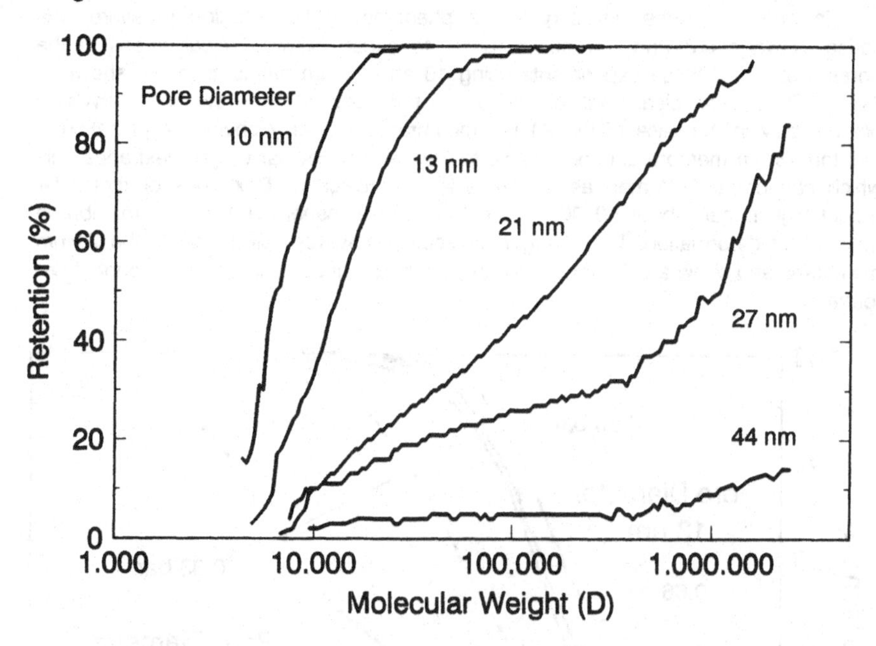

Figure 2 Retention of dextranes by BIORAN porous glass membranes of different pore diameters (pore data were determined by mercury porosimetry, dextrane molecular weight was calibrated with single molecular weight standards). The filtration pressure for all membranes was 1.0 bar.

It is immediately apparent that the shape of these retention curves changes between the 13 and the 21 nm membrane: while they are very steep for smaller pores, they seem to indicate a more diffuse cut-off for larger pore membranes. The pore size distribution was analyzed by mercury porosimetry for all membranes, and it was equally sharp in all cases (i. e. the standard deviation of the distribution from the mean value was less than 5 %). Furthermore, the experiments with the 21 nm membrane were repeated with a second membrane of the same pore size and confirmed.

One likely explanation for this finding is that especially large dextrane molecules are deformed under shear stress in a fluid flow field (e. g. in a pore) and may therefore be able to penetrate a pore through which they could not pass in a globular configuration. de Balmann and Nobrega [11] have partly attributed similar findings in a polysulfone membrane to this effect.

In order to further investigate this phenomenon the filtration pressure was varied between 0.3 and 1.0 bar in order to cause different shear stress in the pores. Results of these experiments using 13 and 21 nm membranes are shown in Fig. 3. It is very clear that an effect of pressure on the retention curves is present only in the case of the 21 nm membrane. The only obvious explanation is that the 13 nm membrane is permeable to the lower molecular weight dextranes only which cannot be deformed as easily as larger molecules. Dextranes of molecular weight higher than about 50 000 D are too large to penetrate a 13 nm membrane even under deformation. These larger molecules, however, will permeate the 21 nm membrane and show a different degree of permeation depending on the amount of deformation.

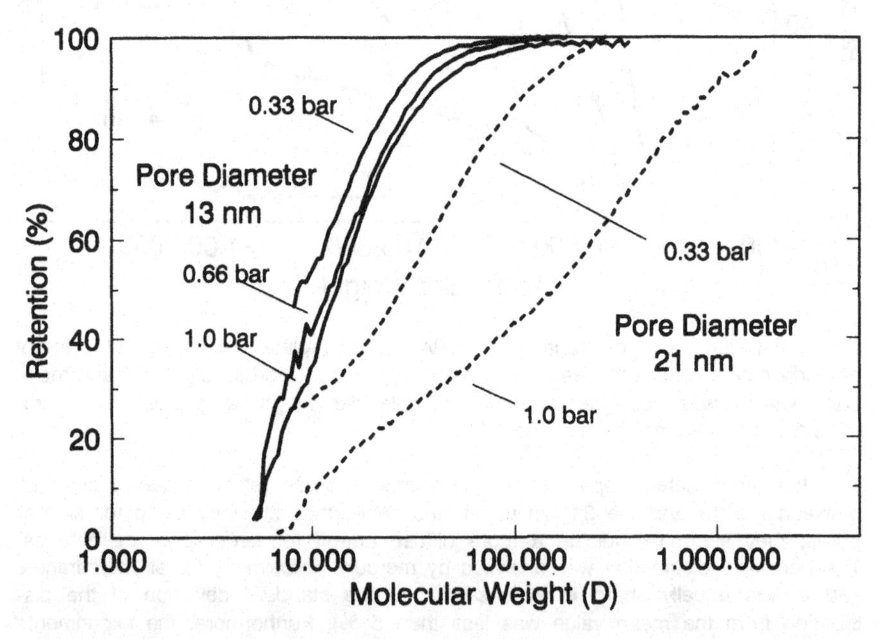

Figure 3 Effect of different filtration pressure on the retention curves for dextranes. VR = 0.1 in both cases.

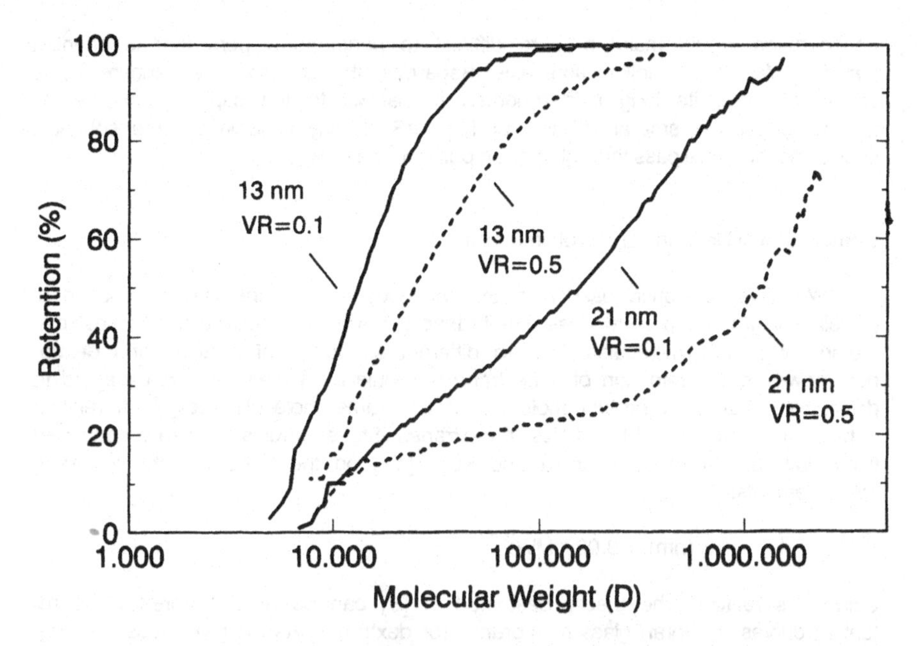

Figure 4 Dextrane retention curves under different volume reduction (ratio of feed to permeate flux) at a pressure of 1 bar. Axial flow velocities are almost a factor 10 higher at VR=0.1 as compared to VR=0.5.

Another cause of shear stress to molecules in a hollow fibre membrane is the axial flux in a fibre. Changing the volume reduction VR at constant pressure affects the axial flow velocity. In the case of the 21 nm membrane, the axial flow velocity is 1 cm/s at VR=0.5 and 8.5 cm/s at VR=0.1. Fig. 4 shows that this parameter has a significant effect on the retention curves for both the 13 and 21 nm membranes.

The effect of decreasing the axial flow velocity in the hollow fibres is a decrease in retention. Concentration polarisation may be partly responsible for this effect. On the other hand, varying the solution concentration between 0.5 and 6 % did not have a very strong influence on the retention curve.

A more plausible explanation could again be the flexibility of dextrane molecules. Decreasing the volume reduction implies increasing the axial flow velocity, thereby "stretching" the dextrane molecules in this direction. This configuration would make it more difficult to penetrate a pore that is oriented perpendicular to the hollow fibre axis. Regarding the deformed dextrane molecule as an ellipsoid, its long axis (oriented in parallel to the capillary axis) would be the longer the smaller VR is. For high VR, it may assume a more spherical shape and therefore pass through a given pore more easily.

Comparison of Dextrane and Protein Results

We have previously used a theoretical model to describe the fractionation of protein mixtures by porous glass membranes [9]. This model used the Stokes' diameters of protein molecules. Due to different structures of proteins and dextranes, however, the retention of a dextrane molecule by a membrane is expected to differ from that of a protein molecule of the same molecular weight. A relation between molecular weight and size of dextranes (Stokes' radius R_s) can be deduced from data published by Granath and Kvist [12] on the basis of light scattering measurements:

$$R_s\,(\text{nm}) = 0.03 * MW^{0.47}$$

Using this relation, the theory described in [9] can be used to predict the retention curves of Bioran glass membranes for dextranes. As the model assumes rigid, spherical molecules and does not account for deformation, it is expected to give the most accurate prediction for the case of low filtration pressure. The resulting curves are shown in Fig. 5 along with the experimental data obtained at a pressure of 0.33 bar.

Considering that the theory does not contain adjustable parameters, the agreement with the experimental results in Fig. 5 is good. This is a clear indication that the model which predicts the retention of a molecule by a porous membrane is perfectly adequate to describe the transport mechanisms prevailing in glass membranes.

This conclusion is even more stringent if protein and dextrane data are plotted on one diagram. As the abscissa the ratio of molecular diameter (Stokes' diameter) and pore diameter is chosen so that the theoretical prediction for all pore sizes is a single curve. Fig. 6 shows this representation with the two dextrane curves from Fig. 5 and the protein retention data for two different membranes as they were published previously [9].

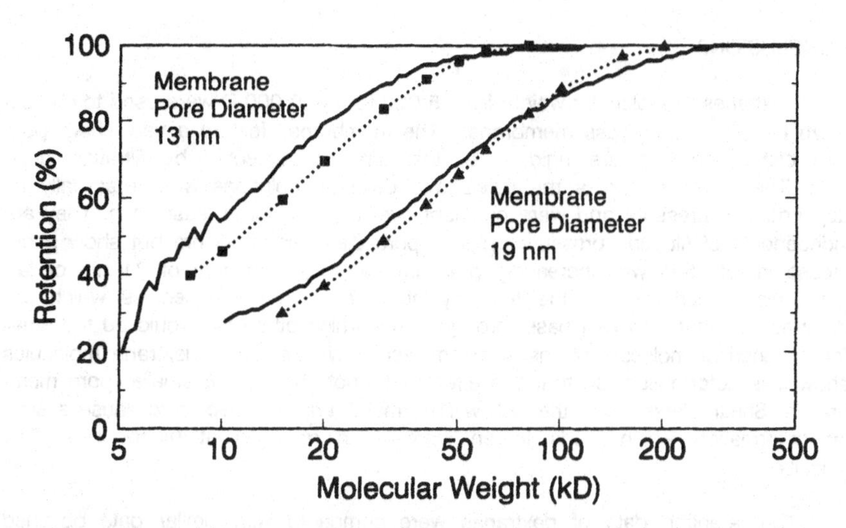

Figure 5 Experimental data of dextrane retention by Bioran membranes of 13 and 19 nm at a filtration pressure of 0.33 bar (solid lines) along with a theoretical prediction (dotted lines with symbols) as given in [9].

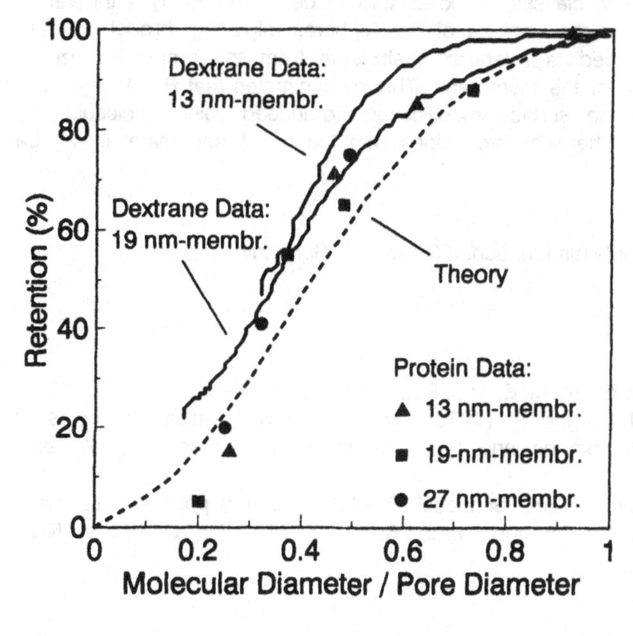

Figure 6 Retention of protein and dextrane molecules by Bioran glass membranes of different pore diameters, plotted over the ratio of molecular to pore diameters.

CONCLUSIONS

Dextranes of molecular weight from 5 000 to 2 000 000 D were used to characterize Bioran porous glass membranes. The membranes feature a very sharp pore size distribution and are rigid, i. e. they are not afffected by filtration pressure. The retention curves measured with dextrane solutions were practically independent of pressure and were constant over a period of at least 24 h. They are independent of filtration pressure up to a pore diameter of 13 nm but show a decrease in retention with increasing pressure for pore diameters of 21 nm or larger. This is attributed to the flexibility of larger dextrane molecules which are deformed so that they can pass through pores which otherwise would be too small for a spherical molecule of the same molecular weight. Small dextrane molecules show less deformation, so that this effect does not show up in smaller pore membranes. Shear stress along the hollow fibre membrane axis seems to cause a similar deformation which leads to an increase in retention at higher axial flow velocities.

The retention data of dextranes were compared with similar data obtained previously for proteins. If instead of molecular weights the molecular diameters of dextranes and proteins are compared it turns out that molecules of identical diameters are retained to the same degree by a membrane of a certain pore diameter. The behaviour of both classes of molecules can be described by a theoretical model which is based on the concept of rigid spheres migrating through cylindrical pores and which predicts retention coefficients from the molecular size in relation to the pore size of the membrane. This demonstrates that Bioran membranes with a low-adsorption surface modification do indeed select molecules by their size and that any chemical interactions with the membrane material can be neglected.

"Bioran" is a registered trade mark of Schott Glaswerke, Germany.

REFERENCES

[1] Hood, H. P. and Nordberg, M. E., U.S. Patent #2,106,744, Feb. 1, 1938

[2] Schnabel, R. and Langer, P. (1989), 'Structural and chemical properties of glass capillary membranes and their use in protein separation', Glastech. Ber. 62, 56-62

[3] Langer, P., Eichhorn, U. and Schnabel, R. (1990), 'Bioran porous glass membranes for solvent purification', Proc. 5th World Filtration Congress, Nice, 134-138

[4] Unger, K.K., 'Porous Silica', Elsevier, Amsterdam 1979

[5] Chang, S. H., Gooding, K. M. and Regnier, F. E. (1976), J. Chromatogr. 125, 103.

[6] Messing, R. A. (ed.), 'Immobilized Enzymes for Industrial Reactors', Academic Press, New York 1975

[7] Langer, P. and Schnabel, R. (1990), 'Separation with porous glass membranes and controlled pore glass (CPG) chromatography', Separations for Biotechnology Vol. 2, Ed. D. L. Pyle, Elsevier, London, 371-380

[8] Schnabel, R., Langer, P. and Breitenbach, S. (1988), 'Separation of Protein mixtures by Bioran porous glass membranes', J. Mem. Sci. 36, 55-66

[9] Langer, P., Breitenbach, S. and Schnabel, R. (1989), 'Ultrafiltration with porous glass membranes', Proc. Int. Techn. Conf. on Membrane Separation Processes, Brighton, 105-119

[10] Gekas, V. (ed.), 'Terminology for pressure driven membrane operations', Lund 1986

[11] de Balmann, Hélène and Nobrega, Ronaldo (1989), 'The deformation of dextrane molecules. Causes and consequences in ultrafiltration', J. Mem. Sci. 40, 311-327

[12] Granath, K. A. and Kvist, B. E. (1967), 'Molecular weight distribution analysis by gel chromatography on Sephadex', J. Chromatogr. 28, 69-81

TITLE: ANTIBODY PRODUCTION IN A MEMBRANE REACTOR - A START-UP COMPANY'S CHOICE

AUTHORS: DR T G WILKINSON, M J WRAITH, P D G COUSINS
BIOCODE LIMITED

ABSTRACT:

Biocode's technology and commercial business is dependent on the use of antibodies for toxin detection and for coding products. Antibody production is therefore crucial. A number of antibody producing systems have been evaluated. In particular, it is important to minimise the use of animals. This paper describes the use of a membrane reactor for antibody production.

The paper concludes that by comparison with other systems, the membrane system can be run efficiently and at high productivities and that the cost of antibodies is economical compared with other systems and the conventional techniques of antibody production using animals.

INTRODUCTION

Biocode, a wholly owned subsidiary of Shell UK, is a company which was set up to commercially exploit biotechnological methods using monoclonal antibodies. Biocode's market is focused on two main areas:

1. Development of a unique patented biological barcode for identifying trace levels of marker chemicals, which have been added to products such as wine, cosmetics, drugs, bulk chemicals, oil products, etc. The system enables the products to be uniquely identified to prevent counterfeiting, as well as marking separate batches for identification and process loss purposes.

2. The development of diagnostic kits to detect trace amounts of contaminants in food, water, air, etc.

The Company's main strengths are its patented portfolio; its ability to develop and select a range of antibodies and link these to simple-to-use test kits; and its quality approach (we have applied for registration for BS 5750 Part 1).

Biocode's main vulnerability is in two areas:

First, whether there is a sizable enough market to exploit; and secondly on the technical front, its ability to produce large quantities of high specification antibodies. The focus of this paper is on antibody production, where there is a need to concentrate on:

- Cost - the market needs to bear the cost of the antibody used in the test.

- Quality - the antibody needs to be produced to a defined specification.
- Ethics - the minimisation of animal use is essential.

- Parent company - antibody production techniques need to be appropriate to a Company belonging to Shell.

Effective Industrial Membrane Processes — Benefits and Opportunities, pp.267–277

CHOICE

An antibody producing cell line (hybridoma) is created by the fusion of a myeloma (cancer) cell with an antibody producing splenocyte (see Figure 1). Once this hybridoma cell has been created there are a number of methods by which the antibody can be produced in reasonable quantities. These are:

1. Ascites production using animals (where hybridoma cells are injected into the peritoneal cavity of an animal and the subsequent tumour produces antibodies).

2. In simple flasks, where the antibody producing cells are grown up in batch flasks and only the temperature is controlled.

3. In a variety of controlled stirred reactors

    i. Using simple stirred vessels where only the initial media composition and temperature and to a certain extent agitation are controlled, or,

    ii. In a fully controlled stirred tank reactor with full pH control, etc. This type of system can be run, in a batch, a batch fill, or continuous mode.

4. Using an air lift reactor - where agitation is maintained by forced aeration.

5. Using a membrane system where the cells are grown on membrane surfaces. Those units which use membranes are many and varied and there is considerable choice to be made.

Although difficult to operate, data suggest that there are higher productivities from membrane systems due to the hybridomas preference for growing on surfaces.

A complete evaluation of these systems is difficult, because one thing that is missing in nearly all cases, is real data which has been generated using commercial antibody producing cell lines. It is not possible to study scientific data to make a rational choice of reactor. The 'reality' is the need to obtain "off the peg" equipment, to trial it for a period and determine whether the system works or not. This naturally requires considerable background knowledge within your own staff, a considerable amount of patience and a great deal of faith. At this point it is worth commenting that when we chose a production systems for trial, the word "membrane" (on which the system being discussed is based) did not come into the equation until very much later. The authors therefore feel something of a fraud discussing membranes at this symposium.

TRIALS WITH A MEMBRANE REACTOR

The system (manufactured by the Endotronics Corporation (USA) and modified from a kidney dialysis unit) that was chosen by us, contained split media and harvesting streams. Media was pushed through the inside of the fibre membranes and cells were grown on the outside. The advantage of this particular system being that the cells could feed off the relatively dilute media and at the same time can produce antibodies which can be harvested separately in a more concentrated solution. A circuit diagram of the reactor is shown in Figures 2, 3 and 4. The system is also set up with a separate "lung" which is used to achieve both aeration and mixing. Clearly the design of the membrane fibres is crucial to ensure separation of the media from the antibodies. The characteristics are shown in table 1 below.

Figure 1

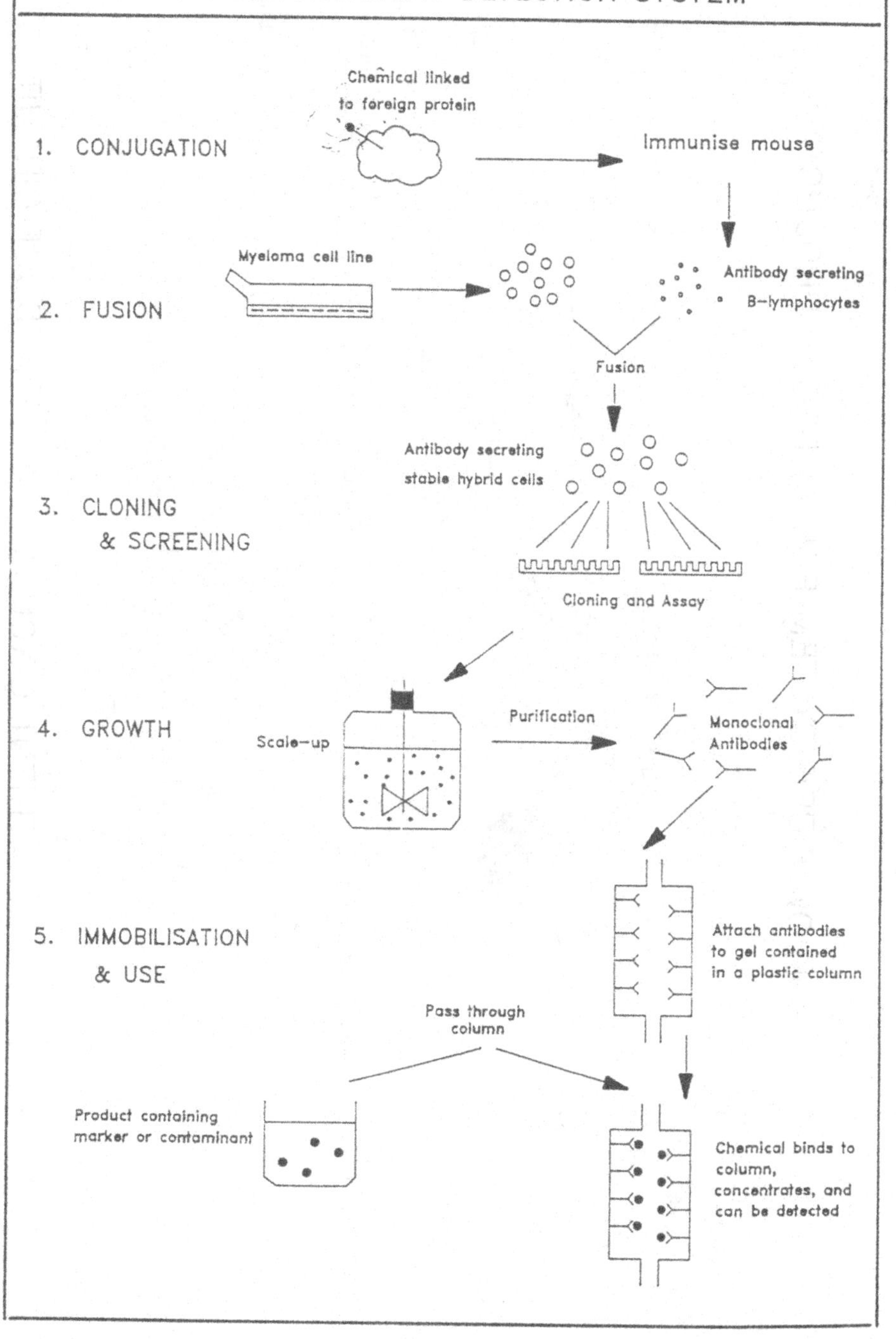
SCHEMATIC REPRESENTATION OF THE DEVELOPMENT OF AN IMMUNOASSAY DETECTION SYSTEM
1. CONJUGATION
Chemical linked to foreign protein
Immunise mouse
2. FUSION
Myeloma cell line
Antibody secreting B-lymphocytes
Fusion
3. CLONING & SCREENING
Antibody secreting stable hybrid cells
Cloning and Assay
4. GROWTH
Scale-up
Purification
Monoclonal Antibodies
5. IMMOBILISATION & USE
Attach antibodies to gel contained in a plastic column
Pass through column
Product containing marker or contaminant
Chemical binds to column, concentrates, and can be detected

Figure 2

HOLLOW FIBRE SYSTEM FOR ANTIBODY PRODUCTION

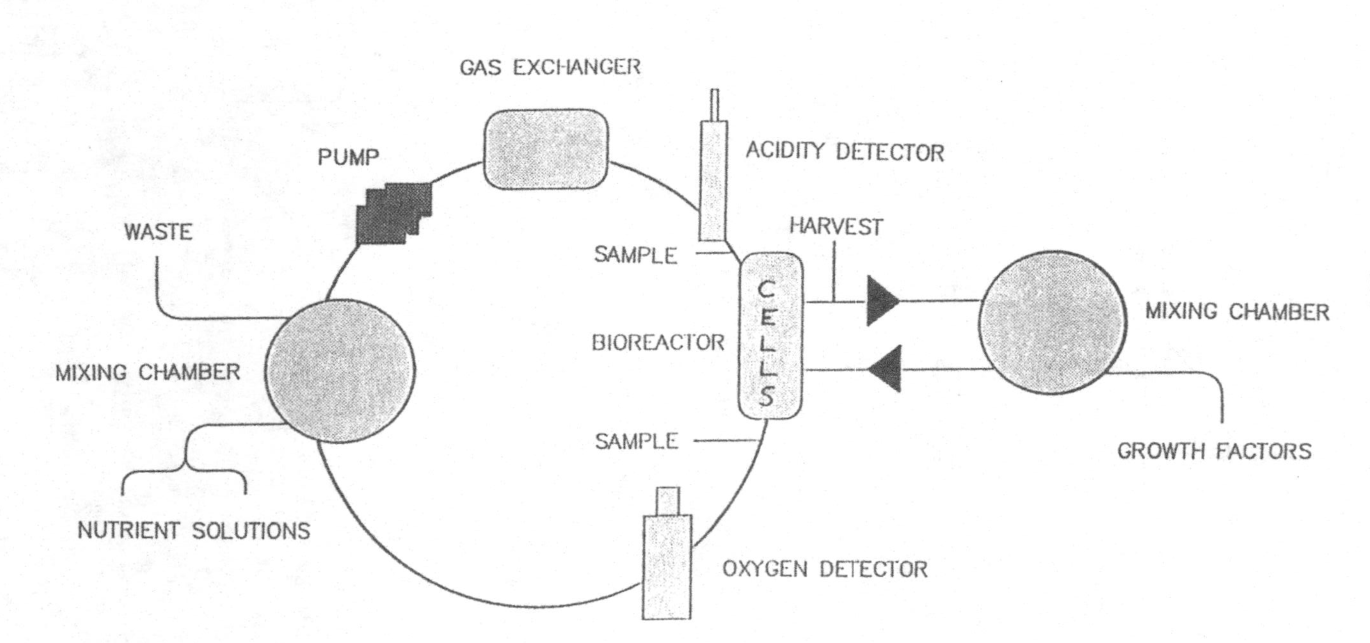

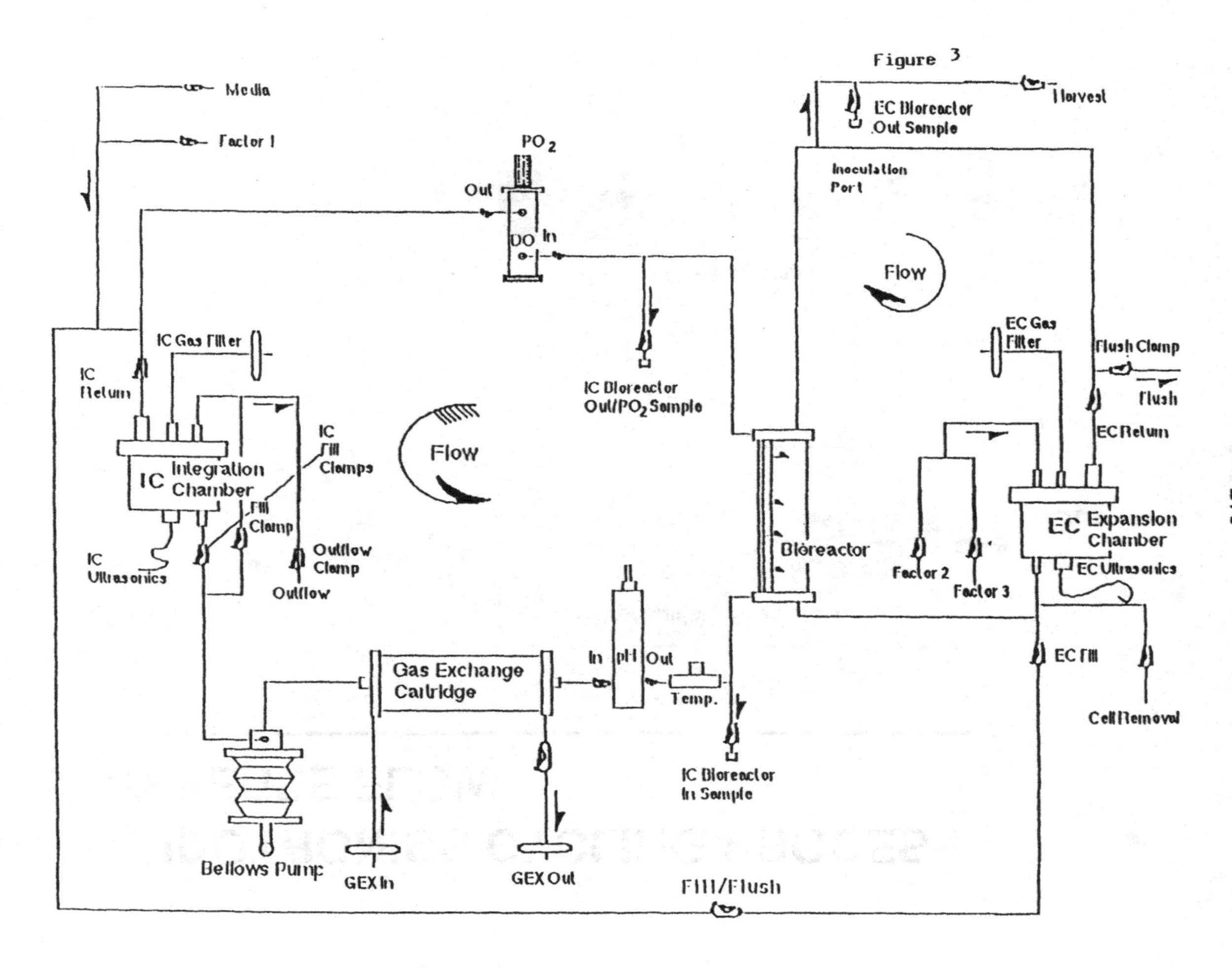
Media
Factor 1
PO2
Out
In
DO
EC Bioreactor
Out Sample
Harvest
Inoculation
Port
Flow
IC Gas Filter
IC
Return
IC Bioreactor
Out/PO2 Sample
EC Gas
Filter
Flush Clamp
Flush
EC Return
IC
Fill
Clamps
IC
Integration
Chamber
Fill
Clamp
Flow
Bioreactor
EC
Expansion
Chamber
EC Ultrasonics
Factor 2
Factor 3
IC
Ultrasonics
Outflow
Clamp
Outflow
In
pH
Out
Temp.
EC Fill
Cell Removal
Gas Exchange
Cartridge
IC Bioreactor
In Sample
Bellows Pump
GEX In
GEX Out
Fill/Flush
Figure 3

ENDOTRONICS CYCLING PROCESS REVERSE FLOW

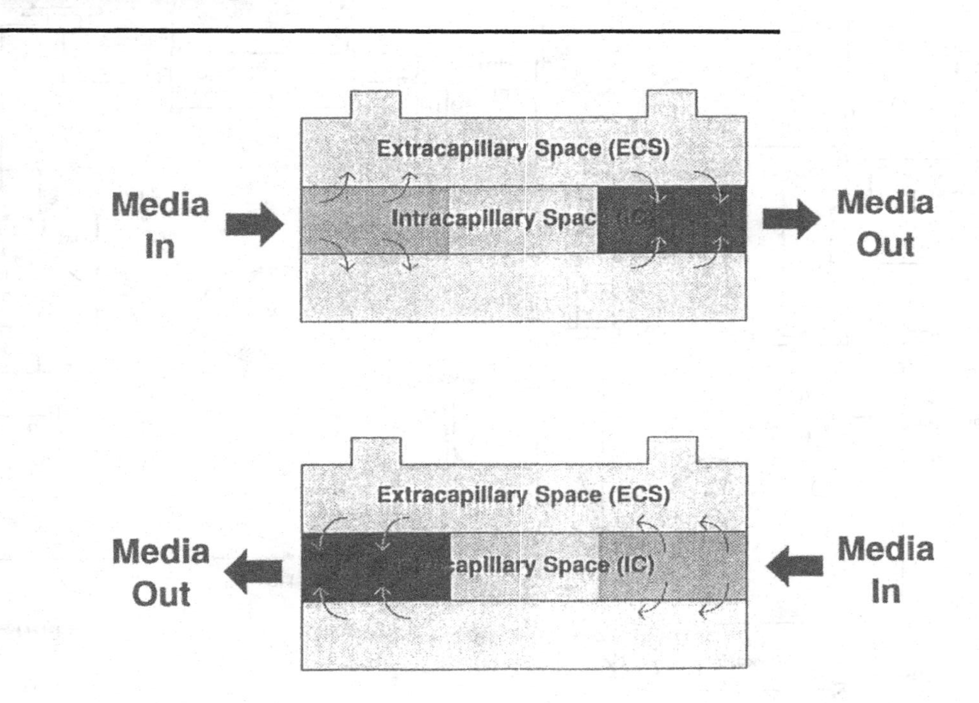

TABLE 1

ACUCELL HOLLOW FIBRE BIOREACTORS:

Fibre material	Regenerated Cellulose (Cuprophan)
Fibre diameter	id 200 um od 220 um
Number of Fibres	8970 per bioreactor
Pore size	6-10K daltons (molecular weight cut off membrane) antibodies are about 150K daltons
Surface Area	1.1 square metres
Fibre Length	230 mm
Extracapillary Volume	110 mL
Intercapillary	70 mL

The main objective of our initial studies were to demonstrate that we could grow the cells in sterile conditions at reasonable productivity. Essentially, the first run was disastrous, with many problems in both understanding the software of the system, breakages and leakages in the hardware of the system and needless to say, contamination. Despite all this, the productivity was reasonable, approaching 100 mg a day which was very much better than other systems we had studied. We therefore made a decision to go ahead with the acquisition of one and later, a second system.
Typical results from what is now regarded as a "day-to-day operation" are shown graphically in Figure 5a and b. Growth takes place in a number of stages.

1. An initial growth stage where the cells grow but produce few antibodies.

2. A long (as long as possible) production phase where antibody productivity increases.

3. The drop-off of antibody production as the cells age.

During this particular run cells were grown on RPMI media, (a media normally used for cell culture) without antibiotics and with pH control. Glucose was the main carbon source. Productivities from the membrane can be compared favourably to other systems in use in Biocode's operations as is shown in Table 2 below.

Figure 5a

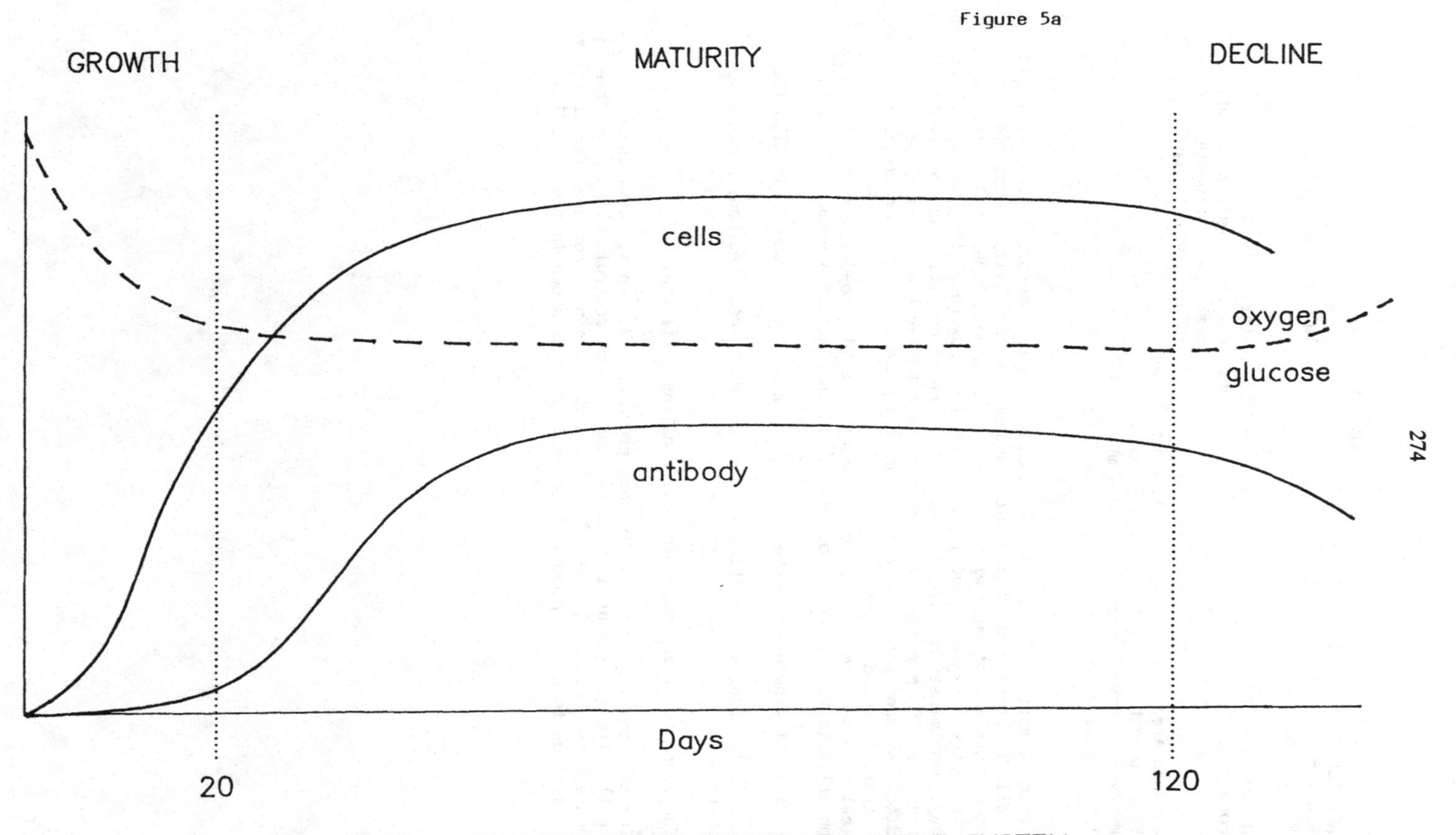

TYPICAL RUN ENDOTRONICS HOLLOW FIBRE SYSTEM

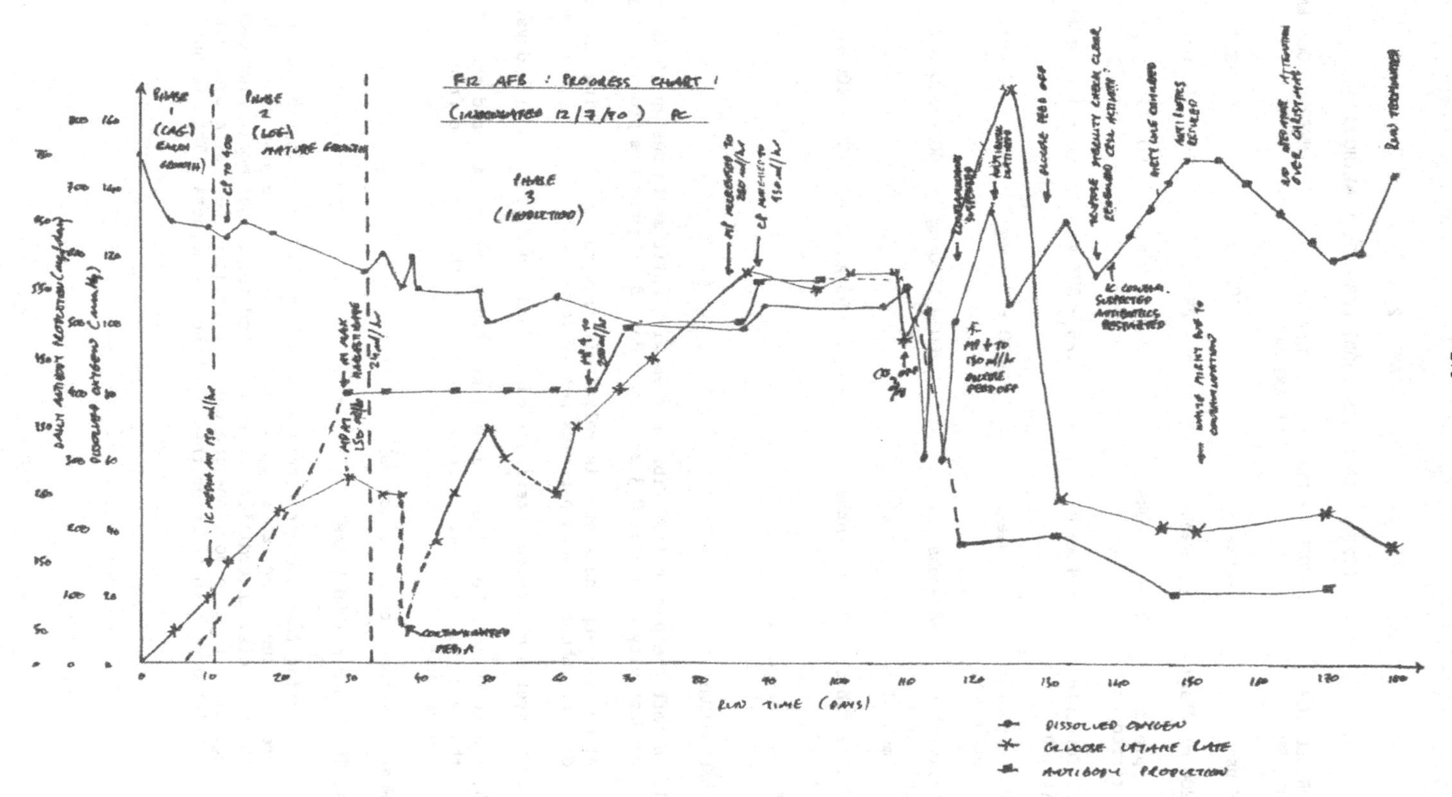
PHASE 3 (PRODUCTION)
RUN TIME (DAYS)
DISSOLVED OXYGEN
GLUCOSE UPTAKE RATE
ANTIBODY PRODUCTION

TABLE 2

COMPARISON OF ANTIBODY PRODUCTION METHODS

PRODUCTION SYSTEM	LENGTH OF RUN	TOTAL YIELD OF ANTIBODY	ANTIBODY CONCENTRATIONS
Mouse	10 weeks	10 - 100 mg	2 - 15 mg/ml
200 ml flask	2 - 4 weeks	6 - 25 mg	50 - 100 ug/ml
simple 2L stirred pot	2 - 20 weeks	125 - 250 mg	50 - 100 ug/ml
2L fermenter (batch/fill)	2 - 4 weeks	125 - 375 mg	50 - 150 ug/ml
2L fermenter (continuous)	4 - 8 weeks	250 - 700 mg	50 - 150 ug/ml
Airlift* fermenter (2L media)	2 weeks	250 - 1000 mg	50 - 150 ug/ml
Hollow fibre system	3 - 6 months	30 - 90 g	200 - 1000 ug/ml

*literature data

CONCLUSIONS

1. An 'off the peg' hollow fibre system with split media design gives a productivity exceeding 0.3 grams per day, during maximum productivity.

2. After start-up the system is relatively easy to run, involving, in total, two days of labour support per week per system.

3. The system can be run sterile - our record run now exceeds 150 days.

4. The system can be run efficiently at high productivities and the antibodies that are produced compete economically with those produced from ascites (animal use).

What therefore are our concerns?

1. The Black Box Software.

2. Monitoring the reaction kinetics. With the limited monitoring which is present, see Figure 2, it is often difficult to determine what is happening in the reactors. In addition there is a need for better monitoring/recording equipment to give more on-line data. Currently the system uses dot matrix tabulated print-out rather than continuous line data. This is hard to follow.

All these points are being tackled.

Finally, you will see from this talk that we are mainly interested in the operating conditions of the fermenter systems and little in the actual design of the membrane, which has barely been mentioned, but which is clearly the "heart" of the system.

The membranes are designed to restrict the transfer of molecules with a molecular weight in excess of 6-10 kilodaltons. In other words, it is possible to separate the media constituents from the antibodies inside the culture vessel. Membranes are also used downstream where both cross-flow and dialysis membranes are used to purify the antibody which is the basis of our diagnostic kits.

By comparison, we have found the system considerably easier to operate than the stirred tank reactor or airlift systems where the cells basically are not nearly so stable and tractable.

Our final conclusion is that, despite intensive scientific study on which system we should be operating, in reality when given a choice, testing 'off the peg' systems is the only approach which is likely to be successful and in our case fortunately worked.

The authors would also like to thank employees of Endotronics Inc, in particular Ken Anderson, without whose support the current system - could not be used.

SESSION H:

APPLICATIONS IN THE CHEMICAL INDUSTRY

INDUSTRIAL APPLICATIONS OF

PERVAPORATION

LE CARBONE - LORRAINE

LE CARBONE LORRAINE, in his separation techniques division, proposes for the chemical industry two main processes :

. PERVAPORATION : for liquid-liquid separation with his daughter company GFT

. CROSS FLOW MICROFILTRATION MEMBRANES and systems developped and produced in LE CARBONE LORRAINE plant in PAGNY-SUR-MOSELLE in FRANCE.

Membranes are tubular and with a carbon activ layer on a carbon fiber support.

This paper is about industrial applications of GFT pervaporation techniques.

Effective Industrial Membrane Processes — Benefits and Opportunities, pp.281–293

INDUSTRIAL APPLICATIONS OF PERVAPORATION

Pervaporation is a very effective and economical technique for the separation of water from organic solvents and solvent mixtures. The heart of the process is a non-porous membrane which either exhibits a high permeation rate for water but excludes and thus does not permeate organic or, vice versa, permeates the organics but excludes water. Although known for more than 70 years this technique was developed to industrial application only during the past 10 years by the invention of a new, highly selective composite membrane by GFT. Numerous plants for different application have been installed around the world ranging from a few m2 of membrane area to more than 2 000 m2 of membrane in one plant, for the dehydration of 150 m3/d of ethanol from a initial ethanol concentration of 93 % b.w. to a final concentration of 99,8 % of ethanol.

For dehydration membranes, developed and produced by GFT are of the composite type and can be used for the removal of water from alcohols, ethers, esters, ketones, hydrocarbons, and halogenated hydrocarbons, and mixtures thereof.

New membranes have been found by GFT in cooperation with other research institutes which also allow the dewatering of acids and amines and the removal of water produced in reaction mixtures. Organophilic membranes, which permeate volatile organics but retain water, are very useful for the recovery of organic solvents from dilute aqueous streams and on the verge of industrial application.

Usually the pervaporation process is separated into three steps :

- Sorption of the permeable component into the separating layer of the membrane,

- Diffusive transport of the substance across the membrane,

- Desorption of the substance at the permeate side of the membrane.

A gradient in the chemical potential of the substances on the feed side and the permeate side acts as the driving force for the process. The best means for keeping this driving force at a maximum is by applying low pressure to the permeate side of the membrane, combined with immediate condensation of the permeated vapors.

The most important advantages of the pervaporation technique are :

- Azeotropic mixtures and mixtures of components with close boiling points can easily be separated,

- As usually the minor component is removed from the mixtures, only this component has to be evaporated through the membrane, requiring significantly less energy than conventional distillation,

- By removal of the minor component from the mixture, any wanted purity of the retentate, the product, can be reached. Limits will only be found by economical consideration.

- Tho-process can be operated at any temperature, allowing an easy treatment of temperature sensitive substances.

- No chemicals, entrainers, or other additives are required, by recirculation and posttreatment of the permeate environmental pollution can be totally avoided.

- Because of the modular design of membrane plants, small installations can operate very economically. The plants are very flexible with respect to partial loads and overloading.

Let us, however, leave the general description and refer to a few plants, which GFT commissioned during the last year.

PRINTING ON ALUMINIUM CANS (CEBAL, F)

In the printing process a solvent mixture with a high content of MEK (methylethylketone) is used. This solvent is evaporated and the solvent laden air is washed with oil and the solvent is separated. The recovered solvent is wet and cannot be reused without drying.

In a GFT pervaporation plant, 6 000 l/d of the recovered solvent mixture are dried from 7 % of water to 0,3 % of water, using 120 m2 of membrane area. Because of its importance with respect to environmental protection, this project was partially funded by the EC.

ETHANOL PRODUCTION FROM GRAIN (UNIGRAIN, F)

In a plant of the group UNIGRAIN a pilot plant is installed for the production of ethanol from grain to be used as fuel additive or in the pharmaceutical industry.

After comparison of azeotropic distillation and dehydration by molecular sieve pervaporation was chosen as the process for final dehydration. GFT delivered at the end of 1989 a pervaporation plant which produces 300 hl/d of ethanol of a final water content of either 2 000 ppm or 500 ppm.

The installation comprises 480 m2 of membrane area and is designed for high flexibility with respect to part load and changing product and feed concentrations.

DEHYDRATION OF ESTERS (FRG)

2 500 kg/d of an ester are dehydrated from 2 % of water to less than 1 500 ppm of water. As the feed contains acid (from partial hydrolyzation in a phase separator), 50 m2 of a high selective, acid resistant GFT pervaporation membrane used. The recovered ester is of high purity and is used as a thinner in a coating process.

The savings through the possibility of solvent recirculation lead to a pay-back period of less than one year for this plant.

The plant is fully automized and operated without personal.

SOLVENT RECOVERY (USA)

A solvent stream which contains too much water for reuse and which cannot be purified by conventional techniques is now incinerated. In cooperation with BASF laboratory tests were performed to develop a suitable membrane and necessary process parameter for the dehydration of this solvent by pervaporation. Economical calculations again lead to a pay-back period of less than one year.

This installation will be operated in a batch mode, dehydrating 2 t/d of the solvent from 45 % of water to less than 2 % of water.

DEHYDRATION OF VARIOUS SOLVENTS (UK)

The client is engaged in the reclamation and purification of spent solvents for the chemical industry. Solvent and solvent mixtures are pretreated by distillation and other techniques and finally dehydrated by pervaporation.

After operation of a pilot plant, a small production plant, incorporating 64 m2 of membrane area, was installed. At the end of 1989, a second industrial plant was added, comprising 150 m2 of GFT-pervaporation membrane. This plant will mainly be used for the dehydration of isopropanol from 13 % to 1 % of water.

The plant was designed for high flexibility in order to allow for the dehydration of different solvents and solvent mixtures.

BETHENIVILLE PLANT

BETHENIVILLE plant is the biggest pervaporation plant with 2 100 m2 of membranes.

The production capacity is 150 000 l/day starting from an alcohol 94°GL to reach a level of 2 000 ppm of water.

This plant is very flexible and produces also an alcohol of a final water concentration around 250 ppm, in that case, the capacity is only 700 l/day.

You will find herewith the energy resultats which are obtained and an economical comparison with an azeotropic distillation.

For an alcohol at 2 000 ppm, the running costs are less than one third of the running cost of the traditionnal process.

List of membranes

Code	1000	1001	1510	1005	2302	1170
Main appl.	Neutral solvent dehydration	Neutral solvent dehydration (high water content)	Neutral Solvent dehydration (IPA)	Organic acid dehydration	Organic amine dehydration	Neutral organic extraction
Active layer	PVA	PVA	PVA	PVA	PVA	Elastomeric polymer

Major operating conditions and limitations

Water in feed	≤ 15 wt-%	≤ 50 wt-% depending on solvent	≤ 20 wt-% depending on solvent	≤ 80 %	under investigation	no known limitations
Neutral solvents concentration	no limitation	no limitation	no limitation	no limitation	pure and mixtures	5 - 25 %, depending on solvent
Excluded solvents	aprotic solvents like DMF, DMSO,... glycols not recommended	aprotic solvents glycols not recommended	aprotic solvents	aprotic solvents	aprotic solvents	aprotic solvents
Organic acid Maximum	≤ 1000 ppm	≤ 1000 ppm	≤ 1000 ppm	no limitation formic acid excluded	limitation by solubility	good resistance, but membrane doesn't separate acids
Mineral acids	excluded	excluded	excluded	H2SO4 ≤ 1 %	limitation by solubility	not known yet
Bases	organic and inorganic excluded	organic and inorganic excluded	organic and inorganic excluded	organic and inorganic excluded	NaOH ≤ 3 %	organic and inorganic excluded
Max. temper. (short time)	105 °C	105 °C	105 °C	100 °C	100 °C	90 °C
Max. operating temperature	100 °C	100 °C	100 °C	100 °C	100 °C	80 °C

Typical performance data

Standard feed solution	95 % EtOH 5 % H_2O	90 % EtOH 10 % H_2O	90 % IPA 10 % H_2O	80 % acetic acid 20 % H_2O	80% amine 20 % H_2O	8 % EtOH 92 % H_2O
Operating temp. °C	80 °C	80 °C	80 °C	80 °C	under investigation	50 °C
Total flux at operating temperature kg/m².h	0.225	0.350	0.700	0.5	under investigation	0.225
% of organic component in permeat	< 5	< 3	< 5	≤ 1	under investigation	≥ 45

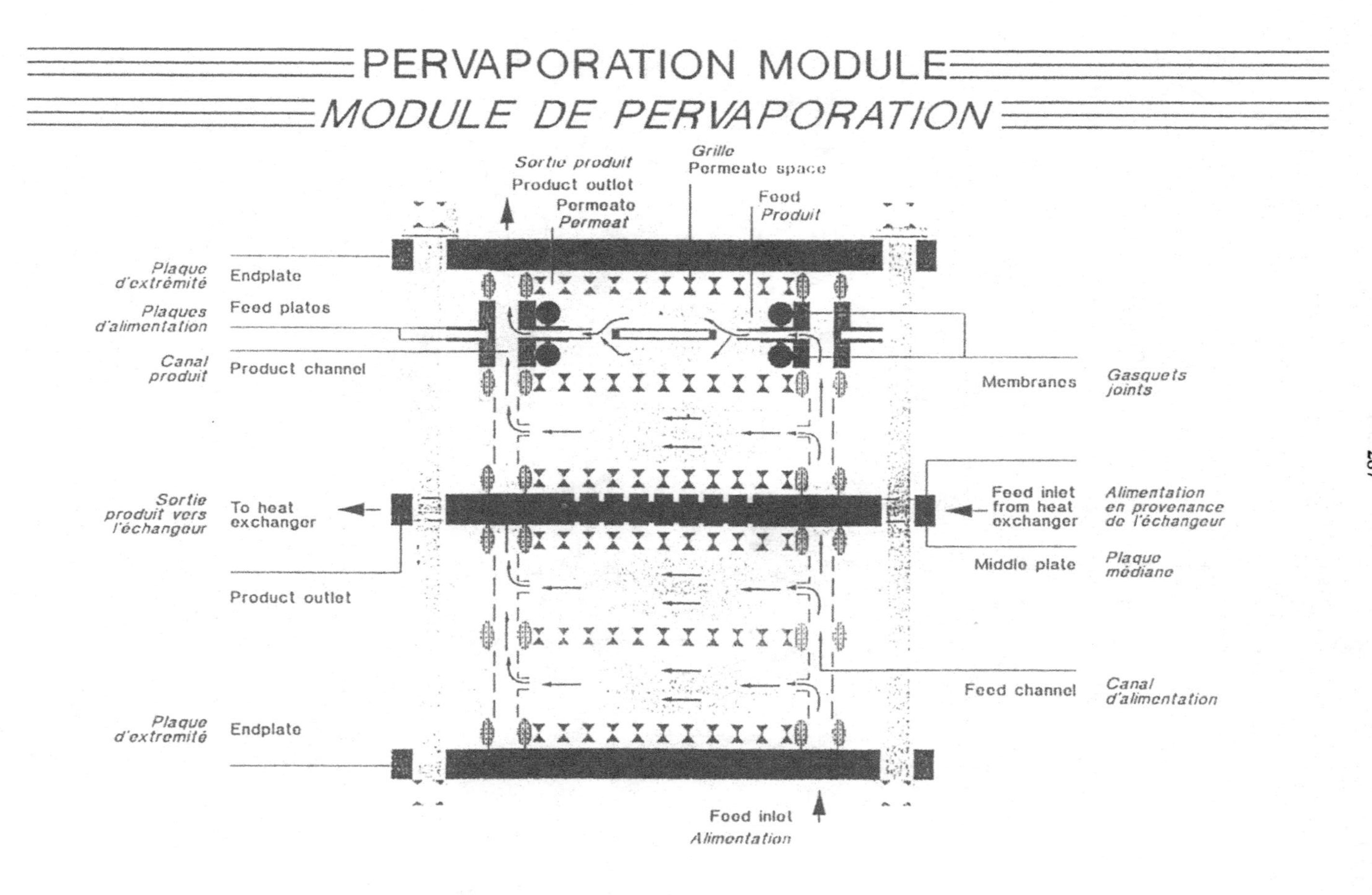
PERVAPORATION MODULE
MODULE DE PERVAPORATION
Sortie produit
Product outlet
Permeate
Permeat
Grille
Permeate space
Feed
Produit
Plaque d'extrémité
Endplate
Plaques d'alimentation
Feed plates
Canal produit
Product channel
Membranes
Gasquets joints
Sortie produit vers l'échangeur
To heat exchanger
Feed inlet from heat exchanger
Alimentation en provenance de l'échangeur
Middle plate
Plaque médiane
Product outlet
Feed channel
Canal d'alimentation
Plaque d'extrémité
Endplate
Feed inlet
Alimentation

PROCESS FLOW DIAGRAM

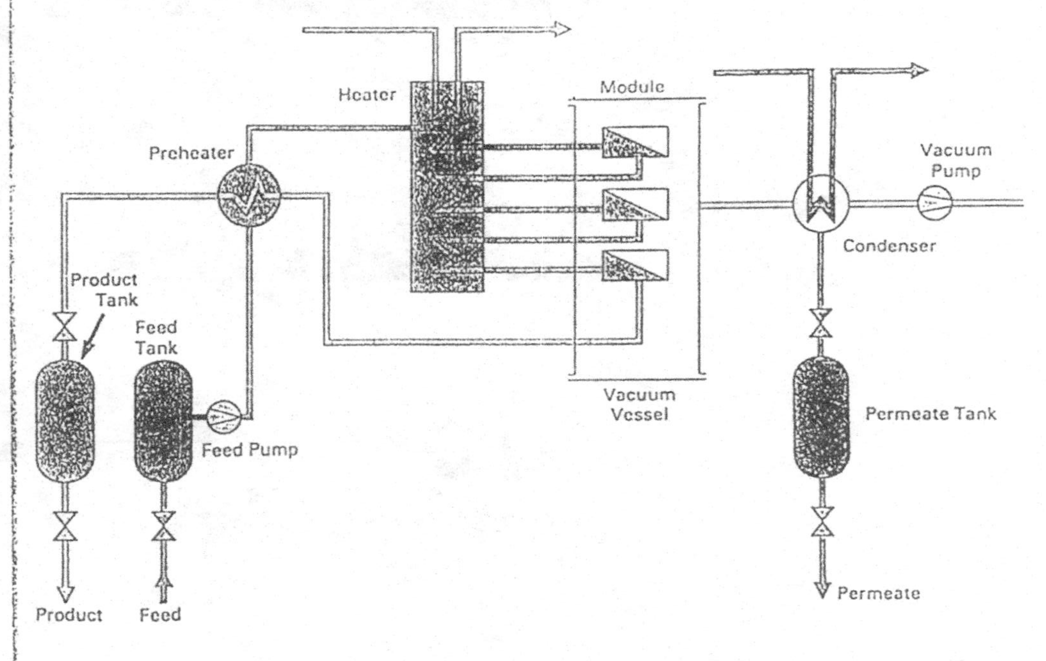

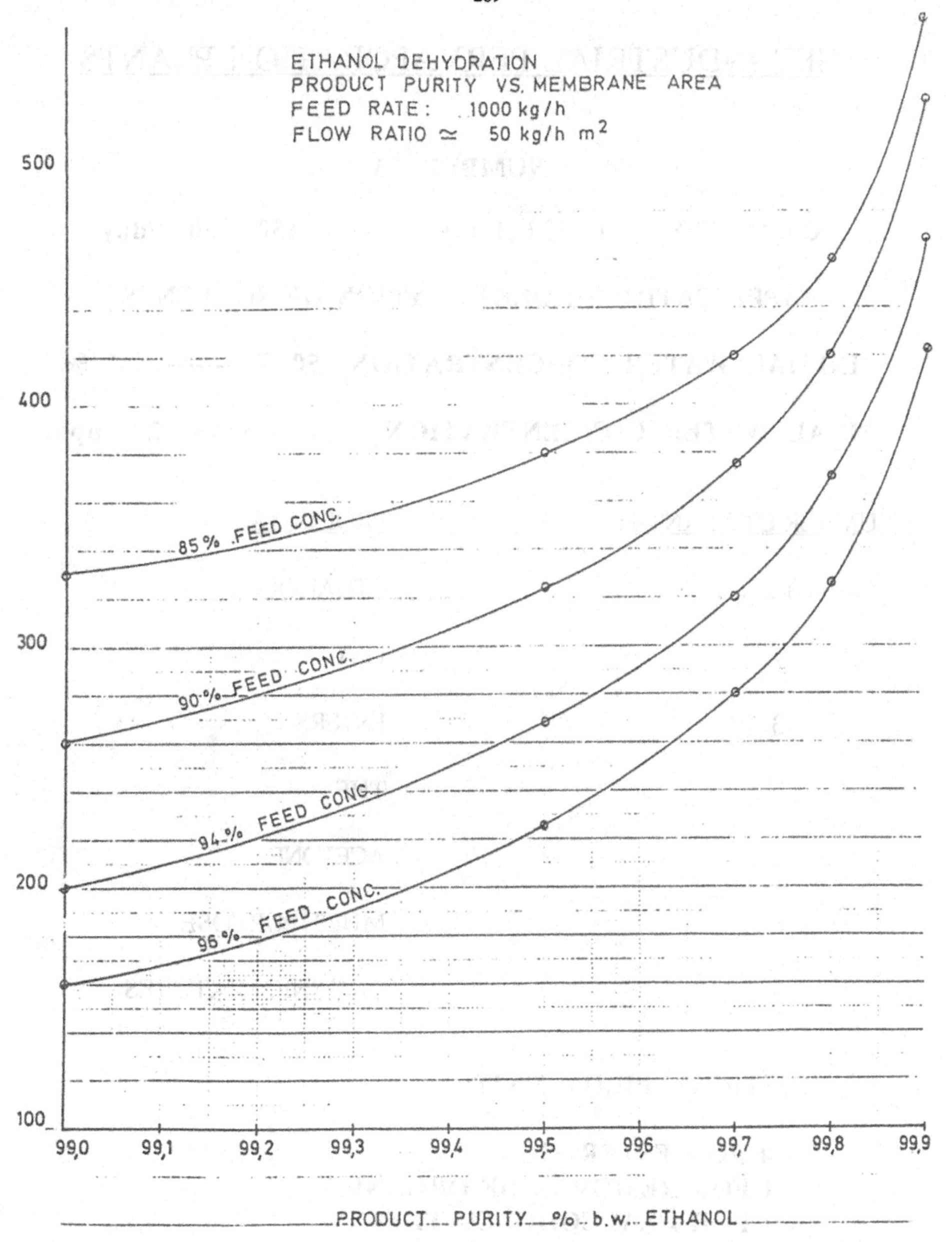
ETHANOL DEHYDRATION
PRODUCT PURITY VS. MEMBRANE AREA
FEED RATE: 1000 kg/h
FLOW RATIO ≃ 50 kg/h m2
85 % FEED CONC.
90 % FEED CONC.
94 % FEED CONC.
96 % FEED CONC.
500
400
300
200
100
99,0
99,1
99,2
99,3
99,4
99,5
99,6
99,7
99,8
99,9
PRODUCT PURITY % b.w. ETHANOL

GFT INDUSTRIAL PERVAPORATION PLANTS

NUMBER 33

CAPACITY : 1 000 l/day -------> 150 000 l/day

APPLICATIONS : DEHYDRATION OF SOLVENTS

INITIAL WATER CONCENTRATION 50 % -------> 1 %

FINAL WATER CONCENTRATION 5 % -------> 200 ppm

NUMBER OF PLANTS :	PRODUCTS :
12	ETHANOL
4	IPA
3	ESTERS
2	THF
1	ACETONE
6	MULTIPURPOSE
5	ORGANIC MIXTURES

20 INDUSTRIAL PILOT UNITS

- **4 FOR ESTERIFICATION**
- **1 FOR REMOVAL OF ORGANICS**
- **1 FOR DEALCOHOLIZATION**
- **1 FOR AMINE DEHYDRATION**

Bétheniville results

	Steam (Kg/Hl) of anhydrous alcohol			Electric Energy (Kwh/Hl) of anhydrous alcohol			TOTAL Kwh/Hl
	Perva-poration	Redistilled permeate	Total	Perva-poration	Redistilled permeate	Total	
Alcohol < 2000ppm from a flegm at 95,5°GL	8,60	2,40	11,00	3,40	0,10	3,50	11,60
Alcohol < 500ppm from a refined alcohol at 96,3°GL	8,30	19,00	27,30	3,40	0,40	3,80	23,70

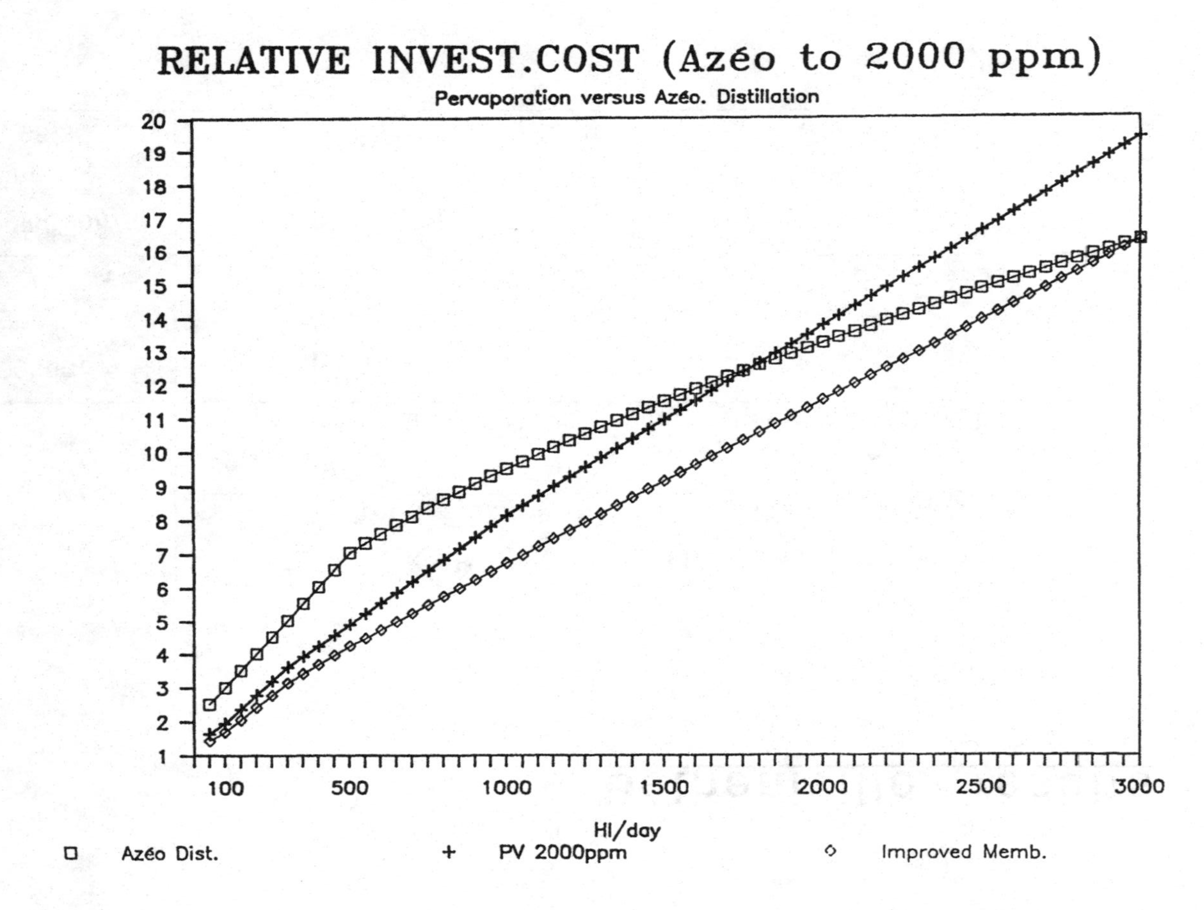
RELATIVE INVEST.COST (Azéo to 2000 ppm)
Pervaporation versus Azéo. Distillation
20
19
18
17
16
15
14
13
12
11
10
9
8
7
6
5
4
3
2
1
100
500
1000
1500
2000
2500
3000
Hl/day
□ Azéo Dist.
+ PV 2000ppm
◇ Improved Memb.

COST COMPARISON FOR DEHYDRATION OF ETHANOL 94°
PERVAPORATION / AZEOTROPIC DISTILLATION

		Pervaporation Consumption	Pervaporation Cost (FF/Hl)	Azeotropic Distillation Consumption	Azeotropic Distillation Cost (FF/Hl)
FINAL CONCENTRATION (2000ppm)					
STEAM (kg/Hl)	(100 FF/t)	11	1,1	110	12,1
ELECTRICITY (kwh/Hl)	(0,30 FF/kwh)	3,5	1,05	1	0,3
MEMBRANE	(2000 FF/m2)		1,5		0
TOTAL			3,55		12,4

FINAL CONCENTRATION (500ppm)					
STEAM (kg/Hl)	(100 FF/t)	15	1,5	110	12,1
ELECTRICITY (kwh/Hl)	(0,30 FF/kwh)	5	1,5	1	0,3
MEMBRANE	(2000 FF/m2)		2,5		0
TOTAL			5,5		12,4

"PERMEATION MEMBRANES" -AN EMERGENCE OF NON-CONVENTIONAL SEPARATION TECHNOLOGY

BY

RAVINDER SINGH

Production Engineer
OIL & NATURAL GAS COMMISSION
INDIA

ABSTRACT:

Membrane separation processes represent an attractive alternative to conventional separation methods.The permeation membrane technology is revolutionising the chemical industry with its far reaching consequences. Because of their potential for both low capital cost and high energy efficiency, membranes are emerging as a viable alternative to the traditional separation methods.

Commercial membranes are available today for specific applications in processes such as reverse osmosis,dialysis,microfiltration,ultrafiltration and gas separations.The earlier disadvantage of membranes,i.e.low fluxes has been overcome in many cases by using hollow fibre modules which pack a large area into a small volume (Ref.5,pg.6).

Techniques like facilitated transport,charged membranes and use of solid absorbents have increased the membrane selectivity and added a new dimension to this technology.On the gas front,installation of a gas treatment membrane sustem is known to reduce treatment costs by as much as 85%.

INTRODUCTION:

In the chemical industry,separation processes are as important as chemical reactions. It is not uncommon to find that more than one-third of the processing cost can be attributed to separation operations.The new developments in this arena can be traced either to discovery of new phenomena or a changed emphasis due to factors such as rapidly escalating energy prices or tighter restrictions on effluent streams.

A membrane separation is a pressure-driven process with no moving parts and without the use of toxic chemicals.Over and above the

Effective Industrial Membrane Processes — Benefits and Opportunities, pp.295–305

advantages in process design, skid-mounted membrane modules offer significant advantages vis-a-vis other conventional technologies. The membrane characteristics are shown in table 1. (Ref. 2).

TABLE-1 - MEMBRANE CHARACTERISTICS

	Membrane M1		Membrane M2	Membrane M3	Membrance M4
	Permeability		Selectivity Change	Selectivity Change	Permeability Change from M1
Component	(md)	Selectivity to C1	from M1	from M1	(md)
H_2O		500	1,250	2,500	
H_2S		30	75	150	
CO_2		20	50	100	
N_2		0.8			
C_1	0.5×10^{-5}	1			5×10^{-5}
C_2		0.6			
C_{3+}		0.5			

Selectivity Scale of An Average Permissible Membrane

C^{3+}	C2	N2	C1	CO2	H2S	H2O
0.5	0.6	0.8	1.0	15 - 25	15 - 30	50 - 500

Ref. 2

Gas separation membranes are thin barriers that allow preferential passage of certain gases when a multi-component feed gas is exposed to the membrane at some pre-determined pressure. The barriers in such separations are predominantly polymeric materials like cellulose derivatives, poly-sulphones, polyamides and polyimides. The present paper presents an overview of various gas-based membranes being used in the petrochemical industry.

Membranes are manufactured as flat sheets or hollow fibres. The flat sheet ones are usually wound as spiral packed elements. The smallest physical membrane is called an element and collection of elements in a pressure housing as a module. Any membrane module may consist of several hundred elements depending on process demand and end-user requirement. They can also be added or removed depending on process requirement. A typical membrane element is shown in fig. 1 (Ref. 6, pg. 126).

Removal of impurities from gas streams or selective removal of one or more components from a multi-component gas feed is one of the major activities in any gas based industry. Traditionally, it is achieved by either cryogenic fractionation, selective solid-bed absorbent procedures, selective physical absorption processes and solvent extraction methods etc.

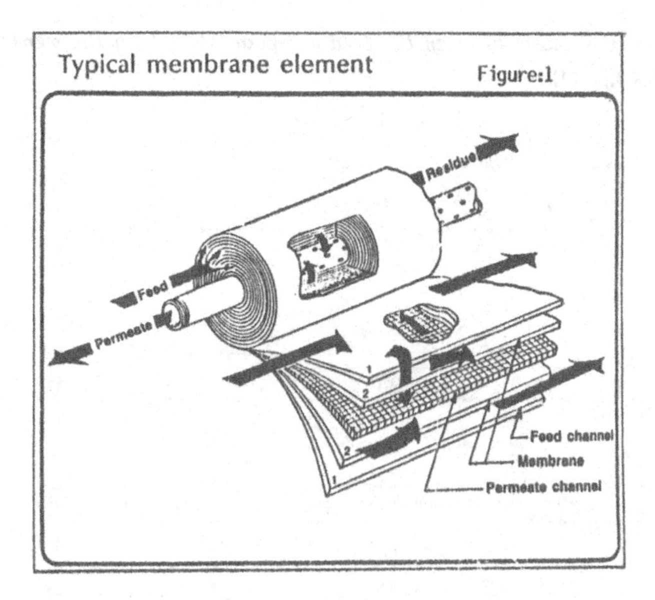

Ref.6,pg.126

Implementation of new permeation processes,where the membranes are gathered in bundles and then introduced in the pressure vessels have been discussed.Performance and economic comparisons between permeation and conventional processes have been presented for some specific cases with emphasis on its possible implications.

MEMBRANE SEPARATIONS-Gaseous Systems:

Membrane separation units operate on the principle of selective permeation.Each gas introduced into the separator has a characteristic permeation rate that is a function of its ability to dissolve in,diffuse through and exit a polymeric membrane of given properties.For example,when a high pressure feed stream of Hydrogen gas and hydrocarbon flows through a membrane,a "fast"or more permeable gas like Hydrogen gas diffuses through the membrane at a higher rate than does a "slow" or less permeable gas like Methane.Once separated,the permeate and residual streams are then channeled through two gas outlets and piped to their respective points.The rate at which a gas permeates across a membrane depends upon the difference *in its partial pressure between the feed and permeate side of the membrane. (Fig.2,Ref.4,pg.71)*

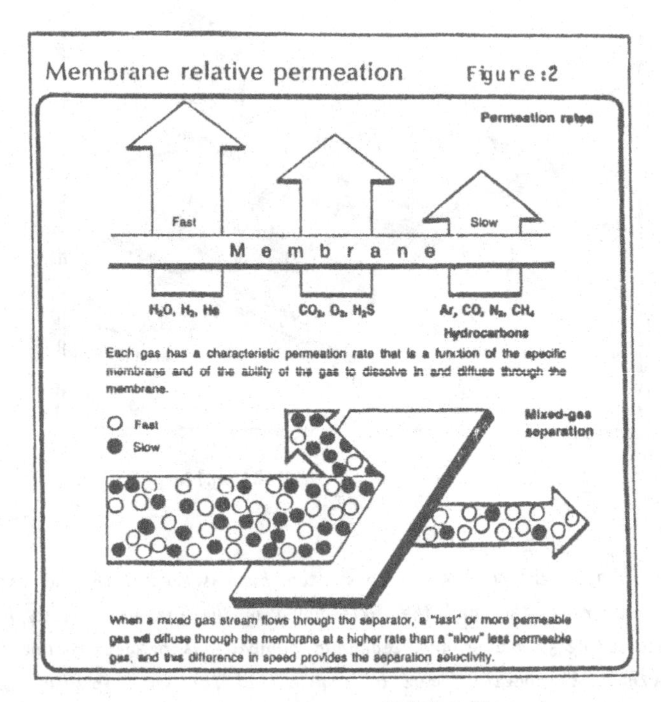

Ref 4,pg.71

DESIGN CRITERIA FOR MEMBRANE SYSTEMS:

The design for gaseous systems involves pressure differential between feed and permeate streams as key factor.The membrane requirement, volume and composition of residue and permeate streams for a specific operation are manifested in the differential pressure.The selection of an optimal membrane process depends upon the flow rate,feed gas composition and desired level of purity in residue gas.The pressure drop of the residue stream is approximately 1.36-3.40 kg/cm^2 whereas the permeate stream exists at a lower exit pressure.

In general, the increase in area for a single stage process,improves residue quality and decrease in area results in purer permeate. In a multi-stage separation process, the permeate stream from the first stage is compressed and passed through second stage membrane module and optionally recycled back to the feed of first stage to improve product recovery. However, optimisation of membrane processes is a difficult exercise due to commercial non-availability of a wide latitude of intricate process design and dedicated software.

The flow behaviour of gases across a membrane is governed by Fick's Law and Henry's Law (Ref.1).

$$J_i \Rightarrow -D\ dC_i/dz$$

Where J_i is the flow rate, D-diffusivity of gases, z is membrane thickness and C_i is the concentration of gas.

The concentration C_i, of the gas is expressed as (for the lower case).

$$C_i = S_i \times P_i$$

Where S_i is the solubility of the gas in a particular membrane and P_i is the pressure of the gas.

Permeability i.e. the rate at which the gases permeate across a membrane is expressed as;

$$P_{gi} = \bar{P}_{gi} / z$$

Where $\bar{P}_{gi}$ is the permeance given by the equation $\bar{P}_{gi} = S_i \times D_i$

S_i being the solubility of gas and D_i the diffusivity.

The permeability of gases and hence separation factor or selectivity, α, are dependent on membrane material and modular design. The permeability is dependent on feed gas composition, pressure and temperature of the system.

For a two - gas system, separation factor, α is expressed as

$\alpha\ [i/j] = P_{gi}/P_{gj}$ (where i and j are two gases)

Therefore, the gas flow rate across a membrane is given as

$$J = P_g \times A[P_p - P_f],$$

Where A is the surface area of membrane, P_f is the pressure of feed gas and P_p is the pressure of permeate gas.

PERMEATION MEMBRANES- Process Developments and Established Fields of Application:

1. Natural Gas Treatment and Processing:

-CO_2/Hydrocarbon for Natural Gas Sweetening.

-CO_2 extraction and concentration from EOR produced gas.

-H_2S/Hydrocarbon for sweetening.

-H_2O/Hydrocarbon for dehydration.

In a typical sweetening process,sour gas is supplied to the membrane system directly from several parallel well systems.The membrane system is designed for 0.5 MMSCMD feed flow,saturated at 80 kg/cm^2 and 95°C.Feed gas is first led to a filter/coalescer for removal of entrained contaminants such as sand,pipe scale,lubricating oil and H_2/water.Filtetred gas passes through a feed pre-heater to superheat the feed.Warm gas enters the membrane elements where the CO_2 content is reduced from 6% to 3% for pipeline distribution.The membrane system is shown in Fig.3(Ref.4,pg.72) whereas the system design parameters in Table 2.

Table 2

Typical Membrane System Parameters

	Feed Gas	Natural gas to pipeline
Pressure,psia	1,195	1,150
Temperature,°F	115	107
Composition,vol%		
H_2O	0.1	0.01
CO_2	6.1	3.0
N_2	0.6	0.6
C1	84.4	86.8
C2	8.8	9.6

The membrane skid consists of four tube bundles arranged in 2x2 matrix. Each bundle consists of five tubes containing 200 mm membrane elements packed inside a pressure vessel.(Fig.4,Ref.6,pg.128)

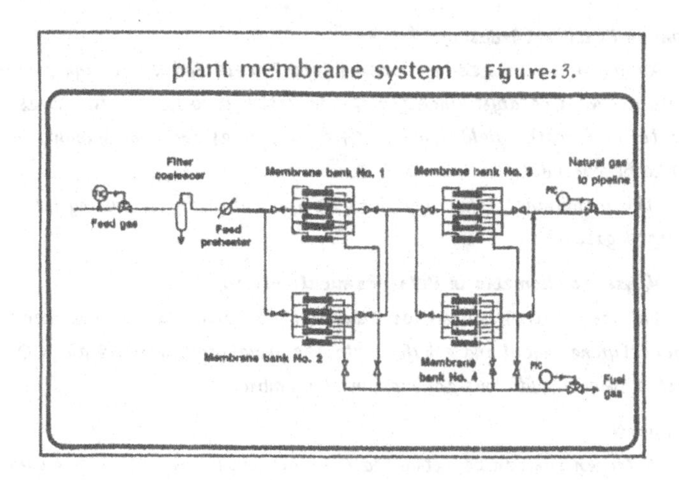

Ref 4,pg.72

By using a similar membrane system, H_2S% of 4 ppm by volume and moisture content of 112 kg/MMSCMD can be achieved. Compared to a conventional Di-ethanol Amine(DEA)system with a two-stage offers 26% savings on capital items and 62% savings on the operational costs.

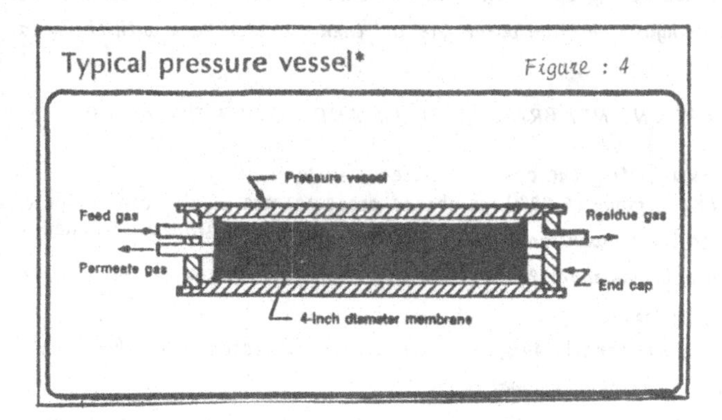

Ref.6pg.128

2.Applications in Fertilizer Industry:

Ammonia is produced by reacting together H2 and N2 at high presure and temperature over a catalyst.Since the % conversion is low,unreacted gases are recycled to improve the yield.However,CH4,inert gases and contaminants build up and need to be purged.

The membrane technology finds its application in recovering H2 from this purged gas.

3. Variation of gas stoichometry in Petrochemical Industry:

Syn gas (H2+CO) is used for industrial production of petrochemicals like methanol,ethylene glycol and ethylene etc.The stoichometry of H2 and CO can be adjusted through judicious application of membranes.

4.Gas Enrichment:

Nitrogen gas can be extracted from air and a purity of 95% easily attained through use of membranes.For O2,a purity level of only 50% is achievable. However,with recycling the % purity can be improved.

Helium recovery from natural gas is technically feasible through use of membranes only when the % content of Helium in the feed gas is substancial. Nevertheless,due to very low Helium content in the natural gas stream,the economics is not established.

5.Refinery Applications:

Hydrogen is consumed in the refineries for hydro-desulphurisation, coking and hydrocracking operations.The membranes would be effectively utilised for the recovery of hydrogen from purge gas and then recycled back with the feed gas.

COST COMPARISONS-MEMBRANE SYSTEMS AND CONVENTIONAL PROCESSES

Various factors affecting the costs for gaseous systems are:
1.For low acid gas range (5-20%),membrane processes are much more competetive as compared to the conventional ones (as far as number of stages are concerned)
2.For medium acid gas range(20-40%),the investments would be the same as the conventional processes.
3.For high acid gas range(40%),the membrane area decreasesand the investment is usually lesser than conventional processes.

The cost comparisons can be used for balancing the advantages and disadvantages of the membrane applications.The cost presented for the existing membrane based processes and conventional processes are drawn from actual production cases and thus incorporate supplier information. For new membranes, the cost will only be roughly estimated because no real change has occured in the membrane prices.

CAPITAL AND OPERATING COSTS:

For gaseous systems, estimates have been prepared for the cost of processing a 0.5 MMSCMD gas stream at 55kg/cm^2 with varying CO2 compositions. Since the cost of recompression is so important, relative to the cost of the permeation process, it has been identified as a separate cost and plotted separately as shown in fig.5 below (Ref.6,pg.132).

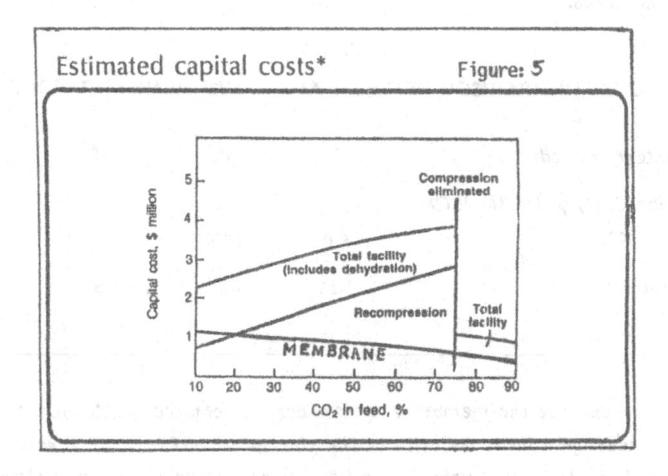

Ref6,pg.132

In general, even with the gas compression and a second stage of separation, the capital cost of membrane system for a comparable gas having 20% CO2 in the feed gas is approximately one-half that of a conventional chemical reaction absorption process.

CONCLUSIONS:

The potential advantages of membrane separation systems vis-a-vis conventional systems can be summarised as follows:

1. Membranes can be utilised for sour gas sweetening and dehydration

and would be particularly attractive for marginal fields.Skid mounted membrane modules would be most convenient to be installed in marginal fields and isolated pools and the sweet gas can be delivered to the consumers.An amine-based gas sweetening plant would probably be highly uneconomical under such conditions.

2. Application of membranes at offshore platforms where sour gas would pose a safety and a health hazard to operating personnel.The membrane system is particularly attractive in offshore platforms since these modules are skid-mounted and light-weight with better space utility for gas sweetening.

3. Removal of bulk amounts of CO2 by conventional procedures is an expensive proposition while a single stage inexpensive membrane module would suffice the needs.

Table : 3

COST COMPARISON BETWEEN AMINE AND MEMBRANE PROCESS

CO_2 content in feed	5%	15%	30%	40%
Processing Cost, $ US./MMSCM				
- Amine Process	6.0	10.60	16.24	19.40
- Membrane Process	3.89	8.20	9.53	9.20

4. Because the membrane system can be operated unattended, the customer can reassign a large majority of the plant personnel to other functions. Elimination of the large fuel volume required for amine and glycol circulation and regeneration drastically reduces operating costs of the facility. The heating value of the membrane system permeate, however, is high enough to be used as a fuel stream.

5. The membrane system is a dry system and thus there atre no storage problems in the plant.

6. The membrane skid is small and compact and requires only 15-25% of the plot area of a conventional separation facility, drastically reducing site preparation costs.

7. The membrane system avoids the operating costs associated with foaming,solution degradation and corrosion complications.

8. Membrane systems greatly reduce capital costs.

9. Membranes eliminate all process heat requirements.

10. They provide a large reduction in electric power and fuel requirements.

11. The addition of modules in stages is possible which accomodates increased sour gas contact in the inlet feed.

ACKNOWLEDGEMENT:

The author gratefully acknowledges his gratitude to the organisation he represents,i.e.,Oil and Natural Gas Commission,for having encouraged and permitted to submit and present this paper.

LIST OF REFERENCES:

1. Banerjee,S and Sinha,N.,"Application of Membranes for removal of impurities from Natural gas and other streams",(1990),Proceedings-Petroleum Conference,Indian School of Mines,Dhanbad,India,January.

2. Fourie,F.J.C.and Agostini,J.P.,"Permeation membranes can efficiently replace conventional gas treatment processes",(1987),J.Petroleum Technology,June.

3. Markiewicz,Gregory,"Membrane system lowers treatment costs at gas plant",(1988),J.Oil and Gas,October 31.

4. Review on "Background papers-Engineering Sciences and Thrust Areas", Chemical Engineering,(1988),Deptt.of Science and Technology,Govt.of India,November.

5. Russell,Fred.G and Coady,A.B.,"Gas permeation process economically recovers CO2 from heavily concentrated stream",(1982),J.Oil and Gas,June.

****

MECHANISM OF THE FACILITATED TRANSPORT OF THE α-ALANINE THROUGH A SULFONIC MEMBRANE

D. Langevin, M. Métayer, M. Labbé, B. El Mahi, University of Rouen, France

ABSTRACT

A facilitated transport process of α-alanine through a sulfonic membrane is studied. A model is propounded assuming in the membrane: 1) the reversible protonation of the zwitterionic form of the amino-acid by the carrier H^+, 2) the interdiffusion of the carrier and of the protonated form.
Experimental investigations have been made in order to check the model: 1) a study of α-alanine sorption by a sulfonic membrane dipped in various solutions, 2) mobility measurements, using self-diffusion of optically active α-L-alanine, 3) transport measurements of α-alanine trough the membrane separating two well-stirred solutions. The comparison of the experimental results with the theoretical calculations allows to discuss the validity of the model.

NOMENCLATURE

Subscipt i refers to component i. Superscripts I and II refer to compartments I and II respectively. Superscript * refers to membrane phase.

A	effective area of the membrane (cm^2)
a_i	molar activity of component i (mol. l^{-1})
C_i	molar concentration of component i (mol. l^{-1})
D_i	diffusivity of the component i ($cm^2.s^{-1}$)
E	electric term in eqn. (9)
F	Faraday constant
f	flow of solution ($cm^3.s^{-1}$)
J_i	Flux density of component i ($mmol.cm^{-2}.s^{-1}$)
K	stability constant of the complex ST^z ($l.mol^{-1}$)
k^*	apparent stability constant of ST^z in the membrane phase ($l.mol^{-1}$)
ℓ	thickness of the membrane (cm)
m_i	mass of component i
N_i	ionic fraction of component i (dimensioless)
P^*_S	permeability coefficient of the membrane to S ($cm^2.s^{-1}$)
R	observed optical rotation
$\mathcal{R}$	universal gaz constant
S, T, ST	Permeant, carrier, complex permeant/carrier
$\mathcal{T}$	absolute temperature
V	volume (cm^3)
X	molar ion-exchange capacity of the membrane ($mol.l^{-1}$)
x	distance parameter (cm)

Effective Industrial Membrane Processes — Benefits and Opportunities, pp.307–319

Y	co-ion
z	electric charge of ions
α	optically active species corresponding to the excess of α-L-alanine with regard to α-D-alanine
δ	thickness of the diffusion boundary layers (cm)
ε	porosity of the membrane (dimensionless)
Φ	electric potential
λ	distribution coefficient (dimmensionless)
σ	selectivity factor (dimensionless)

1 INTRODUCTION

A fundamental property of ionic membranes is permselectivity with regard to counter-ions. An adequate transport reaction coupling can make a membrane permselective with regard to neutral species S [Le Blanc et al, 1980]. Reaction of S with the carrier counter-ion T^z gives selectivity; interdiffusion of carrier T^z and product ST^z controls permeability. Fast formation and dissociaton of a moderatly stable intermediate complex ST^z are quite suitable reactions for this process.
In this way, in a previous communication [Métayer, et al, 1989], we have analysed a facilitated transport system including an ionic membrane between two aqueous solutions as shown in Fig. 1. Assuming fast reactions and taking into account the polarization of the membrane, we have calculated optimal conditions for transport in relation with the stability of the intermediate complex ST^z. Our conclusion was that values of ST^z stability constant between $10^{1.5}$ and $10^{2.5}$ $l.mol^{-1}$ were the most favourable.

CONCENTRATED SOLUTION OF S	BOUNDARY LAYER	IONIC MEMBRANE	BOUNDARY LAYER	DILUTED SOLUTION OF S
S ⟶	⟶	T^z ⟵ ⟶ ST^z	⟶	⟶ S

FIGURE 1: Facilitated transport of S through an ionic membrane using an ionic carrier T^z

α-alanine was found very convenient in order to illustrate this conclusion. In zwitterionic form it behaves like a neutral permeant [Helfferich, 1990] and the stability constant of its protonated form in solution is $10^{2.34}$ $l.mol^{-1}$ [Sillen et al, 1964], which would correspond exactly to the optimal constant range.
In this paper, our purpose is to present a model and experimental confirmations of a facilitated transport process of α-alanine through a cation-exchange membrane with proton as carrier.

2 MODELLING OF THE TRANSPORT.

2.1 The reaction.

The permeant-carrier combination is here the protonation of the α-alanine:

$$\underset{(S)}{COO^{-}\text{-}CH(CH_3)\text{-}NH_3^{+}} + \underset{(T^{+})}{H^{+}} \rightleftarrows \underset{(ST^{+})}{COO^{-}\text{-}CH(CH_3)\text{-}NH_3^{+}}$$

This equilibrium is characterized by a stability constant defined by:

$$K = a_{ST} / (a_S\, a_T) = 10^{2.34}\ l.mol^{-1} \tag{1}$$

where a_i are the molar activities of S, T^+ and ST^+.

When a cation-exchange membrane in T^+ form is dipped in a mixed solution of S, T^+, ST^+, the equilibrium situation can be summarized by Table 1:

species	solution	membrane
neutral S	a_S	
	$C_S \simeq a_S$	$C^*_S = \lambda\, C_S$
counter-ions T^+, ST^+	a_T , a_{ST}	
	$C_{ST} \simeq a_{ST}$ $C_T \simeq a_T$	$C^*_{ST}/C^*_T = \sigma\, C_{ST}/C_T$
co-ion Y^-	a_Y	
	$C_Y \simeq a_Y$	$C^*_Y \simeq 0$

Table 1: Equilibrium membrane/solution

where C_i and C^*_i are molar concentrations, λ is a distribution coefficient and σ is a selectivity factor.

Consequently, in dilute solution, K can be written as

$$K = C_{ST} / (C_S\, C_T) \tag{2}$$

and, in the membrane, we can define an apparent stability constant:

$$k^* = C^*_{ST} / (C^*_S\, C^*_T) \tag{3}$$

which can be related to K by:

$$k^* = \sigma\, C_{ST}/(\lambda\, C_S\, C_T) = K\, \sigma/\lambda \tag{4}$$

Finally, assuming the electrical neutrality in the membrane:

$$C^*_T + C^*_{ST} = X \tag{5}$$

with X = molar ion-exchange capacity

and defining the overall concentration C^*_R of the permeant:

$$C^*_R = C^*_S + C^*_{ST} \tag{6}$$

Eqn.(3) allows to relate C^*_R, C^*_S, C^*_T and C^*_{ST} everywhere in the membrane as soon as the equilibrium constant k^* is known.

2.2 Interdiffusion in the membrane.

2.2.1 Hypothesis. The protonation of α-alanine is instantaneous so that the equilibrium is reached everywhere in the membrane. The different species diffuse in the stationary state in a direction Ox perpendicular to the plane of the membrane, without charge transfer:

$$J_T + J_{ST} = 0 \tag{7}$$

where J_i are flux densities

The diffusion of the neutral permeant follows the first Fick's law:

$$J_S = -D^*_S \; dC^*_S/dx \tag{8}$$

and the interdiffusion of the charged species T^+ and ST^+ follow the Nernst-Planck equation:

$$J_i = -D^*_i \; (dC^*_i/dx + C^*_i E) \qquad \text{with } E = F/\mathcal{R}\mathcal{T}.d\Phi/dx \tag{9}$$

where D^*_S, D^*_i are apparent diffusion coefficients in the membrane and E is an electric term
with F = Faraday's constant, $\mathcal{R}$ = universal gaz constant
Φ = electric potential and $\mathcal{T}$ = absolute temperature

2.2.2 Relations between local flux densities. Eqns. (5), (7) and (9) lead to:

$$J_T \; (1/D^*_{ST} + 1/D^*_T) = X \; E \tag{10}$$

and eqns. (8) and (9) give:

$$J_S/(D^*_S C^*_S) + J_T/(D^*_T C^*_T) - J_{ST}/(D^*_{ST} C^*_{ST}) = d/dx \; [\mathrm{Ln} \; (C^*_{ST}/C^*_S C^*_T)] \tag{11}$$

From definition (3), Ln $(C^*_{ST}/C^*_S C^*_T)$ is Ln(k^*) and its derivative is zero, so that, considering relation (7), we can write:

$$J_S/(D^*_S C^*_S) = J_{ST}[1/(D^*_{ST} C^*_{ST}) + 1/(D^*_T C^*_T)] \tag{12}$$

or $$J_S = J_{ST}(D^*_S C^*_S D^*_T C^*_T + D^*_S C^*_S D^*_{ST} C^*_{ST})/(D^*_{ST} C^*_{ST} \; D^*_T C^*_T) \tag{13}$$

Using the coefficients

$$p_1 = D^*_S C^*_S D^*_T C^*_T; \; p_2 = D^*_S C^*_S D^*_{ST} C^*_{ST}; \; p_3 = D^*_{ST} C^*_{ST} \; D^*_T C^*_T;$$
$$p_{12} = p_1 + p_2; \; p_{13} = p_1 + p_3; \; p_{123} = p_1 + p_2 + p_3; \; r_1 = p_1/p_{123}; \; r_2 = p_2/p_{123};$$
$$r_3 = p_3/p_{123}; \; r_{12} = p_{12}/p_{123}; \text{ and } r_{13} = p_{13}/p_{123}$$

and defining the overall permeant flux density as:

$$J_R = J_S + J_{ST} \tag{14}$$

we obtain the relations between the individual fluxes:

$$J_{ST} = r_3 \; J_R \tag{15}$$
$$J_S = r_{12} \; J_R \tag{16}$$
$$J_T = - r_3 \; J_R \tag{17}$$

and the expression of the electric term:

$$E = B \; J_R \qquad \text{with } B = (1/D^*_T - 1/D^*_{ST}) \; r_3/X \tag{18}$$

2.2.3 Overall permeant transport. The overall flux density J_R can be also related to the overall concentration C^*_R by the first Fick's law:

$$J_R = - D^*_R \, dC^*_R/dx \quad \text{with} \quad dC^*_R/dx = dC^*_S/dx + dC^*_{ST}/dx \qquad (19)$$

where D^*_R is an overall apparent diffusion coefficient of the permeant in the membrane.

From eqn. (8), $dC^*_S/dx = -J_S/D^*_S$

and, from eqn. (9), $dC^*_{ST}/dx = -J_{ST}/D^*_{ST} - C^*_{ST}E$

so that the expression of D^*_R is given by:

$$1/D^*_R = r_{12}/D^*_S + r_3(N^*_T/D^*_{ST} + N^*_{ST}/D^*_T) \qquad (20)$$

where N^*_i are ionic fractions C^*_i/X

2.3 Transport curves calculation.

Fig. 2 shows a schematic concentration profile of the overall permeant through the membrane ($C^*_R = C^*_S + C^*_{ST}$) and through the diffusion boundary layers ($C_R = C_S$) in a solution/membrane/solution system and in the stationary state. Fig. 3 presents a diagram of an experimental stationary state transport measurements device.

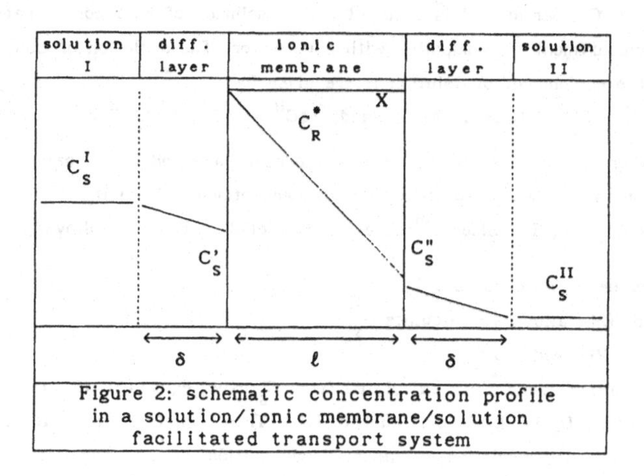

Figure 2: schematic concentration profile in a solution/ionic membrane/solution facilitated transport system

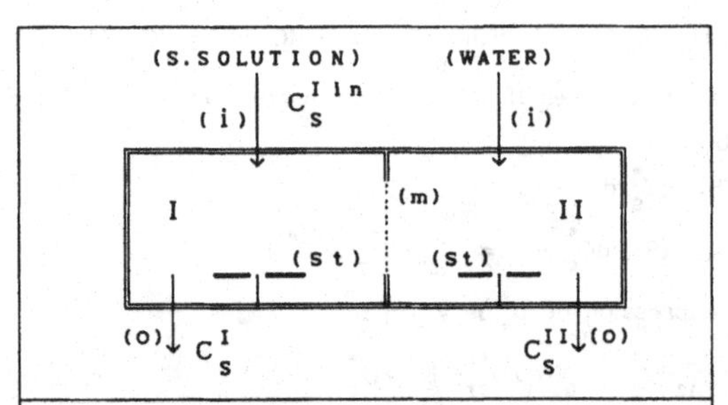

Figure 3: Diagram of a stationary state transport measurement device: i: inlet, o: outlet, m: membrane, st: stirrer

In Fig. 2 the boundary concentrations, C_S' and C_S'', are related to the concentrations C_S^I and C_S^{II} of the upstream and downstream solutions by:

$$J_R = D_S(C_S^I - C_S')/\delta = D_S(C_S'' - C_S^{II})/\delta \tag{21}$$

where D_i are the diffusion coefficients in solution and δ is the thickness of the boundary layers.

C_S^I and C_S^{II} correspond to the outlet concentrations of compartments I and II of Fig. 3. Compartment I is supplied with a solution of S of concentration C_S^{Iin}, compartment II is supplied with pure water. The depletion of solution I and the enrichment of solution II are given by:

$$\Delta C^I = C_S^{Iin} - C_S^I = J_R\ A/f^I \quad \text{and} \quad \Delta C^{II} = C_S^{II} = J_R\ A/f^{II} \tag{22}$$

where A is the effective area of the membrane and f^I, f^{II} are the constant flows of solution through compartments I and II.

To calculate for instance C_S^{Iin} as a function of J_R needs the following steps:

-1) A value is assigned to J_R

-2) Eqn. (22) allows to calculate C_S^{II}

-3) Eqn. (21) leads to C_S''

-4) The interfacial $C_i^{\bullet}$ are calculated from C_S'' using eqns. (3), (5), (6).
Consequently $D_R^{\bullet}$ is obtained from eqn. (20). Then Eqn. (19) allows to make correspond fixed small increments of concentration, $dC_R^{\bullet}$, to variations of the distance parameter, dx. For each new increment, $C_i^{\bullet}$, $D_R^{\bullet}$ and a new position x are calculated, so that the upstream interface of the membrane is reached step by step (when $\sum dx = \ell$, the thickness of the membrane) and the interfacial concentration C_S' can be calculated.

-5) C_S^I is calculated using eqn. (21)

-6) Finally, C_S^{Iin} is calculated by eqn. (22).

3 EXPERIMENTAL DETERMINATION OF PARAMETERS

In order to check the model, values must be assigned to the parameters involved in the equations. Some of these like the diffusivities $D_{S,}$ D_T and D_{ST} are known from the literature [Robinson et al, 1959] but, in our particular system, others has to be determined . Experimental measurements allowed us to estimate the thickness δ of the boundary layers, the constant k^* and the mobilities of the species S, T^+, ST^+ in the membrane.

3.1 Thickness of the diffusion boundary layers.

The thickness δ has been determined by using an experimental method already described in our previous papers. In a "Facilitated Extraction " system [Langevin et al, 1986], when the membrane separates a permeant solution from a carrier solution, the flux density has an upper limit proportional to $1/\delta$. This allows an indirect estimation of δ. In our device, δ has been determined from flux measurements in [α-alanine/membrane/HCl] systems.

3.2 Sorption of α-alanine by the membrane.

Three series of equilibrium [membrane(T^+ form)/solution(S, T^+, ST^+)] were performed by immersing samples of H^+ form membrane in α-alanine solutions partially neutralized by HCl ($C_Y = C_{Cl} = 0 - 10^{-2}$ and $5\ 10^{-2}$ mol.l^{-1}). After equilibrium, the α-alanine was determined in the membrane and in the solution using a potentiometric method or ionic chromatography.

3.3 Diffusivity of S in the membrane

In order to determine the diffusivity of the zwitterionic form of the α-alanine in the membrane phase, permeation measurements of the amino-acid have been made through the membrane in the non-reactive Na^+ form. The experiments were performed in the stationary state using the device of Fig. 3. Compartment I was supplied with an α-alanine solution (concentration C_S^{IIn}) and compartment II was supplied with water as for facilitated transport experiments. In this situation, the concentration profile of the permeant can be represented by Fig. 2 with, in the membrane, $C_R^* = C_S^*$ (ST^* does not exist) and, for the interfacial concentrations, $C_S^{*\prime} = \lambda\ C_S'$ and $C_S^{*\prime\prime} = \lambda\ C_S''$ (see table 1).

An equation similar to eqn. (21) allows to express the flux density J_S of the permeant through the boundary layers:

$$J_S = D_S(C_S^I - C_S')/\delta = D_S(C_S'' - C_S^{II})/\delta \tag{23}$$

and through the membrane:

$$J_S = D_S^*\ (C_S^{*\prime} - C_S^{*\prime\prime})/\ell = D_S^*\ \lambda\ (C_S' - C_S'')/\ell \tag{24}$$

The product $D_S^*.\lambda$ is also a Permeability coefficient, P_S^*, of the membrane to the α-alanine.

From eqn. (26) we can write: $(C_S' - C_S'') = C^I - C^{II} - 2\delta\ J_S/D_S$ (25)

so that:

$$J_s = (C_s^I - C_s^{II})/(\ell/P_s^* + 2\delta/D_s) \tag{26}$$

where C^I and C^{II} are the outlet concentrations of compartments I and II. J_s is obtained from the amino-acid depletion of compartment I or from the enrichment of compartment II. The slope of the experimental curve J_s vs. $(C_s^I - C_s^{II})$ allows to calculate an average value of P_s^*.

3.4 Diffusivity of T^+ in the membrane.
The diffusivity of T^+ in the membrane phase has been estimated by using the Nernst-Einstein equation which relates the diffusion coefficient of an ionic species to its electric mobility. Conductivity measurements were made on samples of membrane in H^+ form, following a method previously described [Métayer et al, 1988].

3.5 Diffusivity of ST^+ in the membrane.
The apparent diffusion coefficient of the protonated permeant through the membrane has been determined using self-diffusion of the protonated optically active α-L-alanine.
As shown in Fig. 4, the membrane, initially in H^+ form, was placed between a finite volume V^I of an α-L-alanine solution (concentration C_T) and a finite volume V^{II} of a racemic α-DL-alanine solution (concentration C_T). In this system, without overall concentration gradient, the optically active α-L-alanine diffuses through the membrane, in the protonated form ST^+, from compartment I to compartment II and, vice-versa, the α-D-alanine diffuses from compartment II to compartment I. The rotation in a compartment is proportional to the difference of concentration between the two forms L and D. For instance in compartment II:

$$R^{II} = R_M (C_L^{II} - C_D^{II}) \tag{37}$$

where C_L^{II} and C_D^{II} are the concentrations of the L and D forms and R^{II} is the observed rotation.
R_M is a molar rotation of the α-L-alanine, defined as $[\alpha]_\lambda^t$ M/1000, where $[\alpha]_\lambda^t$ is the specific rotation and M is the molecular weight of the α-alanine [Greenstein et al, 1961].

It is easy to show that this system can be described by the quasi-stationary diffusion of an optically active species, α, on an optically inactive backround, with $R^{II} = R_M C_\alpha^{II}$ and $C_\alpha^{II} = (C_L^{II} - C_D^{II})$.

The flux density of α through the boundary layers and through the membrane is given by:

$$J_\alpha = D_S(C_\alpha^I - C_\alpha')/\delta = D_{ST}^*(C_\alpha^{*\prime} - C_\alpha^{*\prime\prime})/\ell = D_S(C_\alpha'' - C_\alpha^{II})/\delta \tag{28}$$

Assuming, for the interfacial concentrations,

$$C_\alpha^{*\prime}/C_\alpha' = X/C_T \quad \text{and} \quad C_\alpha^{*\prime\prime}/C_\alpha'' = X/C_T \tag{29}$$

we obtain:

$$J_\alpha = Q\, D_{ST}^*\, X\, (C_\alpha^I - C_\alpha^{II})/\, (\ell\, C_T) \tag{30}$$

with $Q = 1/[1 + 2\delta\, D_{ST}^*\, X\, /\, \ell\, D_S C_T)]$

The evolution of the concentrations is given by:

$$-dC_\alpha^I/dt = dC_\alpha^{II}/dt = J_\alpha\ A/V = (C_\alpha^I - C_\alpha^{II})\ A\ Q\ D_{ST}^*\ X\ /(V\ \ell\ C_T) \qquad (31)$$

with $V = V^I = V^{II}$, the volumes of the solutions I and II, and A, the effective area of the membrane.

Considering the assumption $V^I = V^{II}$, the balance $C_\alpha^I(0) - C_\alpha^I = C_\alpha^{II} - C_\alpha^{II}(0)$ and the initial concentrations $C_\alpha^I(0) = C_T$ and $C_\alpha^{II}(0) = 0$, the difference $(C_\alpha^I - C_\alpha^{II})$ can be written as $(C_T - 2C_\alpha^{II})$, so that eqn.(31) becomes:

$$dC_\alpha^{II}/dt = G + H\ C_\alpha^{II} \qquad (32)$$

with $G = A\ Q\ D_{ST}^*\ X\ /(V\ \ell)$ and $H = 2\ A\ Q\ D_{ST}^*\ X\ /(V\ \ell\ C_T)$

The integration of this differential equation leads to:

$$\mathrm{Exp}(-H\ t) = 1 - 2C_\alpha^{II}/C_T \quad \text{or} \quad H = -\ 1/t\ \mathrm{Ln}(1 - 2C_\alpha^{II}/C_T) \qquad (33)$$

Using the expressions of H (eqn.(32)) and Q (eqn.(30)), this last relation allows to calculate finally D_{ST}^*:

$$D_{ST}^* = \ell\ C_T\ H\ V\ D_S/\ [2X\ (D_S A - H\ V\ \delta)] \qquad (34)$$

Experimental measurements were made using the diffusion cell of Fig. 3, without the pumping device used for the stationary state transport experiments. The evolution of the concentration C_α^{II} versus the time was followed by determining the rotation R^{II} on small samples of solution, using a Perkin Elmer 241 polarimeter at 365 mμ and 25 °C, in the presence of HCl.

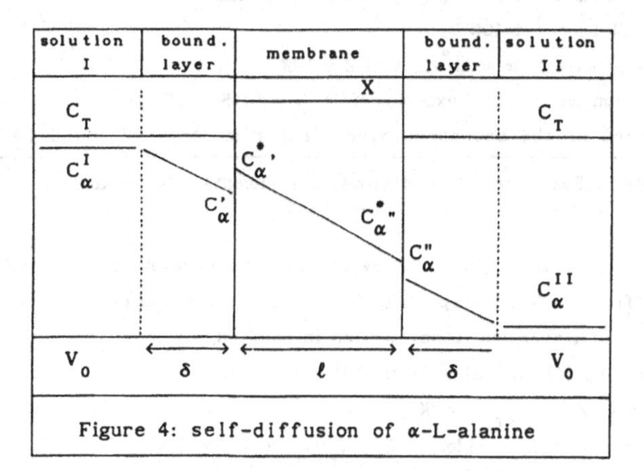

Figure 4: self-diffusion of α-L-alanine

4 EXPERIMENTAL RESULTS.

Table 2 summarizes our experimental conditions, experimental results and all the data necessary to compute facilitated transport theoretical curves.

Membrane type: Rhône Poulenc sulfonic cation exchange membrane
Thickness: $\ell = 5.8\ 10^{-2}$cm ± 1%
Molar ion exchange capacity (experimental): $X = 1.57$ mol.l^{-1}± 1%
Experimental conditions:
Effective area of the membrane: $A = 4.15$ cm^2± 1%
Volume of solution I and II (self diffusion):
$V^I = V^{II} = 100$ cm^3 ± 1%
Flow of solution I and II (permeation, facilitated transport):
1.8 cm^3.s^{-1} ± 1%
Diffusivities in solution (literature):
$D_S = D_{ST} = 9.1$ cm^2.s^{-1}
Diffusivities in the membrane:
$P^*_S = 8.46\ 10^{-7}$ cm^2.s^{-1} and $D^*_s = 1.88\ 10^{-6}$ cm^2.s^{-1} ± 5%
$D^*_T = 3.14\ 10^{-6}$ cm^2.s^{-1} ± 10%
$D^*_{ST} = 2.0\ 10^{-7}$ cm^2.s^{-1} ± 5%
Stability constant of ST^+ in the membrane phase:
$\lambda k^* = 264$ l.mol^{-1} ± 10%
Selectivity factor: $\sigma = \lambda k^*/K = 1.2$ ± 10%
Distribution coefficient (experimental): $\lambda = 0.45$ ± 2%
Thickness of the boundary layers (experimental): $\delta = 4.9\ 10^{-3}$cm ± 5%

Table 2: Experimental conditions, experimental results and literature data

The results of α-alanine sorption by the membrane are shown in Fig. 5 where the ionic fraction N^*_{ST} is represented as a function of the decimal logarithm of the overall amino-acid concentration in solution, C_R.

From eqn. (3), (4) and (5) we can write:

$$\lambda k^* = N^*_{ST}/C_S(1-N^*_{ST}) = \sigma K \qquad (35)$$

with, from eqn. (2), considering $C_Y = C_T + C_{ST}$ and $C_R = C_S + C_{ST}$

$$C_S^2 + (C_Y - C_R + 1/K)\ C_S - C_R/K = 0 \qquad (36)$$

The experimental data allows to calculate an average value, $\langle \lambda k^* \rangle = 264$

$l.mol^{-1}$, not far from the value of the constant K in solution: 218.7 $l.mol^{-1}$. That shows, in agreement with previous results [Zaikov <u>et</u> <u>al</u>, 1988] and taking into account the accuracy of the measurements, a very weak selectivity ($\sigma \simeq 1.2$) of the membrane between T^+ and ST^+.

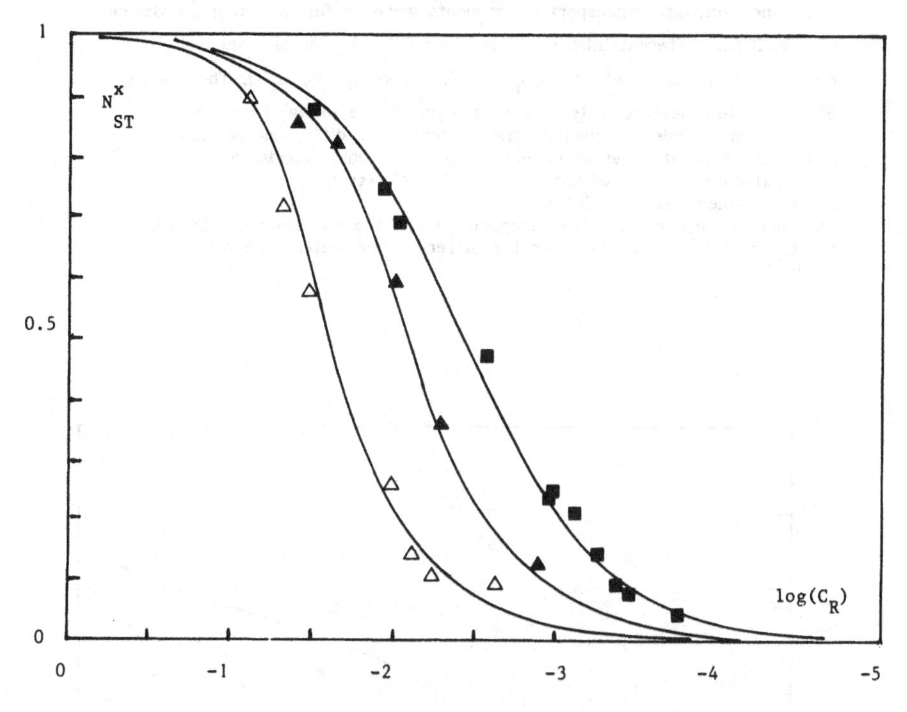

Figure 5: Absorption of α-alanine by the membrane, initially in H^+ form, dipped in various solution of amino-acid partially neutralized with HCl. Ionic fraction of ST^+ in the membrane vs. $\log(C_R)$. $C_{Cl} = 0$ (■), $C_{Cl} = 0.01$ (▲) and $C_{Cl} = 0.05$ $mol.l^{-1}$ (△).

The three curves drawn in Fig. 5 have been calculated using the average value $\langle \lambda k^{*} \rangle$.

Experimental measurements of water uptake by the H^+ form membrane dipped in α-alanine solutions give an estimation of ε, the effective porosity of the material:

$$\varepsilon = V_P/(V_P+V_M) \simeq (m_S-m_D)/m_S \qquad (37)$$

where V_P is the porous volume, V_M is the matrix volume, m_S is the mass of the swollen material and m_D is the mass of the dry (H^+ form) material.

The results lead to $\varepsilon \simeq 0.45$. Neglecting salting-in or salting-out effect for the zwitterionic form of the α-alanine, the distribution coefficient λ can be assimilated to the porosity coefficient ε, so that $C_s^* \simeq \varepsilon\, C_s$ and $k^* \simeq 119$ mol.l^{-1}.

α-alanine facilitated transport experiments were performed using the device of Fig. 3 for different inlet concentrations C_s^{Iin} .The results are represented in Fig. 6 where $\log(J_R)$ is plotted versus $\log(C_s^{Iin})$. The results of permeation measurements through the membrane in Na^+ form are also represented in order to compare the diffusion and diffusion-reaction processes. A third series of experimental points corresponds to the "facilitated extraction" of the α-alanine by HCl (system S solution/membrane/T^+solution).
The theoretical curve has been computed using the equations of the model presented above, considering for the different parameters the values of Table 2.

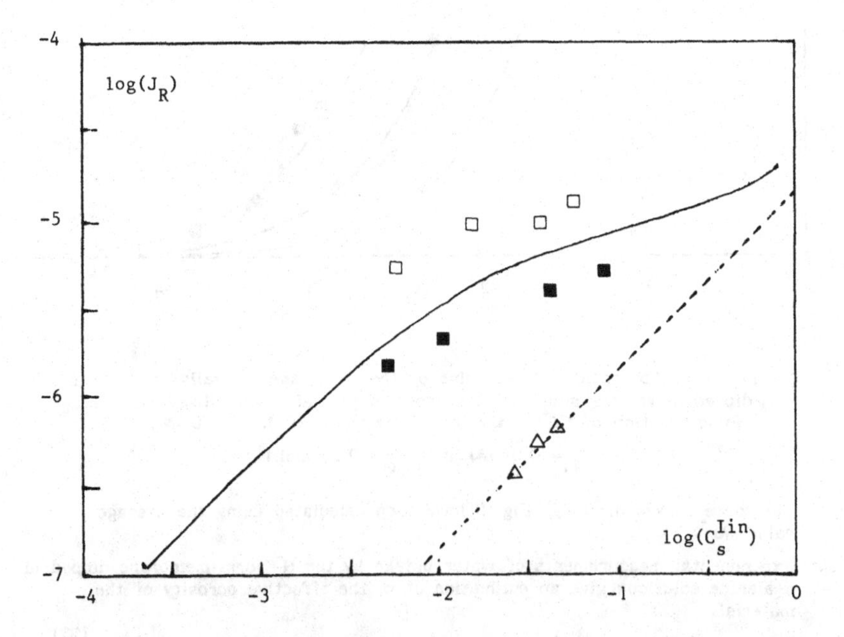

Figure 6: Experimental results of facilitated transport (■), permeation (△) and facilitated extraction (□). $\log(J_R)$ vs. $\log(C_s^{Iin})$. Theoretical curves have been computed using the values of Table 2.

CONCLUSION.

The stability constant of the protonated α-alanine in the membrane phase has been experimentally determined and was found practically equal to the constant in solution, in agreement with other authors. According to our previous theoretical analysis, [Métayer et al, 1989] this value is quite favorable to a facilitated transport mechanism through an ionic membrane. Actually Fig. 6 shows a high facilitation factor (10 to 20) with regard to the permeation transport rate which repesents the lower limit in such a system. In addition, the facilitated transport results are not far from the upper limit corresponding to the facilitated extraction process, this is predicted by the model when the stability constant k^* is in the optimal range.

In conclusion, the facilitated transport of α-alanine using the proton as carrier is a good illustration of facilitated transport processes through ion-exchange membranes.

The difference between the facilitated transport theoretical curve and the experimental results observed in Fig. 6 could be reasonably attributed to kinetic factors. These are not taken into account in our model and, in order to verify this hypothesis, the next step of this work will be to introduce a kinetic parameter in the equations and to compute its influence on the discrepancy between theoretical curve and experimental points.

REFERENCES

Greenstein, J.P. and Winitz, M. (1961) *Chemistry of the amino acids,* Vol. 2, John Wiley and Sons,Inc., New York., Chapt. 18.

Helfferich, F.G. (1990) 'Ion-exchange equilibria of amino acids on strong-acid resins: theory', *Reactive Polymers,* vol. 12, pp. 95-100.

Langevin, D., Métayer, M., Labbé, M., Pollet, B. Hankaoui, M and Sélégny, E. (1986) *Carrier facilitated transport and extraction through ion-exchange membranes, illustrated with ammonia, acetic and boric acid.*, in Membranes and membrane processes, Drioli, E. and Nakagaki, M. Eds., Plenum Press, N. Y. and London, pp. 309-318.

Leblanc, O.H., Ward, W.J., Matson, S.L. and Kimura, S.G. (1980) 'Facilitated transport in ion-exchange membranes', *J. Memb. Sci.,* vol. 6, N°.3, pp. 339-343.

Métayer, M., Langevin, D., Hankaoui, M. Pollet, B., Labbé, M. and Roudesli, S. (1988) *'Reaction in ion-exchangers, apparent stability constants of complexes and salting-in (out) effects',* Reactive Polymers, vol. 7, P. 111.

Métayer, M., Langevin, D., El Mahi, B., and Pinoche, M. (1989) 'Facilitated extraction and facilitated transport of non-ionic permeants through ion-exchange membrane. Influence of the stabilities of the permeant/carrier complexes', *conference presented during the 6th International Symposium on Synthetic membranes in Science and Industry,* Tubingen, RFA., to be published in *J. Memb. Sci.*

Robinson, R.A. and Stokes, R.H. (1959) *Electolyte solutions,* Butterworths, London.

Sillen, L.G. and Martell, A.E. (1964) *Stability constants of metal ion complexes,* The Chemical Society, London., p. 295.

Zaikof, G.E., Iordanskii, A.P. and Markin V.S. (1988) *Diffusion of electrolytes in polymers,* VSP, Utrecht, p. 99.

PHOSPHAZENE POLYMER MEMBRANES: HAZARDOUS CHEMICAL REMOVAL FROM GAS AND LIQUID FEEDSTREAMS.

M. L. Stone (EG&G Idaho), E. S. Peterson (EG&G Idaho),
M. D. Herd (EG&G Idaho), D. G. Cummings (Argonne),
R. R. McCaffrey (Ciba-Corning)

SUMMARY

Phosphazene polymers consist of nitrogen-phosphorus backbones which have high thermal and chemical stabilities. Experiments at the Idaho National Engineering Laboratory demonstrate that phosphazene polymer membranes are capable of removing: (1) chlorinated hydrocarbons from water and air; and (2) acid gases such as SO_2, from nitrogen. These results point to a wide range of potential industrial and environmental applications.

INTRODUCTION

The increased industrial and governmental efforts to develop and promote energy efficient processes, reduce environmental degradation, and remediate existing pollution problems have accelerated membrane research for chemical separations. Energy inefficient processes decrease competitiveness and usually are significant contributors to environmental pollution. Recent studies conducted by the Department of Energy indicate that a substantial reduction in energy consumption is possible by using membranes (Baker 1990, Leeper 1984). The Idaho National Engineering Laboratory (INEL) has been conducting membrane research sponsored by the Department of Energy's Office of Industrial Technologies to develop energy efficient membranes based upon phosphazene polymers.

One problem delaying a more widespread implementation of polymer membrane technology has been that many polymers degrade in the harsh thermal (>100°C) and chemical environments often encountered in these processes. To address this problem, the INEL research has been directed at a class of polymers, polyphosphazenes, known to have good thermal and chemical properties. Another important feature of phosphazene polymers is the rich variety of end products that can be made from a single precursor (ACS Symposium 1988). The research conducted has shown that polyphosphazene membranes are suitable for liquid and gas separations in harsh environments (Allen 1989, Allen 1987, McCaffrey 1988, McCaffrey 1987, McCaffrey 1986).

This paper briefly introduces two more applications for phosphazene polymer membranes: (1) chlorinated hydrocarbon removal from water via pervaporation, and (2) gas separations involving SO_2, N_2, H_2S, CH_4, CO_2, and some binary mixtures of some of these gases. The results indicate that phosphazene polymers may be well suited for removing chlorinated hydrocarbons from water and air and good candidates for removing acid gases and performing natural gas sweetening.

Effective Industrial Membrane Processes — Benefits and Opportunities, pp.321–326

METHODS/EXPERIMENTAL PROCEDURE.

Polymer synthesis.
Phosphazene polymers consist of alternating phosphorus and nitrogen atoms joined by alternating single and double bonds. Figure 1 depicts the polymers used in this study as well as some others being investigated. The polymers were synthesized by following published methods.

Figure 1 Polyphosphazene structures

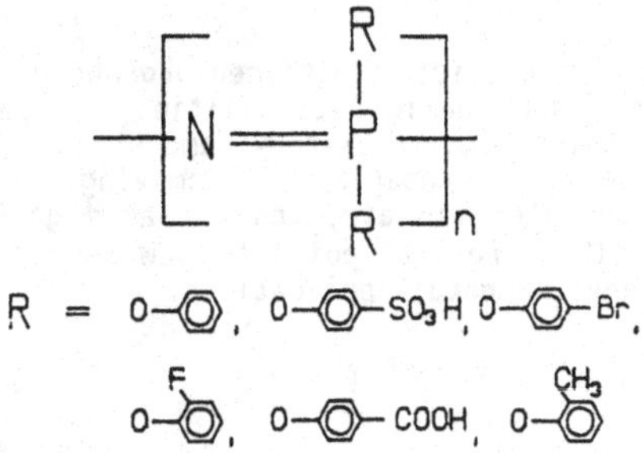

Polybis(phenoxy)phosphazene. Synthesis of this polymer is detailed by Allen (1987) and Singler (1974). Briefly, it was performed by ring-cleavage polymerization of hexachlorocyclotriphosphazene at 250°C under vacuum, and subsequent substitution of the chlorines by the desired side groups.

Polybis(8% p-bromophenoxy)phosphazene and polybis(8% p-carboxyphenoxy)phosphazene (8% COOH-PPOP). These two polymers were synthesized utilizing a lithiophenoxy intermediate procedure described by Allcock (1980).

Membrane Casting.
Solutions of 1% to 5% of the polymer in tetrahydrofuran (THF) were used to solution cast films onto glass plates. Evaporation of the solvent yielded thin dense films of from 1 to 20 microns thick. The edges of the film were scored. When the plate was slowly lowered into a water bath, the membrane floated off the glass plate. Finally, a suitable support was lowered into the water bath, positioned under the membrane, and raised yielding a final supported membrane for subsequent testing.

The same casting solutions used above have also been dropped onto thin, rapidly spinning ceramic supports to form membranes. The excess polymer was spun off leaving a uniform thin film in intimate contact with the ceramic support. Spray casting, and knife casting have also been utilized to yield phosphazene polymer membranes (McCaffrey 1986).

Membrane Testing.
Pervaporation. When a permeate changes phase across a membrane (from a liquid to a gas) the process is called pervaporation. The experiments were performed by passing a liquid mixture containing the material to be separated across the

membrane on the feed side. A vacuum was pulled on the permeate side and the permeate was collected by a cold trap. The feed side was a closed loop consisting of a glass container serving as the reservoir, a small liquid pump (2-20 ml/min), and the cell containing the membrane. The permeate side consisted of a tube connecting the cell to the cold trap and another tube connecting the cold trap to the vacuum pump. The cold trap was cooled with liquid nitrogen to ensure collecting all materials coming through the membrane. When a sample had been collected the trap was closed and a sample of the permeate was analyzed using a GC. In some cases the feed side was heated and pressurized to increase fluxes.

Mixed Gas. Mixed gas tests were run on equipment developed at the INEL. The commercially prepared gas mixture consisted of 10% SO_2 in nitrogen. An 8% COOH-PPOP membrane was cast on the inside of a tubular ceramic support. The support was mounted in a stainless steel test fixture, placed in an oven, and attached to the system.

In the course of the experiment, the feed gas entered the cell containing the membrane, passed over the membrane, and exited the cell as the rejectate. The permeate was swept from the cell with a helium purge gas and all three gas streams (feed, rejectate, and permeate) were transmitted to a gas chromatograph (GC) for analysis. The pressure of the feed gas was held at 20 psi with the permeate side at ambient pressure. The flows for the feed and purge were 20 cc/min and 2 cc/min respectively. The thicknesses of the membranes were measured after the experimental runs by freeze breaking the membrane and averaging the thickness values determined at several positions in a scanning electron microscope (SEM). From these tests three parameters were determined: (1) the separation factor which is the ratio of the concentrations of the pair of gases in the permeate divided by the ratio of the concentrations of the pair of gases in the rejectate; (2) the relative permeabilities; and (3) the selectivities which are the ratios of permeabilities for the pair of gases being tested.

RESULTS AND DISCUSSION

Pervaporation.
In early proof-of concept experiments, two binary gas mixtures (CH_2Cl_2/N_2 and CCl_4/N_2) were tested utilizing phosphazene polymer membranes. Excellent selectivity for the chlorinated hydrocarbons over the nitrogen was observed. The separation factor for CH_2Cl_2 was 84 and for CCl_4 was 20. The above observation prompted the pervaporation experiments involving chlorinated hydrocarbons in water. Recently, coworkers reported enrichment factors for chloroform from water of 1500 and fluxes (l/m^2-hr) in the range of 0.003 to 0.05 using the PPOP polymer. These fluxes were obtained from solutions containing 0.6% chloroform in water at room temperature. The membrane surface area was 13.8 cm^2, the permeate vacuum was 200 Torr, and the membrane thicknesses were 2 to 17 microns (Gianotto 1990). These flux values were in the range reported for other pervaporation separations which utilized thinner membranes and higher vacuums (Blume 1990).

Initial results with temperature and pressure enhanced pervaporation appear very encouraging. Experimental conditions include: 200 psig; 24°C; with the permeate being collected by cryotrapping into a liquid nitrogen cooled trap. Under these conditions 1 ml of liquid was collected in 6 hours. Increasing the temperature to 97°C yielded 2.5 ml of liquid over 8 hours. Thus a doubling of the flux was observed with the increased temperature. The experiment has been repeated several times to give similar results. These fluxes would be approximately 0.24 l/m^2-hr; an order of magnitude greater than previous results and is comparable with published results.

Gas Separations.
One of the goals of this research program has been to determine if polyphosphazene membranes could perform separations in harsh conditions at temperatures above 100°C. McCaffrey (1988) reported separation performance data for a number of gases including He, N_2, CO, Ar, CH_4, SO_2, H_2S, and CO_2. The results showed that among the gases tested SO_2, H_2S, and CO_2 had high permeabilities. He concluded that these were more soluble in the polymer than the others. Based on these results, an SO_2/N_2 mixture was selected for use in these studies. The SO_2 mixture is representative of a useful acid gas separation and the 5 and 10% mixtures represent a harsh chemical environments.

The test results from a representative run using the hollow tube configuration is given in Table 1.

Table 1 High temperature hollow tube test data

Temp. (°C)	Separation Factor[1]	SO_2 Conc. (%)	Permeability (barrers)[2]		Selectivity[3]
8%COOH-PPOP					
	SO_2/N_2		N_2	SO_2	SO_2/N_2
80	14.0	9.85	5.4	108.9	19.9
130	15.6	9.85	17.2	374.8	21.7
130	12.3	4.58	19.8	132.7	6.7
160	--	4.58	30.1	141.9	4.7
190	4.8	4.58	43.7	149.4	3.4
210	5.8	4.58	54.8	154.9	2.8
240	4.6	4.58	76.1	163.8	2.2
270	3.2	4.58	118.3	170.7	1.5

[1]Separation factor = $[SO_2]/[N_2]_{permeate}$ / $[SO_2]/[N_2]_{rejectate}$.

[2]Barrer = $[10^{-10}$ cc(STP) cm]/[cm^2 sec cm(Hg)]

[3]Selectivity = ratio of permeabilities.

The selectivity results were very encouraging. The membrane not only survived temperatures over 100°C but actually gave slightly improved values at 135°C. The temperature was increased up to 270°C at which point the membrane slowly failed. It is not known if the polymer simply melted into the support or if the failure was caused by some other means. It appears that the present polyphosphazene membranes can survive harsh conditions at temperatures over 100°C. It is felt that by varying the crystallinity, extent of crosslinking, and types of functional side groups that membranes can be developed to operate at temperatures well in excess of 100°C.

CONCLUSIONS

Membranes have been formed from a variety of phosphazene polymers derived from polybis(phenoxy)phosphazene. These membranes have been tested in pervaporation experiments involving chlorinated hydrocarbons and in mixed gas tests involving binary gas mixtures which are industrially and environmentally important. The results have led to the following conclusions:

- Phosphazene polymers are excellent film (membrane) formers.
- Pervaporation experiments indicate that polyphosphazene membranes may be well suited for separating hazardous chlorinated organics from water (and air).
- Initial experiments show that pervaporation fluxes can be increased significantly by increasing temperatures and pressures.
- Polyphosphazene membranes show excellent selectivities and permeabilities toward SO_2.

ACKNOWLEDGMENT

The work described in this paper was supported by the Department of Energy's Office of Industrial Technologies Contract No. DE-AC07-76ID01570.

REFERENCES

ACS Symposium Series 360, (1988) Inorganic and Organometallic Polymers, M. Zeldin, K. J. Wynne, and H. R. Allcock, editors, American Chemical Society, Chapters 19 through 25 deal with polyphosphazenes.

Allcock, H. R., Fuller, T. J., and Evans, T. L., (1980) Macromolecules, vol. 13, no. 6, pp. 1325-1332.

Allen, C. A., Cummings, D. G., and McCaffrey, R. R., (1989) 'Separation of Cr Ions from Co and Mn Ions by Poly[bis(trifluoroethoxy)phosphazene] Membranes', J. Membrane Science, vol. 43, pp. 217-228.

Allen, C. A., McAtee, R. E., Cummings, D. G., Grey, A. E., and McCaffrey, R. R., (1987) 'Separation of Ions from Co and Mn Ions by Poly[bis(phenoxy)-phosphazene]', J. Membrane Science, vol. 33, pp. 181-189.

Baker, R. W., et. al., April 1990, Membrane Separation Systems: A Research Needs Assessment, U. S. DOE Contract No. DE-AC01-88ER30133.

Blume, I., Wijmans, J. G., and Baker, R. W. (1990) 'The Separation of Dissolved Organics from Water by Pervaporation', J. Membrane Science, vol. 49, pp. 253-286.

Gianotto, A. K., Bauer, W. F., and Cummings, D. G., (1990) 'Application of Polyphosphazene Pervaporation Membranes For Separation of Chlorinated Hydrocarbons from Water', paper presented at the 45th Northwest Regional Meeting of the American Chemical Society, Salt Lake City, Utah.

Leeper, S. A., et. al., (1984) Membrane Technology and Applications: An Assessment, EG&G-2282, U. S. DOE Contract No. DE-AC07-76ID1570.

McCaffrey, R. R. and Cummings, D. G., (1988) 'Gas Separation Properties of Phosphazene Polymer Membranes', Fifth Symposium on Separation Science and Technology for Energy Applications, Knoxville, Tennessee, October 26-29, 1987, Separation Science and Technology , vol. 23, no. 12&13, p. 1627.

McCaffrey, R. R., McAtee, R. E., Cummings, D. G., Grey, A. E., Allen, C. A., and Appelhans, A. D. (1986) 'Synthesis, Casting, and Diffusion Testing of Poly[bis(trifluorethoxy)phosphazene] Membranes', J. Membrane Science, vol. 28, pp. 47-67.

McCaffrey, R. R., Allen, C. A., Appelhans, A. D., McAtee, R. E., Grey, A. E., Wright, R. B., Jolley, J. G., and Cummings, D. G. (1985) 'Inorganic Membrane Technology', Fourth Symposium on Separation Science and Technology for Energy Applications, Knoxville, Tennessee, October 20-24, 1985, Separation Science and Technology, vol. 22, no. 2, p. 873.

Singler, R. E., Hagnauer, G. L., Schneider, N. S., Laliberte, B. R., Sacher, R. E., and Matton, R. W. (1974) J. Polym. Sci., Polym. Chem. Ed., vol. 12, pp. 433-444.

SESSION J:

GAS SEPARATION IN THE PETROCHEMICAL INDUSTRY I

AN OPERATORS OVERVIEW OF GAS MEMBRANES

W.R. Gurr
BP Engineering

1 INTRODUCTION

Separation technologies have a significant role in petrochemical plants. They are both capital and energy intensive. In various forms they require from 40 to 70% of the capital for new plants and the distillation operation alone consumes some 29% of the U.K. industrial energy. Consequently, advances in this area have much to offer the industry.

Membrane technology has emerged over the last decade or so as one of the most dynamic areas of separation technology. It has now reached an initial stage of maturity in several important separations: it has the potential to stretch much further. The drawbacks of polymer performance offer development potential for other material systems to expand areas of application, these include:- permeate losses; limited operating temperature range and moderate to high pressure requirements, offers the potential for other material systems to expand areas of application. Various aspects of these are discussed.

2 BENEFITS TO THE OPERATOR

The major benefit to the operator of using membranes will always be in the process benefits which they can create. There are secondary engineering benefits associated with their simplicity, low weights and small physical size. They have no moving parts neither have they cyclical valve operations. Coupled to this they can, in a combined step, replace complex unit operations such as gas/liquid absorption for dehydration and acid gas removal. Not only will specific benefits be obtained from the improved performance of the individual units and columns within a unit, but also indirectly, from interrelated process units. These could include for example, a cat. reformer production can be increased, or alternatively a steam reformer can be cut back.

Against this there are always permeate losses. These increase with increased purity requirement and with lower feed pressure. However, the potential adverse economic effects of permeate losses can be mitigated by integrating the membranes into the overall process arrangement.

Effective Industrial Membrane Processes — Benefits and Opportunities, pp.329–336

3 APPLICATIONS AND OPERATING SYSTEMS

BP companies have owned and operated membrane systems since 1981. The three plants and their performances are outlined in the following table.

1 Hydrogen Recovery from Ammonia Purge

Permea unit installed in 1981

Feed Gas	8800 Nm3/hr	126 bar g	61%H2
Product	83.7%H2	Recovery	92%

2 CO2 Removal from Natural Gas

Grace unit commissioned in 1988

Feed Gas	910 Mscfd	630 psia	14% CO2
Product	4% CO2		

3 Hydrogen and Carbon Monoxide Hybrid System

Union Carbide/UOP unit commissioned in 1990

Feed Gas	20MMscfd	420 psig	63.4%H2
Products	99.999% H2	Syn gas 49.1% H2 49.1 CO	

Operational performance and plant histories will be presented visually.

4 PROCESS APPLICATIONS

4.1 Applications Within Refineries and Chemical Plants

Authorities are now exerting severe pressure for reductions in the sulphur content of a wide range of oil products and the availability of higher purity hydrogen will assist in achieving these lower levels. As crude slates tend to get heavier, it will become advantageous for refinery operators to have more flexibility to process the cheaper sour crudes, and this can often be constrained by the lack of hydrogen. Membrane separators can readily remove this constraint by recovering hydrogen for improved utilisation.
The principal beneficiaries of upgrading of hydrogen content are the hydrocracking and desulphurisation units. Depending on the design and equipment in these units, the location chosen for a membrane system can vary to achieve the optimum benefit: this could be in the gas feed to the unit; the purge from the reactor loop or within the reactor loop itself. Specific consideration must be given to the effects on compressors in the system.
Process benefits can also be obtained by removing the hydrogen and lighter hydrocarbons from cracking operations in chemical plants and refineries.

4.2 Integrated Systems

It is not only the simple "unit product" which should be considered in the credit side of the equation, the consequential effects within a complex can be important. In seeking to obtain the maximum benefit from the introduction of membrane technology it is considered that the complete hydrogen system should be reviewed. In this way the membrane separators can then be located such that a balance can be achieved, in both composition and flow rate, between the emergent streams from the producer units and the requirements of the consumer units. This would then minimise the loss of a valuable feed stock commodity to the fuel system and maximise the potential for recovery of saleable hydrocarbons. Hydrogen can be sold as a by-product if a consistent excess is produced.

5 GENERAL DESIGN CONSIDERATIONS

5.1 Pretreatment Requirements

Pretreatment considerations are important when considering the application of a membrane process. The extent to which feed pretreatment is required effects equipment cost, operating flexibility and ease of operation.

Membrane materials can be damaged by the impingement of liquid droplets, and some, notably the composite polymers, will suffer if the coated layers are soluble in gas condensates. Mist droplets in the feed, even from relatively short-term upsets, can damage the elements, and this effect is worsened when large amounts of gas are processed per module. Generally, polymeric membranes cannot process saturated feeds directly, because the non-permeate remains at feed pressure and the condensables are concentrated. Consequently, the dew point of the residue stream is higher than that of the feed gas. A typical pretreatment system for handling saturated feedstocks for membrane applications consists of a knock-out drum with mist eliminator, high performance coalescing filter, and a preheater. The feed is normally heated to approximately 10 °C above the tail gas dew point. The extent of superheat required depends on the design recovery and purity in addition to the feed properties. Good feed temperature control is important, because operation at higher temperatures increases permeability at the expense of selectivity, and the membrane itself can be damaged by high temperatures.

Feed contaminants must be identified and pre-treatment included if they are not acceptable in the product stream, or if they will effect the separator. Important examples are:- fine solids from catalysts or absorbents, ammonia, hydrogen sulphide, organo-sulphur compounds, aromatics and higher cyclic compounds, hydrogen chloride, and acetylenes. These components will partition between the product streams, but may be detrimental to membrane life.

5.2 Turndown

The modular nature of membranes is a major benefit when deciding how to meet turndown requirements. Isolation can be varied as required. Membrane systems are also normally capable of maintaining product purity at feed rates of 30-100+% of design. However, recovery will be sacrificed at turndown to a varying extent. Turndown is accomplished by either reducing feed pressure, increasing permeate pressure or by isolating modules from the system. The first two methods can be used for short-term operation, and the latter when operating at significantly reduced capacity for extended periods.

5.3 Reliability

Reliability of the separation processes is an important consideration, particularly if the separator is a primary source of feedstock to a mainstream process. Membrane systems are extremely reliable with high on-stream factors. The process is continuous and has few control components which can cause a shutdown. On-stream factors, year on year of 100% availability with respect to unscheduled shutdowns are quite common. However, the ability to produce usable product is also a reliability consideration since off-specification product effects the downstream processes even if a shutdown of the separation process is avoided. Consequently, the separator performance must be thoroughly evaluated for all expected variations in feed conditions. When required, changing membrane elements is a simple operation. Downtime is minimised.

5.4 Ease of Future Expansion

Membrane systems are ideally suited for expansion, since this generally, only requires the addition of identical modules. In cases where future expansion is contemplated, a minimum preinvestment is required to provide the tie-ins.

6 ADVANCED SEPARATION SYSTEMS

6.1 Condensation Membranes

New polymers have recently been commercialised that will separate higher hydrocarbons from gas mixtures. Early uses of these were aimed at vapour recovery systems, but further development has improved their operating pressure range and they are being used for the recovery of liquids from natural gases and are currently being evaluated for corresponding applications in refineries and chemical plants.

Early performance indicated recovery of ca.60% C3, 85% C4, 100% C5+. But improved materials now achieve C3 recovery increased to over 90%. These materials will find application in reducing hydrocarbon losses via the fuel and flare systems and the improvement of performance by removing light ends in distillation operations, e.g. debottlenecking de-ethanisers or absorbers.

This type of separation can also be achieved by non-polymeric materials. This results from the use of modern engineering ceramics. These have the capability to have controlled variations created in their structure, which can readily be made asymmetric. The selectivity can be high at very low pressure differentials. This offers prospects of creating a range of special separators with a wide range of operating temperatures.

Applications can be expected to include: organics from inerts; hydrocarbon fractionations; acid components from reactor/flue gases or natural gas; iso/normal paraffins; aromatics from alkanes; olefins from paraffins.

6.2 Hybrid Systems and By-product Recovery

In many refinery hydrogen upgrading processes, the residue stream is largely hydrocarbons, which, if recovered as liquid products, can have value significantly above fuel value. This is particularly true for olefin-containing streams from catalytic crackers. The relatively high value of the hydrocarbons will very often provide attractive paybacks; the hydrogen has, at best, feedstock value.
A secondary recovery process can be used in these applications. In the past PSA and cryogenics having been applied at the appropriate scale. A recent development now being considered is a hydrocarbon dewpointing membrane. Since the hydrocarbon-rich residue remains at high pressure, the new membrane materials are

capable of bulk LPG/NGL separation, but are not capable of providing highly selective separate streams rich in specific hydrocarbon fractions. This makes the dewpointing membrane attractive in two applications: 1) in series with a hydrogen membrane, in recovery applications from offgas streams, and 2) in association with a distillation column, processing a sidestream or overhead vapour.

6.3 Facilitated Transport Systems

Microporous structures of engineering polymers can be utilised as an ultra thin support for reactive liquids. Consequently, it is feasible to enhance the selectivity of membrane separators (200-500 compared with 5-130 for polymers) by employing specific reagents which perform the separation, in combination with the membrane system which gives the large surface area and permeability for economic transport rates.
One such system which has been heavily researched is the reaction between olefins and silver. Here either aqueous solutions of silver salts or even coatings of silver metal are contained in a polymeric pore structure to achieve the separation of olefins from hydrocarbon mixtures. However, there are no known commercial systems available. When such a system is on the market it will undoubtedly find many analogues since the concept should be usable for many different separations.

The use of various salt systems with melting points from near ambient for general gas treatments, up to 2-300°C for process applications, are being investigated. These are reported to show selectivities of 30 CO_2/methane and 130 for H_2/methane, i.e. of the same order as polymers.
The use of engineering ceramics as substrates to contain the reactive medium will overcome the limitations on operating temperature imposed by the use of polymers. Since the transport mechanism involves a gas/liquid surface contact it is reasonable to expect that these systems will operate at substantially lower pressure differentials than polymer separators. Similarly, since the pore structure is physically closed to the bulk gas, other than by relatively slow processes such as absorption, the high selectivity indicates that permeate losses should be minimal.

Applications could potentially encompass all the reactive gas treatment processes: acid gas removal; dehydration; olefin separations. It also opens the way for the development and production of specific chemical complexing reagents to achieve these separations. Potential candidate separations may include - oxygen from air, and nitrogen from methane.

Areas of development that have to be considered in each case include: the long term compatibility of the solution and the membrane material; stability of the solution and effect on cost of solution losses or degradation; recovery of the product from the solution.

6.4 High Temperature Separators

There are developments in progress in many research organisations on advanced concepts which utilise metal or alloy membranes supported on ceramic substrates. Other systems involve metal composites.
These developments are heavily focused on the separation of hydrogen from hydrocarbon mixtures, where the separation is not part of the reaction stage. This method has the significant advantage that at temperatures ranging from 200 to 1000°C fluxes are achieved without requiring high pressure drops across the membrane. Several metals have effectively infinite selectivity for the hydrogen. Economic levels of hydrogen flux have been demonstrated at the small scale. Other possible applications may arise, where due to the selectivity of the membrane, processes could be supplied from impure feed gases as an alternative to expensive conventional processes such as steam reforming.

It is anticipated that the systems could become commercial within the next 5 years and that their introduction could result in useful yield improvements in catalytic processes. The main potential benefit from these separators is shift in equilibrium reactions, such as butane dehydrogenation to butene and to butadiene. Potential drawbacks include the catalytic nature of the metals with the probability of coke laydown, and mechanical and thermal property differences of the materials involved.

The development of commercial scale units will present substantial engineering challenges which will only be given sufficient motivation if the economic case is sufficiently attractive to the end-users.

As research into the properties and coating techniques for thin material films progresses it is feasible that other separations for e.g. oxygen and nitrogen will be developed.

6.5 Membrane Reactors

With the capability to perform gas separations using thin metal films, the removal or addition of gaseous reactants within reactions systems that use conventional catalytic technology becomes a possibility. The ceramic substrate can be used to support both the separating coating and catalyst, or an appropriate metal membrane could perform both functions. In some cases a mixed composition ceramic can also perform this function. In the most advanced concepts, systems have been proposed which have separate reaction systems on each side of the membrane such that mass and thermal balances can be achieved.
Where applications are proved feasible, the technical and engineering development required to commercialise these systems will be lengthy and expensive, and may ultimately only be progressed through joint industry ventures. The goal, not yet clearly defined, could be a new generation of process technology.

7 CONCLUSIONS

Membrane technology is entering a phase of widening application for the existing polymeric systems, with considerable additional potential for "technical stretch" when and if, inorganic materials are specifically applied.
The areas of advanced applications indicated offer much wider potential for development into specific chemical, high temperature and catalytic applications. The potentially high development costs will require considerable economic justification and a co-operative approach to find the solution.

The role of separations is considerably subordinate to the principal objectives of optimising the chemistry and reactor performance. The solution to many separation problems is seen to be located in reaction selectivity and yield.

Acknowledgements

Technical operation staff at Erdolchemie; Oxochemie and BP Alaska

BP International for permission to publish this paper.

Mr.K. Doshi - UOP and Mr.T. Cooley - Grace Membranes for their various advice and assistance.

wrg/3.91

EVOLUTION OF A GAS SEPARATION MEMBRANE 1983 - 1990

By
Robert J. Hamaker
THE CYNARA COMPANY
Houston, Texas

Abstract
New technologies, improvements in design and application of membranes are discussed tracing their evolution from the world's first commercial large scale CO_2 separation plant through present day applications including gas removal from a liquid feed.

Effective Industrial Membrane Processes — Benefits and Opportunities, pp.337–343

"Evolution of a Gas Separation Membrane 1983 - 1990"

A Brief History:

The use of semi-permeable membranes in industrial gas separations dates back to the mid 1970's. The initial development efforts were primarily an offshoot of liquid membrane separation technologies. This work had led to the identification of suitable polymers which had potential for commercial gas separations in the following areas:

a) Air separation to provide a 98%+ N_2 product.
b) Hydrogen separation for purification of process streams in ammonia plants and refinery applications.
c) Helium separation from hydrocarbon gas streams.
d) CO_2 and H_2S separations from hydrocarbon gas streams.
e) General gas dehydration.

During the past 15 years all of these technologies have been commercialized but with varying degrees of success. This paper covers an area where gas separation membrane systems have been successfully deployed—the separation of CO_2 and H_2S from hydrocarbon gas streams. The membrane systems were developed and commercialized through a joint effort between The Dow Chemical Co. and The Cynara Co. The focus of this paper is on the technological improvements which have been made in this area since its initial application in 1983. While specific to CO_2/hydrocarbon separation technology, the improvements described are typical for most new technologies that evolve into accepted products and systems.

While initial research on semi-permeable membranes for hydrocarbon gas separations dates back to the early 70's, the first serious efforts to develop a product for CO_2 separation from hydrocarbon gas streams began in 1976 to 1977. By 1979, prototype products were in field test conditions. Successful operation was first obtained on dry produced gas streams and subsequently, in early 1982, on associated gas streams. By the summer of 1982, commercial size membrane devices were in field operation in a demonstration plant and by late fall of 1983, the first commercial facility was placed in operation at the SACROC Unit in West Texas, where Chevron, USA is Unit Operator.

SACROC Unit Design & Operation 1983 - 1988

Much has been published about this facility, but it is worthwhile to review some of this data in order to understand how it laid the foundation for further technology improvements.

First, The Cynara Co. approached the introduction of membrane separation technology to the Oil & Gas Industry on a service basis wherein Cynara took the technology risk and SACROC paid only for product separated. If the technology failed, SACROC had the option to remove Cynara from the site and install more conventional technology. Realizing that the introduction of new technology is of high risk, Cynara did all possible to insure that such failure would not occur. The results were a 99.5%+ availability from the date of start up through the first five years of operation.

Second, with the inlet CO_2 composition varying between 40% to 60%, it was necessary to design for flexibility of operating conditions. On top of this, inlet capacity was to vary by over 50%. This required the CO_2 separation plant be designed to bring banks of membranes on and off line automatically to keep the system in balance. Initially 8 parallel trains with up to 20 membrane modules per train were incorporated. Each train was instrumented to provide the ability of analyzing train as well as overall plant performance. It was, therefore, possible to analyze banks of modules and their performance as a function of different variables. Throughout the first 5 years of operation performance of the overall system remained within the specified range. The attached set of operating conditions show an approximate material balance with the per cent of the inlet components permeating. In most cases, the original membrane modules installed were in operation 5 years later. Capacity decline was, in other cases, sufficient to make replacement of modules economically feasible. To give an example, capacity decline for one of the plant operating trains during this period is shown.

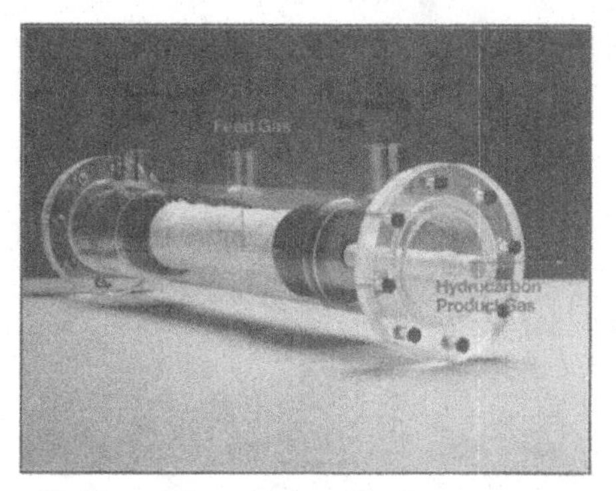

The Cynara/Dow modules are of the hollow-fibre design with double-ended tube sheets. This basic configuration was selected for gas separations in order to provide the maximum gas contact area per unit volume. The feed gas is introduced into the fibre bundle with the CO_2-rich stream permeating to the inside of the fibre and the hydrocarbon rich gas flowing to central core from where it is piped out of the module. In excess of 2 dozen process steps were involved with the manufacture of the final product. Many of these process steps were manual or subject to variable process conditions which affected the performance of the final product. Product performance characteristics were established but, within the acceptance range, performance of any given module varied significantly and it proved necessary to match modules for optimum train and plant performance. This matching was done on the basis of the module's flow characteristics. This allowed for each module in a train to operate in the same range of its performance map.

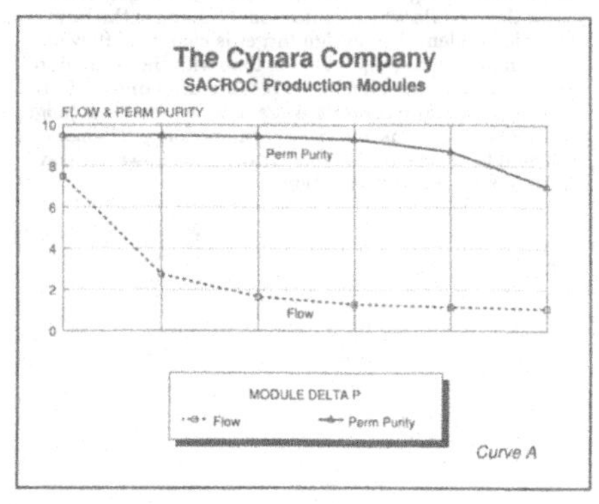

Curve A

Revised Contract and Operating Conditions:

During the five years of initial operation, oil and gas production from the SACROC Unit declined significantly, resulting in changes in the requirements of the CO_2 separation systems. As hydrocarbon capacities declined, CO_2 composition increased and it was projected that it could reach as high as 80%+. This would mean that even with the declining hydrocarbon production, overall gas handling requirements would increase. To handle these changes required an increase in the capacity of the existing CO_2 separation facility and an increase in the number of membrane stages to remove the additional CO_2.

Along with these process system modifications, the decision was made to install new membranes with improved capacity and separation capabilities. To reach the performance goals of this improved membrane separation technology, Dow R&D, Dow manufacturing, and Cynara engineering collaborated to establish product requirements and the necessary membrane manufacturing systems to meet these performance goals.

Membrane Technology Improvements:

Before undertaking the improvement of the existing semi-permeable membrane design, it was necessary to translate field performance and operating requirements to product design requirements. It must be noted that this program was designed to **improve existing technology** as opposed to developing a new product.

Further, it was important to make these product improvements, where possible, applicable to a broad range of applications, some of which were only economically marginal applications and some technically infeasible.

The final product design improvements requirements were:

1) Increased capacity capabilities per membrane device to reduce system cost.
2) Improved separation performance to minimize hydrocarbon product loss and auxiliary system requirements.
3) Improve consistency of the membrane module performance
4) Improve membrane acceptance to start ratio.
5) Maintain product reliability.
6) Increase range of product application.

Basic polymer technology establishes theoretical limits to flow and separation characteristics, but packaging concepts and manufacturing techniques control just how closely these limits may be approached. While basic polymer technology may be similar for different membrane module configurations, (i.e., spiral-wound modules and hollow-fibre modules), the module manufacturing processes are entirely different.

In the case of a Cynara/Dow hollow-fibre module configuration, overall performance is primarily governed by controlling the following variables:

1) Flux rates, the rate at which different gas components permeate through a unit area
2) Relative flux rates or separation factor, the relative rate at which different gas components permeate through the membrane.
3) Flow dynamics which affect (a) by-pass or effective flow distribution (b) pressure drop for the non-permeate gas stream and (c) fibre bore pressure drop.

To further complicate product improvement efforts, the selection of the proper range for these variables is dependent on the type of application under consideration. For instance, in applications where there is a high percentage of CO_2 in the inlet gas stream, it is often desirable to permeate a large percent of the inlet stream. In applications where there is a small amount of CO_2 in the inlet, only a small percentage of the inlet is permeated. While the difference in the flow dynamics of these two examples may be intuitively obvious, the impact on overall module design is complex and significant.

Taking all of these factors into consideration the design team selected the following factors for product design:

1) Maximum flux rates consistent with structural integrity and extended product life.
2) Maximum separation factors, again consistent with structural integrity and extended product life.
3) Two different fibre sizes, the largest for high CO_2 applications and the smaller for low CO_2 applications.
4) Two different module sizes, one for small capacity projects and one for larger capacity projects.

With these selected, we then concentrated on a manufacturing process that would give the desired consistent results and high acceptance rates.

Modern manufacturing facilities throughout the world are designed to essentially meet these same goals: high performance, reliability, consistency, and high acceptance rates. Likewise, the basic steps to obtain these goals are also widely accepted: careful control of feed stock quality, accurate control of process variables, and minimization of the required number of manufacturing steps. The first step must include the cooperation of the vendors who provide raw materials used in the manufacturing process. The second step is established by automated control of all critical manufacturing steps. The third step is established by careful systems analysis of the manufacturing process, material flow, and available manufacturing technologies.

Basic flux rates are controlled by polymer and polymer morphology, discriminating layer thickness, and support structure diffusion capabilities. Theoretical models and laboratory and field testing were used to establish the target configuration which would provide maximum flux rates consistent with reliability and extended operating life while operating in real field conditions. Once this target configuration was established, the manufacturing process had to be designed to consistently and reliably produce the desired results. While basic fibre flux rates and separation factors are established in the early stages of module manufacturing, virtually every manufacturing step has the potential to alter these properties. To the extent that consistency can be obtained throughout the manufacturing process, acceptance rate is high and the number of modules required for a particular application can be minimized. Since the start up of our new manufacturing facility over 400 modules of one specific design have been produced. By reviewing Curve B, it is possible to evaluate just how successful we were in meeting our target. We were able to increase our average flux rate over our original module production by greater than 5%.

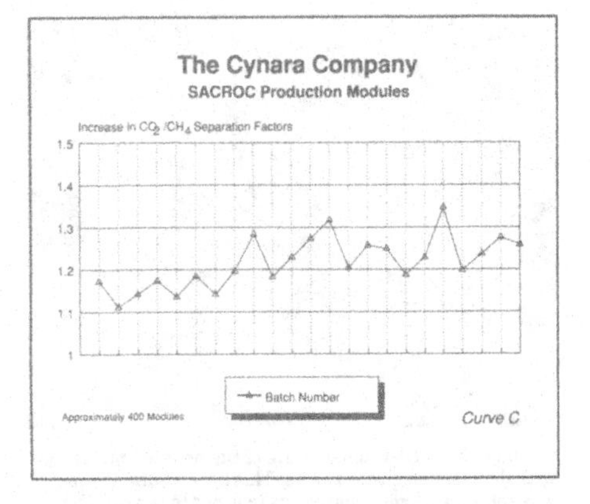

The story for relative flux rates or separation factors is much the same. The target configuration was established concurrently with the overall flux using the same evaluation procedures. Here again, the results were equally encouraging as can be seen on Curve C. Average separation factor increased by over 20%.

The design requirements for flow dynamics are somewhat different than for flux and separation factors. Those familiar with catalyst bed configurations have a reasonable understanding of a hollow fibre bundle's feed to reject flow dynamics. Permeate flow dynamics, flow down the fibre bore, is closely related to laminar pipe flow. In addition, inlet composition changes at varying depths as the gas flows across the module which continuously changes the flow dynamic problem. The design target is even feed flow distribution to provide proper contact time with the entire fibre surface and minimum intrinsic pressure drop in both fibre bed design and fibre bore. Here again, we were able to judge our success by evaluating theoretical module performance vs. actual performance. The attached curve shows the consistency obtained and deviation.

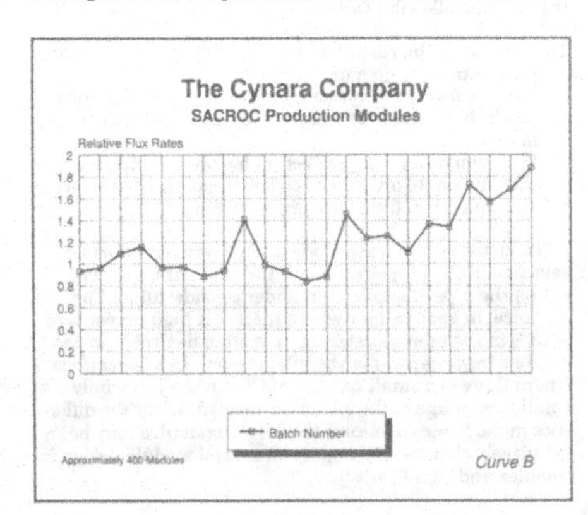

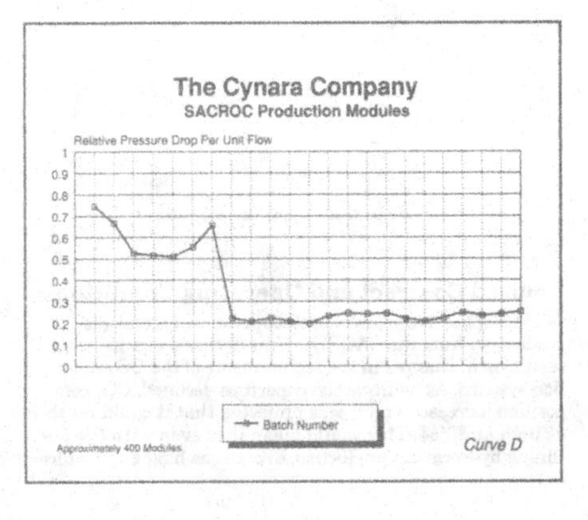

It should be noted that prior to shipment all modules are tested for mechanical integrity and performance characteristics. If these modules do not pass design criteria they are either reworked where possible or rejected. This provides the final and all encompassing evaluation of the success of our new manufacturing facility acceptance rates. In this instance, we are extremely proud of our results which substantially exceed our target. At the end of the production run for this particular module configuration our acceptance rates were consistently greater than 95%.

The final test, of course, is provided by just how well these modules perform in actual field applications.

SACROC, 1988 - 1991:

In late September of 1988, the SACROC CO_2 Separation Plant was down for a period of two weeks to make the necessary modifications to handle the revised operating conditions. The plant was restarted in early October with the first series of new modules installed. Over the succeeding 5-month period, all of the modules in the plant were changed.

Since the start-up of the modified facility, the Plant has continued to operate at over 99% availability, separate CO_2 at greater than contract removal rates, and remained within the product quality specifications. In total, this CO_2 separation facility has operated for over 7 years with continued performance of this quality, separating over 55 BCF of CO_2, and one major membrane change.

Did the new membranes perform as anticipated and have they maintained their performance? Under the modified Plant configuration, the membranes are installed in a total of 12 individual operating trains again having the capability of performance analysis on each of these trains. The following graphs of flow rates, composition, percent of inlet permeated, and product purities are recorded from the operating log. They are for the train that contains the first set of modules to be replaced at the time of the modification. While data points show typical scatter, the overall trend is remarkably consistent. All of the other 11 operating trains show the same consistency and this has been reflected in overall plant performance. In addition, with the exception of some performance enhancement experimentation in one train, all modules changed during the period of modification continue to operate.

Another measure of module performance is the specific rate of CO_2 separation or the contract CO_2 removal rate per unit volume of inlet gas. After all of the new modules were in place, this rate of CO_2 removal has exceeded the contract rate by an average of approximately 10%. When exact performance matches occur (i.e., capacity and composition matching surface area installed), monthly average CO_2 removal rates have reached as high as 20% in excess of contract.

The data which we have been able to collect indicates that the improvements in module manufacturing have led to

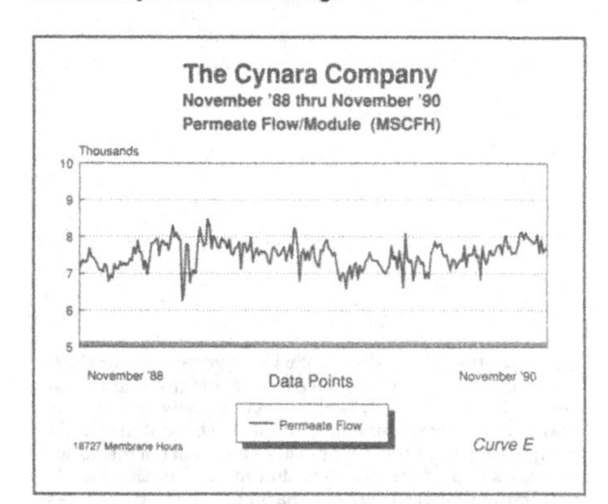

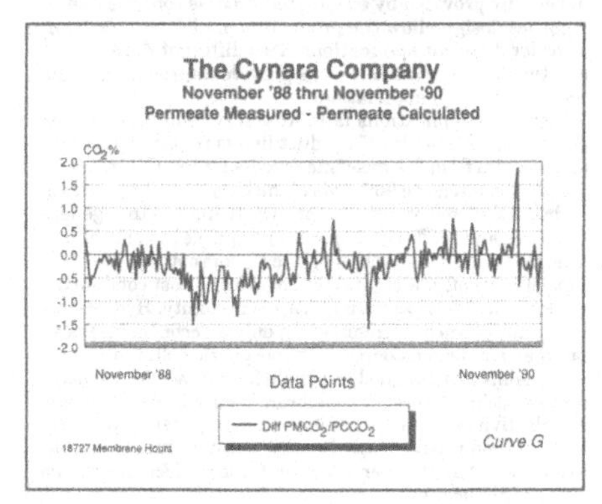

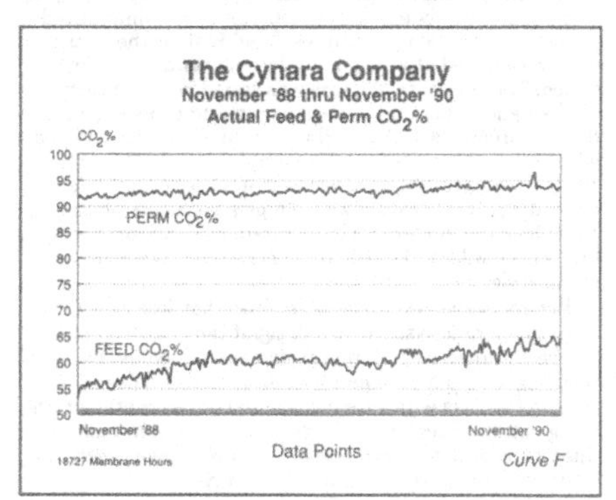

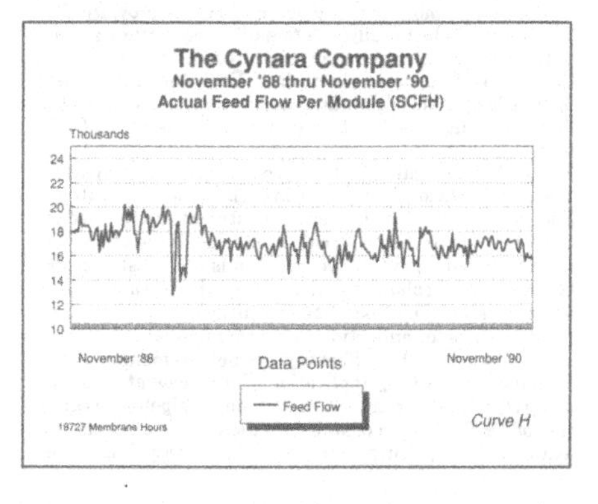

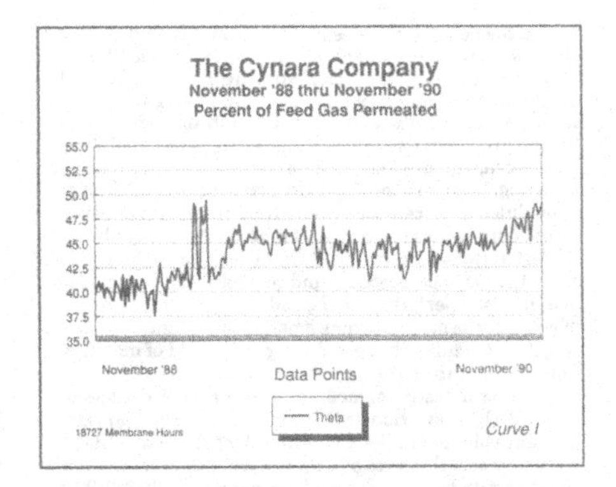

Curve I

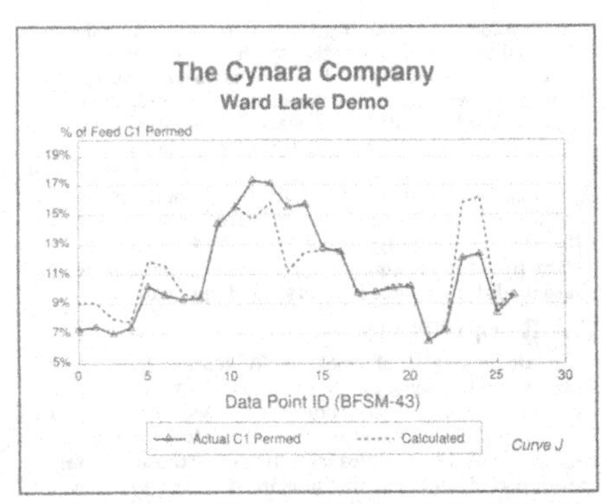

Curve J

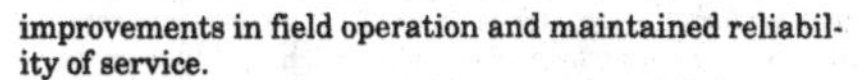
improvements in field operation and maintained reliability of service.

Low CO_2 Applications:

Consistent improved module performance was but one of the benefits of our new manufacturing facility. Additional benefits provided by automated process controls and facilities design allow the production of different configurations for different applications. Two different fibre sizes and two different module sizes provide improved performance for low CO_2 applications.

Low CO_2 applications fall into a service or product category where a "standard" product line is required. Amine technologies form the base line process for low CO_2 removal and they can remove 90% plus of the CO_2 in the inlet stream, hydrocarbon consumption is primarily limited to regeneration fuel, and capital and operating expenses are not excessive. For membrane technology to be competitive in this type of market, low capital and operating cost combined with minimal hydrocarbon loss are a necessity. Hydrocarbon loss to the permeate stream is not only an economic penalty but may also be considered an atmospheric pollutant.

A membrane designed to handle high flow rates reduces system packaging cost. A membrane with consistent improved selectivity and flow distribution is required to reduced hydrocarbon product loss. Finally, the membrane module needs to withstand operating conditions typical for the majority of applications.

The new membrane manufacturing facility provided the consistent product quality. A "small" fibre configuration accompanied by an increase in tube sheet diameter provided the increased surface area desired. Each one of these new devices is capable of handling approximately 10 times the flow of the initial devices that were installed in the SACROC unit.

Just how critical is hydrocarbon loss to the permeate? The answer obviously depends on each specific application. For instance, if all of the permeate stream can be used as fuel, then the permeate stream is of value. If the permeate must be flared, then the hydrocarbon is truly lost and may be considered a pollution problem. Also, there is a wide variation in value of hydrocarbons permeated, ranging from pipeline sales to reinjection to "no commercial value."

To understand the effects of commercial membrane performance on low CO_2 applications, let us look at an inlet gas stream that contains 10% CO_2 and a pipeline specification of 2% CO2. With commercial membrane systems available today, a single stage system (no recycling of the

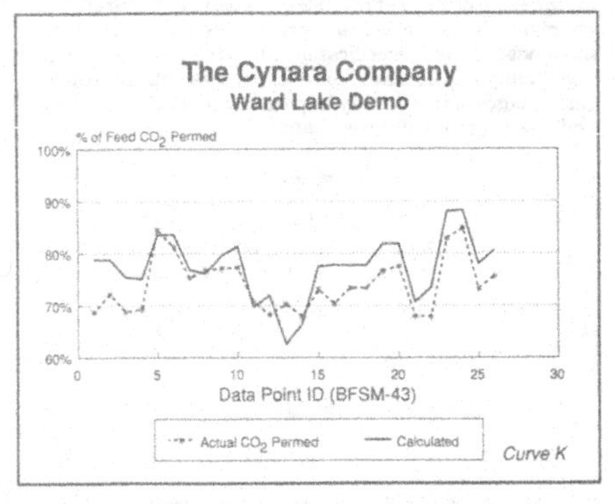

Curve K

permeate stream) to deliver a 2% CO_2 hydrocarbon product would permeate from between 8% to 15% of the inlet hydrocarbon. This reasonably broad range not only reflects the quality of the membrane modules available but also different operating conditions. Our goal as a supplier of membrane systems is to produce a product that minimizes this loss. In installations where there is a market for hydrocarbon gas, this loss is usually the largest contribution to annualized membrane system cost. Curves J and K show the results of data collected at a field installation over a two month period. The permeate pressure was varied to provide actual performance wherein the permeate might be used for site fuel requirements. Calculated vs. actual performance varied within the accuracy of the gas samples taken. Throughout the operating range, however, performance consistency existed. While operating with a 5 psig permeate pressure, a single-stage membrane system removed approximately 80% of the inlet CO_2 with 10% of the inlet hydrocarbon permeating.

For applications of this type, hydrocarbon loss in a single-stage membrane system are still significantly greater than can be obtained with Amine systems which would be in the range of 1% of the inlet hydrocarbon gas. This will remain the case with existing commercial membrane technology and although there is great room for improvement in this technology, membrane separation will have a very difficult time approaching this 1% figure.

With hydrocarbon losses in the 8% to 15% range, a permeate hydrocarbon recovery unit (PHRU) can provide an economical method of reducing overall hydrocarbon losses to levels equal to or less than those experienced with Amine systems. The better the performance of the primary CO_2 system the smaller the size of the permeate hydrocarbon recovery unit (PHRU). Permeate hydrocarbon recovery units (PHRU's) are composed of one to two additional stages of membrane separation and recycle compression. For the conditions shown in this example and depending on the efficiency of the primary membrane separation stage and on the amount of hydrocarbon in the final vent, the compression requirements will vary in a range of 50 to 100 BHP/MM. Since a PHRU incorporates compression, the availability of this secondary system is less than that anticipated for the primary separation system. However, maintenance on this secondary system can be performed separately without affecting the primary system product purities or availability. On land-based installations wherein this gas can be sold, these systems will almost invariably provide an adequate return on investment and operating cost. Further, PHRU's can reduce hydrocarbon emissions by a factor of 10 to 20 times.

Can a primary membrane system with a permeate hydrocarbon recovery unit (PHRU) be competitive with an Amine system? A qualified yes. Much has been printed on this subject. Combinations of membrane systems and Amine units have been considered. From our review of land-based systems, the indication is that with the improved performance we obtain from membranes manufactured in our new facility, membranes can indeed be competitive. It is not however, an overwhelming difference and site specific requirements, capacities, and operating preference all affect the final process choice.

CO_2/Hydrocarbon Liquids Membrane Separations:

In the processing of hydrocarbon gas streams for NGL products, CO_2 that has not been removed from the plant inlet stream concentrates in different areas of the NGL recovery process. Commonly, this can occur in the de-methanizer bottoms and the de-ethanizer overhead or reflux streams. Further, CO_2 in these areas of the process can limit hydrocarbon liquid recovery efficiencies.

During the 3rd quarter of 1990, in cooperation with a major oil company, Cynara conducted a successful demonstration of the separation of CO_2 from de-methanizer bottoms. The inlet hydrocarbon liquid stream contained approximately 6.4% CO_2 and this was reduced to 3% in the hydrocarbon product stream. Feed pressure was approximately 1000 psig and permeate pressure varied from between 100 to 200 psig. The demonstration took place over a 30-day period and the membrane showed essentially stable performance. Several process upsets occurred in the NGL facility during the duration of the test and the membrane system returned to prior performance characteristics after each of these occurrences. For the purpose of the demonstration, the permeated gas was flared, but it is envisioned that in a full capacity facility this gas would be added to the plant fuel system.

In addition to the stability of the membrane system, the separation performance exceeded modeled figures. The following table shows the percent of the inlet liquid stream that permeated:

Component Permeated	Percent of Inlet
CO_2	64.88
C2	1.57
C3	0.11
IC4	0.01
NC4	0.04
IC5	0.00
NC5	0.00

As a result of this successful demonstration, a full capacity facility is in the process of being installed at the NGL facility and should be on stream during the month of April, 1991.

Conclusion:

While the selection of proper membrane materials and configurations are indeed important and set the basic limitations of a membrane system, the actual manufacturing process is essential to insure that the finished product consistently meets the potential of the membrane material. This, along with conservative system design, insures long and reliable membrane operating life. Neither membrane manufacturing technology or system design fall into the "glamor" category of membrane material morphology or design concepts. However, the success of a membrane installation depends to a large extent on their proper execution.

This paper is reproduced with the kind permission of GASTECH.

NITROGEN PRODUCTION BY MEANS OF MEMBRANE TECHNOLOGY FOR SAFETY ENHANCEMENT IN MARINE AND ON-SHORE PETROLEUM AND NATURAL GAS FACILITIES

JOHN THURLEY, CHAIRMAN
RIMER-ALCO INTERNATIONAL LIMITED
CARDIFF, UK

FAZJULLA G GAINULLIN, GENERAL DIRECTOR
VNPO SOJUSPROMGAS
MOSCOW, USSR

VALERY A IRKLEY, DIRECTOR
ALL-UNION SCIENCE-RESEARCH INSTITUTE
OF POLYMER FIBRES, VNIIPV
KIEV, USSR

SUMMARY

Inert gas is required for purging of equipment and for blanketing of potentially dangerous materials and processes. Initially combustion fired inert gas generators were used in process plant and on ocean tanker ships, but acidic exhaust gases produced in these systems caused severe corrosion problems.

Nitrogen is now used for many of these purging and blanketing applications due to the benefits of clean dry nitrogen. Most nitrogen and oxygen is produced by cryogenic air separation techniques in large industrial plants. Distribution is then by pipeline, road or rail tankers or by cylinder. This makes the nitrogen very expensive in remote locations and off-shore.

Effective Industrial Membrane Processes — Benefits and Opportunities, pp.345–380

On-site cryogenic air separation to reduce transport costs can only be considered for very large outputs due to the high capital cost of smaller equipment and cannot therefore be considered for on-site nitrogen generation in most industrial applications.

However the raw material for manufacture of nitrogen and oxygen is air, which is freely available everywhere. This makes the on-site generation of nitrogen by air separation well worth consideration.

Pressure Swing Adsorption (PSA) and Membrane processes offer the operators the ability to produce their own clean, dry nitrogen at low cost.

In the last 10 years many offshore platforms have installed PSA Nitrogen units for purging, well stimulation and gas lift.

More recently membrane air separation units for nitrogen generation have been installed on offshore platforms and on board ship due to their simplicity of operation and their lightness in weight. On-shore the number of membrane Installations have been limited due to their high initial cost, even though the membranes have been proven to be reliable in operation. Recent developments and increased production, have now reduced the cost of air separation membranes for nitrogen generation. This is particularly so with Soviet hollow fibre air separation membranes.

The Soviet Chemical, Oil Refining and Gas Industries have experience in the development and manufacture of flat-plate, spiral-wound and hollow fibre membranes covering a period of approximately 10 years.

After the successful operation of Hollow Fibre Nitrogen Membranes in the USSR Process Plants and on board Ships. Arrangements have been made for VNIIVPROEKT and VNPO Sojuspromgas to collaborate with Rimer-Alco in the packaging and skid-mounting of these Soviet Membranes on an international basis. Hollow fibre membranes have been selected due to their high packing density, approximately 20,000 square metres of surface area for mass exchange/cubic metre

A historical background is given on the international development of gas separation membrane processes. Information is included on hollow fibre membrane technology and comparisons made with other types of membrane gas separation processes.

INTRODUCTION - ONSITE NITROGEN GENERATION.

Many purging and blanketing applications do not require pure nitrogen, as inert gas with an oxygen content not greater than 5% or in some cases 3% is quite acceptable for adequate safety. This enabled other air separation processes to be considered for on-site nitrogen generation as an alternative to expensive delivered high purity nitrogen from remote cryogenic plants.

During the last decade Pressure Swing Adsorption (PSA) Nitrogen Generators were installed in many Offshore and Onshore Hydrocarbon Installations, due to the savings offered over delivered nitrogen. A packaged skid-mounted PSA Nitrogen Generator is shown in the Rimer-Alco Cardiff UK Works, prior to shipment to the Middle East (Fig 1).

FIG. 1. PSA NITROGEN UNIT IN RIMER-ALCO WORKS.

Recently Membrane Nitrogen Generators have been installed in Marine and Offshore Locations due to their simplicity of operation and their lightness in weight. A packaged Nitrogen Membrane Unit, incorporating Dow Generon Hollow Fibre Membrane Modules is shown in the Rimer-Alco Cardiff Works, prior to shipment to a North Sea Floating Platform (Fig 2). Another typical packaged Nitrogen Membrane unit, incorporating A/G Technology Hollow Fibre Membrane Modules is shown in the Rimer-Alco Cardiff Works, prior to shipment to the Norwegian Offshore Oilfields (Figs 3 & 4).

Nitrogen purities above 99% are possible with membranes, but lower purities greatly reduce membrane size and cost. Nitrogen quantity produced by a membrane module increases with the oxygen content permitted in the nitrogen. So although higher purity nitrogen can be achieved, 5% maximum oxygen content is often specified, for low pressure blanketing or purging applications, as satisfying both the safety requirements and minimum equipment cost. The graph (Fig 5) shows the variation in amount of nitrogen produced with both permitted oxygen content and operating pressure, for a medium size individual membrane module producing approximately 10 cubic metres/hour of nitrogen.

MEMBRANE HISTORY

Membrane separation was first proposed over 100 years ago. Active laboratory research in the 1960 was mainly due to the fact membrane separation technology is potentially energy efficient.

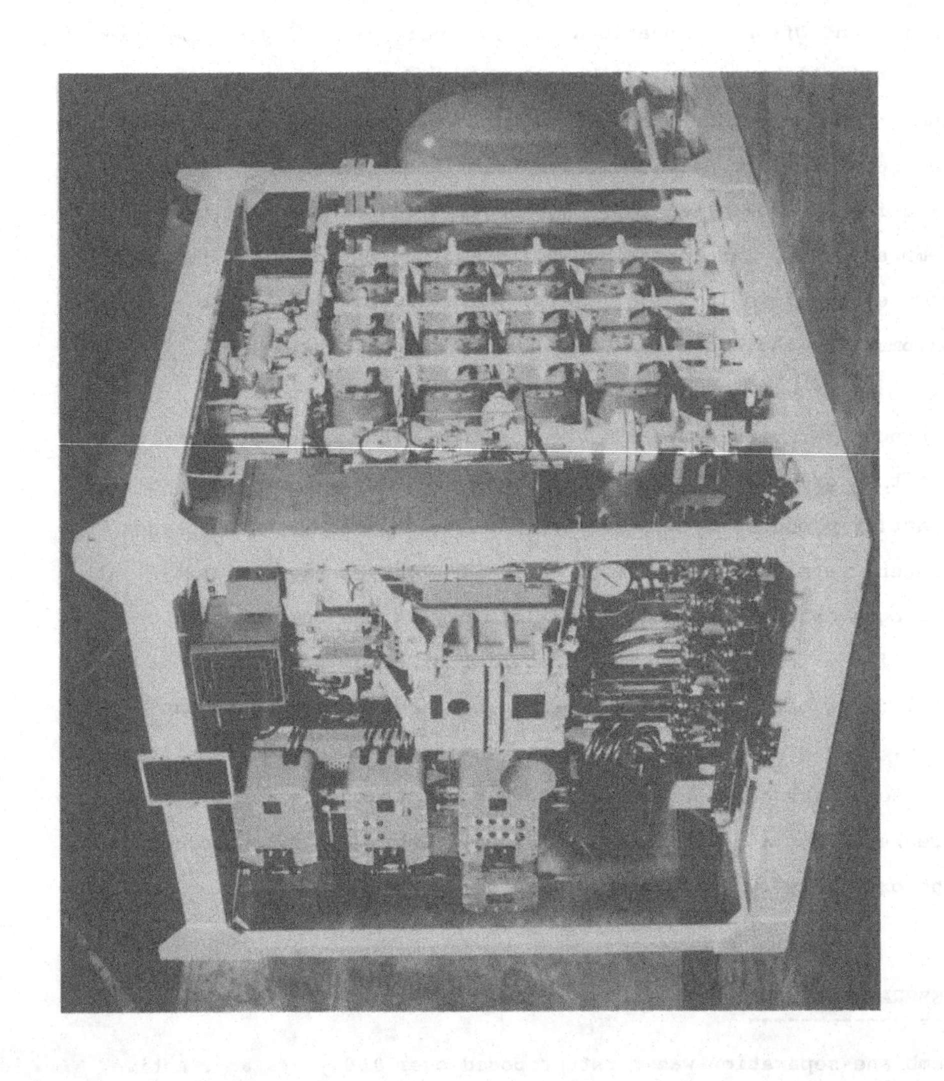

FIG. 2. PACKAGED MEMBRANE NITROGEN UNIT.

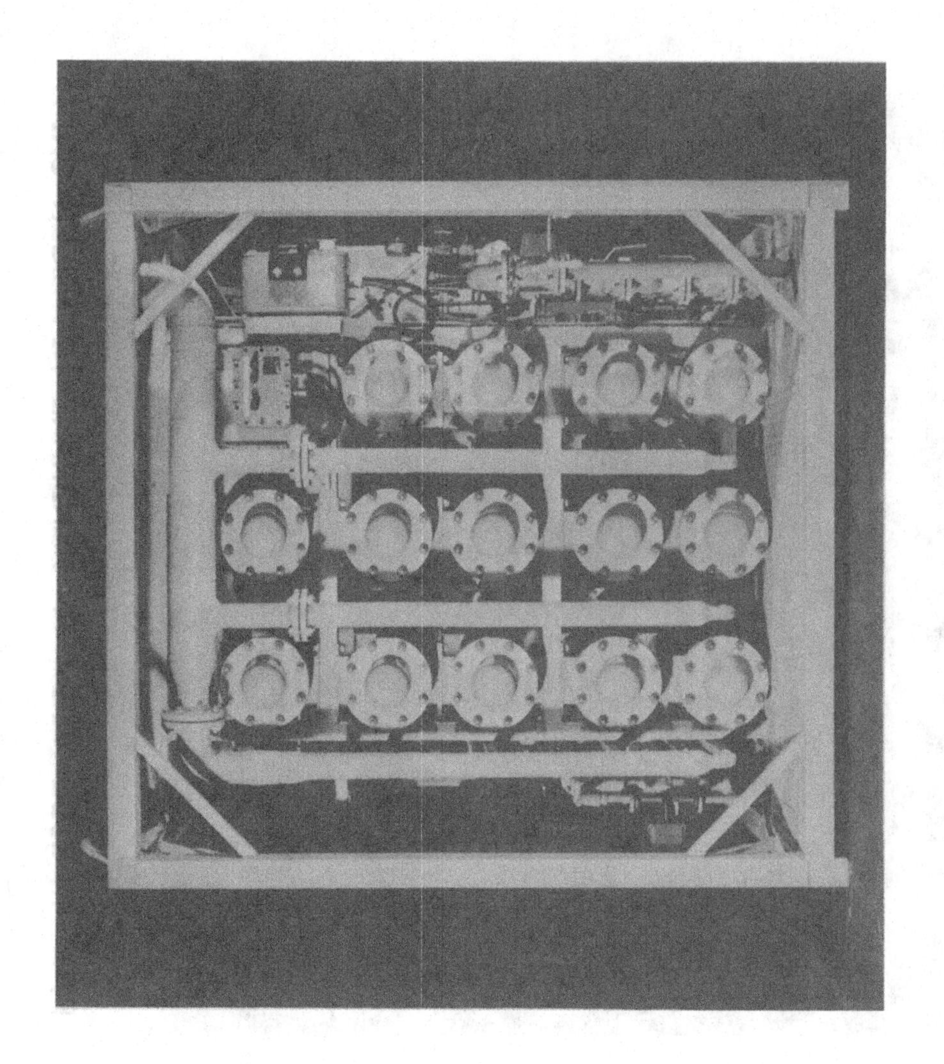

FIG. 3. END VIEW OF MEMBRANE NITROGEN UNIT.

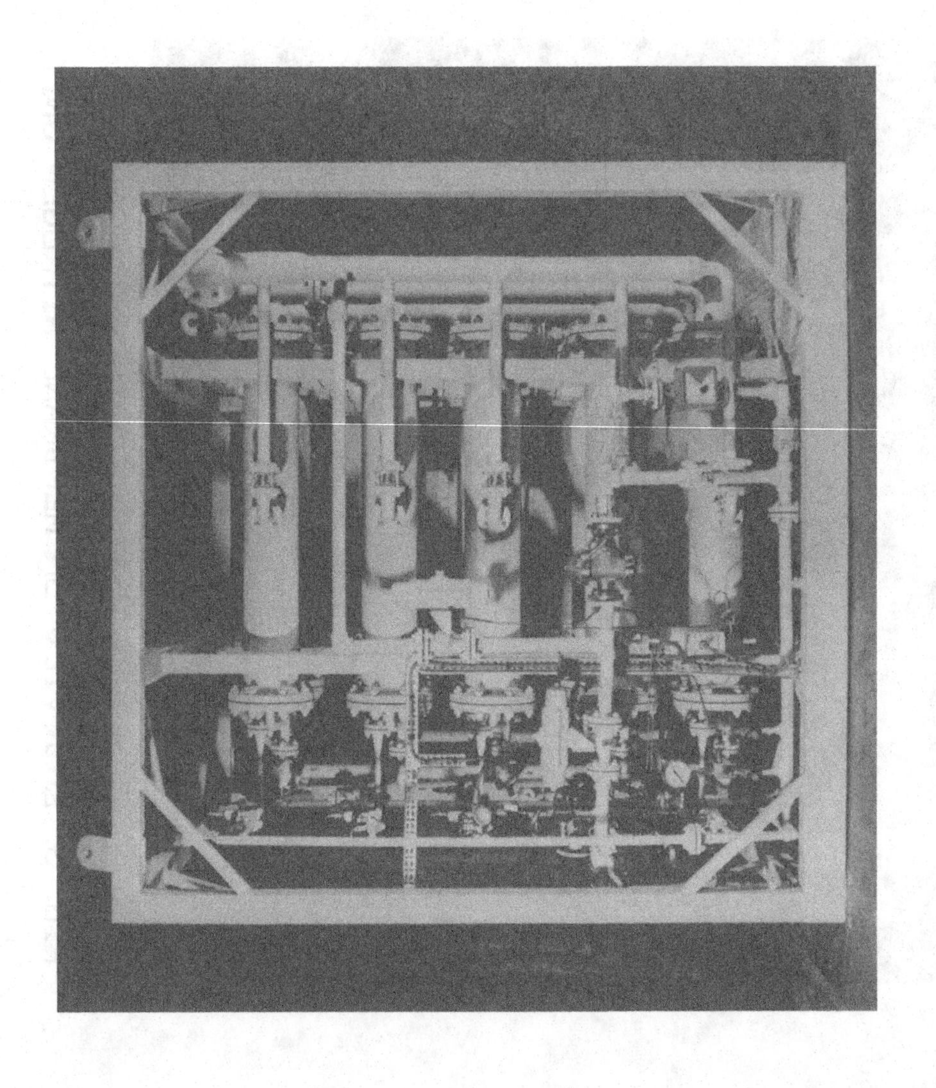

FIG. 4. SIDE VIEW OF MEMBRANE NITROGEN UNIT.

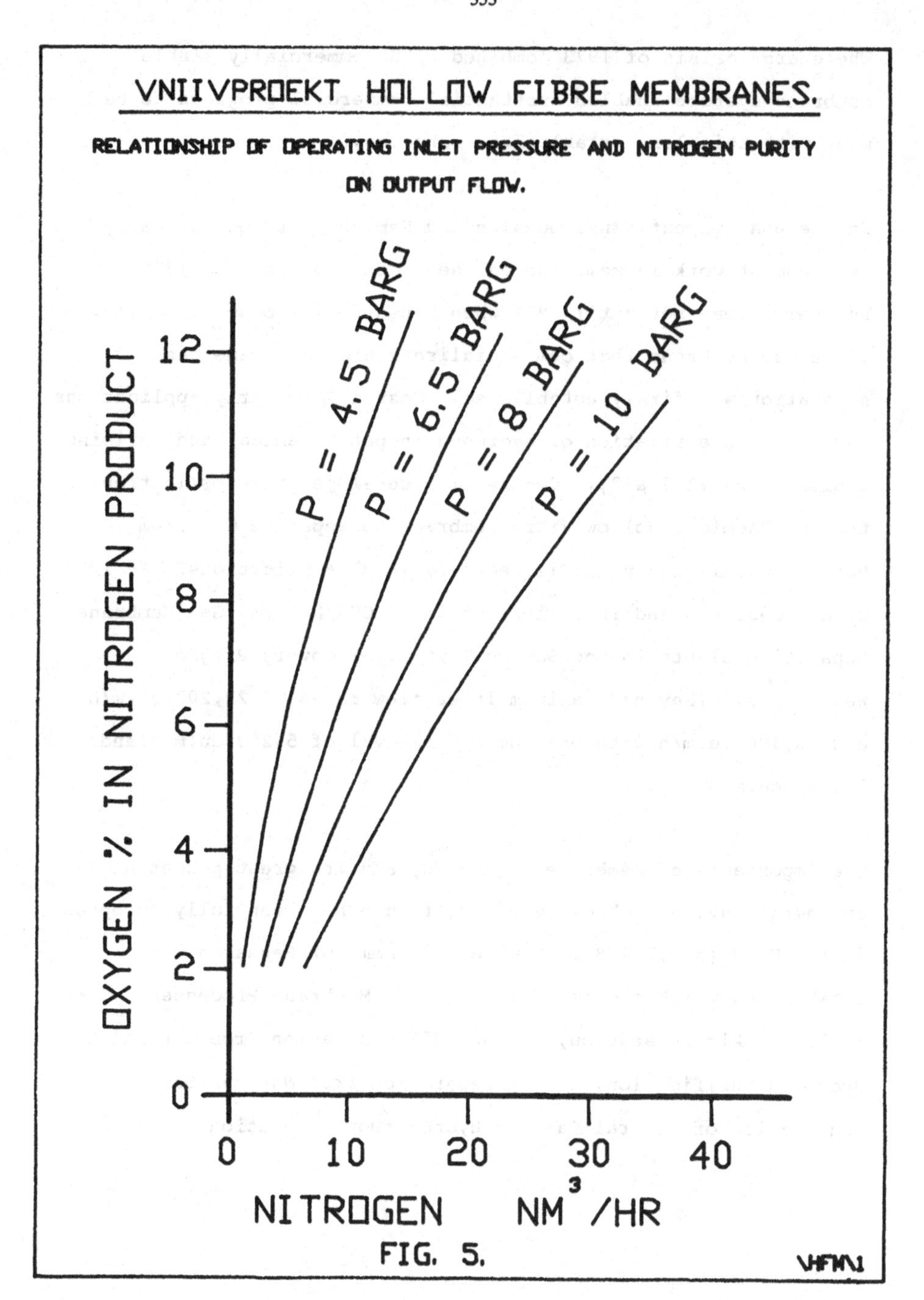
VNIIVPROEKT HOLLOW FIBRE MEMBRANES.
RELATIONSHIP OF OPERATING INLET PRESSURE AND NITROGEN PURITY ON OUTPUT FLOW.
P = 4.5 BARG
P = 6.5 BARG
P = 8 BARG
P = 10 BARG
OXYGEN % IN NITROGEN PRODUCT
0
2
4
6
8
10
12
0
10
20
30
40
NITROGEN NM3 /HR
\HFM\1

FIG. 5.

The energy crisis of 1973 combined with commercially viable membrane systems enabled substantial commercial progress to be made with this long established technology.

In the USA, Dupont, Envirogenics and Separex carried out early development work in membrane gas separation during the 1970's. However it was not until 1979 when Monsanto introduced a hollow fibre gas membrane that commercialisation of membrane gas separation was firmly established. Most of the early applications were for the separation of hydrogen in petrochemical and refining applications (1,2 & 3). Cynara also developed the application of the Dow Chemical Hollow Fibre membrane to separate C02 from Natural Gas in Enhanced Oil Recovery (E0R) applications. In 1983 Cynara designed and installed two large CO2/Natural Gas Membrane Separation plants in the Sacroc Tertiary Recovery Project, West Texas. They had maximum inlet flow rates of 22,200 cu.m/h and 76,400 cu.m/h with maximum CO2 removal of 5,250 cu.m/h and 22,300 cu.m/h (4).

The importance of membrane technology and its great potential as an energy-saving technology of the future has been fully accepted in the USSR (5,6,7 & 8). A wide programme of Research was established which created Membranes and Membrane Processes. These included Air Separation, CO2 and H2S Separation from Natural Gas, Hydrogen Purification, Helium Separation from Natural Gas, Dehydration of Natural Gas and Hydrocarbon Separation.

Soviet membranes have been widely used in Air Separation for generation of Nitrogen for purging and safety protection in the transport and storage of petroleum and other hazardous and sensitive products. Another important application in the USSR is for food storage (fruit, vegetables and seed storage). Initially Soviet membranes were of the plate and spiral-wound types.

Soviet Air Separation Membranes have also been extensively used for the production of oxygen enriched air up to concentrations of 35% and capacities of 1,500 cu.m/h. Two of those plants were installed in 1979 at a chemical enterprise in the USSR (5). Another major area for membrane oxygen enriched air up to 40 per cent oxygen concentration, has been in medicine for respiratory purposes. Other important applications have been in fish farms and in biological purification of city sewage and plant effluent.

The three manufacturers of Air Separation Membranes in the USSR are :-

(a) Kryogenmash - Plate type

(b) Polymersintez - Spiral Wound type

(c) VNIIVPROEKT, Khimvolokno - Hollow Fibre type

Since 1984 in the USSR, The Kiev Branch of VNIIVPROEKT Scientific-Production Association (SPA) "KHIMVOLOKNO" have carried out trial installations of Hollow Fibre Air Separation Membranes for creation of Safety and Inert Environments on Marine and Onshore applications. These are listed in Fig 6.

KIEV BRANCH VNIIVPROEKT NPO "KHIMVOLOKNO"

HOLLOW FIBRE MEMBRANE NITROGEN UNITS IN OPERATION

APPLICATIONS	YEAR UNITS COMMISSIONED	NITROGEN FLOW Cu.m/h
Creation of controlled gas medium for storage of fruit and vegetables.	1984	5
	1986	7
Nitrogen purity 96%. Operating pressure 5 bar.	1987	20
	1989	25
	1990	25
Creation of protective medium in holds of ships containing flammable cargo.	1988	30
	1989	30
Nitrogen purity 94-96%. Operating pressure 6 bar.		
Creation of protective medium in flame hazardous process equipment.	1987	20
	1990	20
Nitrogen purity 94-96% Operating pressure 6 bar.	1990	40
Creation of protective medium for purging of natural gas lines and compressors.	1990	5
Nitrogen purity 93%. Operating pressure 4 bar.		
Medical unit for creation of "artificial" mountain climate.	1987	20
Nitrogen purity 85%. Operating pressure 3 bar.		

FIG. 6.

MEMBRANE TECHNOLOGY

(a) GAS SEPARATION MECHANISM

Gases permeate through nonporous polymer membranes by a SOLUTION - DIFFUSION MECHANISM (9). Application of gas pressure on one interphase of a membrane results in :-

(a) Solution (adsorption) of the gas into the membrane at the interphase.

(b) Molecular diffusion of the gas in and through the membrane.

(c) Release of the gas from solution (desorption) at the other interphase.

PERMEATION is the overall mass transfer of the penetrant gas across the membrane. DIFFUSION is the movement of gas molecules only inside the membrane.

The adsorption and desorption are generally so fast that solution equilibrium is achieved at the membrane interphases. Molecular diffusion is much slower and therefore controls the rate of permeation. A membrane concentration profile is shown in Fig 7 Fick's Laws predict the flow rate (J), whilst Henry's Law governs the concentration (c) of the gas. The rate of Permeation under steady state conditions (Js) is equal to $\bar{P}(p_h - p_l)/\delta$ where $\bar{P}$ is the permeability, p_h & p_l are the penetrant pressures at the two interphases and δ is the thickness of the membrane.

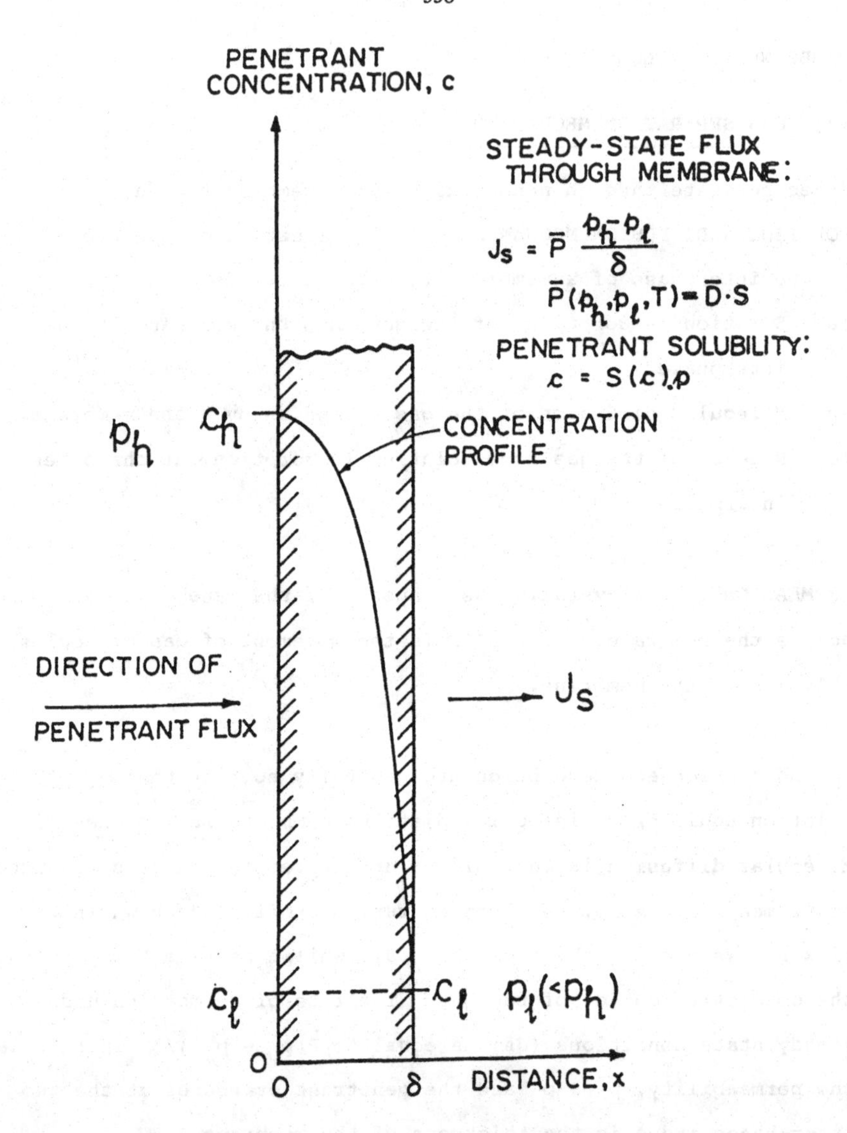

FIG. 7. GAS/AIR PERMEATION.

In the case of a membrane gas separation process where p_h is higher than p_ℓ, the Permeability $\bar{P}$ is equal to the product of the mean Diffusion coefficient ($\bar{D}$) and the Solubility coefficient (S_h), where :-

$$\bar{D} = \frac{\int_{c_\ell}^{c_h} D(c)dc}{(c_h - c_\ell)}$$

and $S_h = c_h / p_h$, c_h & c_ℓ being the equilibrium concentrations of the penetrants dissolved at the two membrane interphases when penetrant pressures at these interphases are p_h & p_ℓ respectively and $D(c)$ is the mutual diffusion coefficient for the penetrant and polymer membrane.

(b) MEMBRANE MATERIALS

The Oxygen Permeability and Separation Factor of Oxygen and Nitrogen for various membrane materials is shown in Fig 8. The most suitable membranes are those nearest to the upper right hand corner of the diagram (i.e. those with both high permeability and high separation factors). It should be noted that membranes that are highly gas selective exhibit a low permeability. Generally membranes with high permeability are in the rubbery state and membranes with high selectivity are in the glassy state.

Many membrane materials have been developed for packaging rather than gas separation so it is possible that they can in future be synthesised with both high selectivity and high permeation.
Use of additives with the polymer can also improve permeability of membrane materials. Resistance of the membrane against deteriation with the gas being handled is also important.

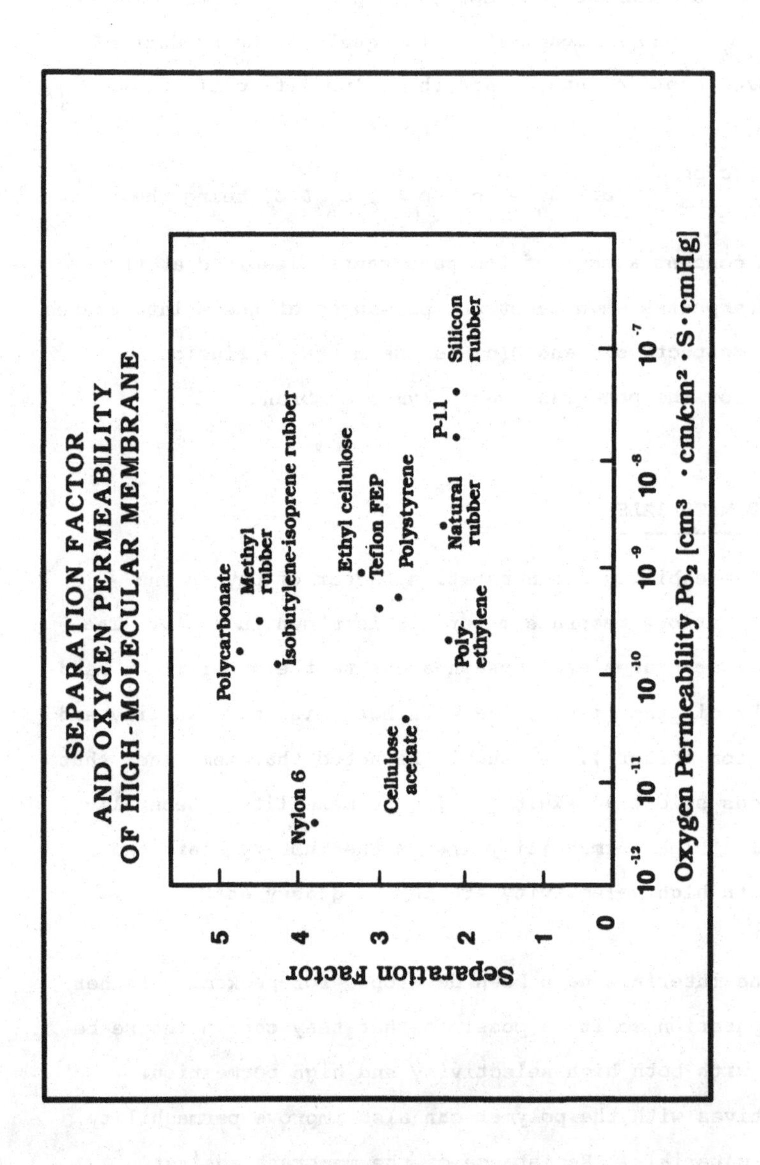
SEPARATION FACTOR
AND OXYGEN PERMEABILITY
OF HIGH-MOLECULAR MEMBRANE
Separation Factor
0
1
2
3
4
5
Nylon 6
Cellulose acetate
Polycarbonate
Isobutylene-isoprene rubber
Methyl rubber
Poly-ethylene
Polystyrene
Teflon FEP
Ethyl cellulose
Natural rubber
P-11
Silicon rubber
10⁻¹²
10⁻¹¹
10⁻¹⁰
10⁻⁹
10⁻⁸
10⁻⁷
Oxygen Permeability Po₂ [cm³ · cm/cm² · S · cmHg]

FIG. 8.

Minimum membrane thickness is of great importance in the design of membrane units, as gas permeation is dependent on the thickness of the membrane. One method of achieving nonporous thin film membranes that exhibit both high permeability and high selectivity is by the use of asymmetric membranes. Three different types are illustrated in :-

Fig 9 - ASYMMETRIC MEMBRANES (9)

(i) SKINNED MEMBRANE (Fig 9a) This consists of a microporous layer (100 - 200μm thickness) one side of which is covered with a very thin (0.1 - 1.0μm) nonporous skin. The separation of the gases occurs in the skin whilst the microporous substrate provides the membrane with mechanical strength. This gives very high permeability due to the thin membrane separation layer. The selectivity can be high, particularly if the membrane is in the glassy state. Newer developments in hollow fibres with graded density skins give dramatic increases in permeation rates with maintenance of selectivity (10).

(ii) THIN FILM LAMINATE (Fig 9b) This consists of an ultra thin (0.05 - 0.1μm) nonporous film laminated to a much thicker (100 - 200μm) microporous backing. Gas Separation takes place only in the ultrathin nonporous film to achieve a very high permeability. The operation is similar to that of the skinned membrane. The difference is that in this case the ultrathin film and its backing are of different materials whereas the skin of a skinned membrane is an integral part of its substrate.

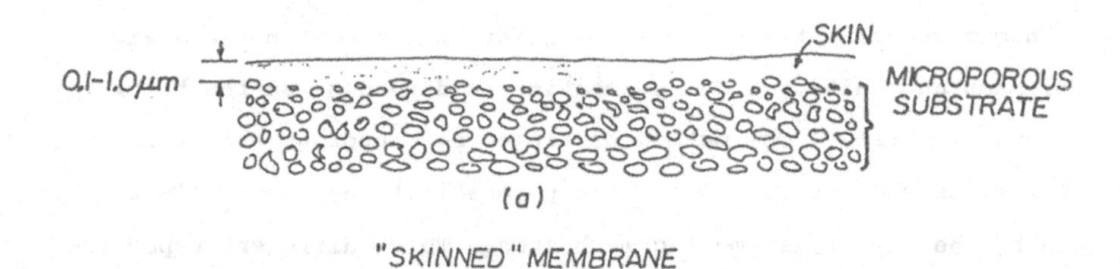

(a)

"SKINNED" MEMBRANE

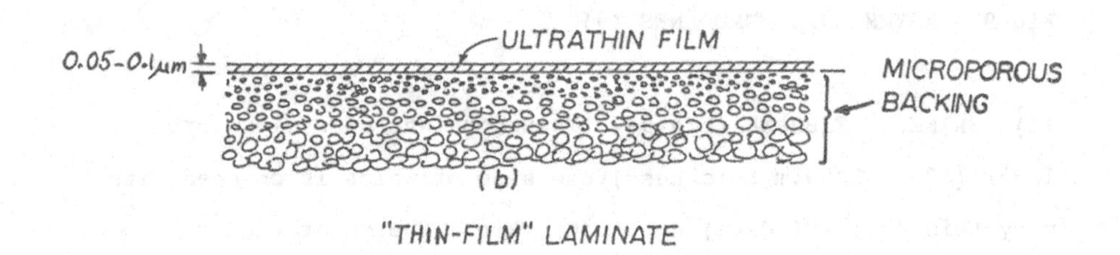

(b)

"THIN-FILM" LAMINATE

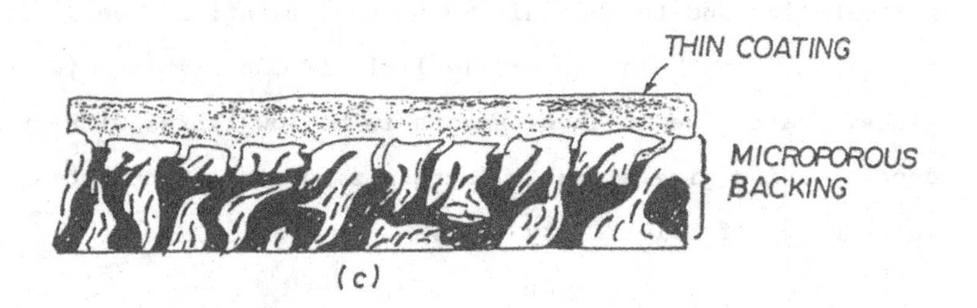

(c)

COATED-FILM COMPOSITE

FIG. 9. ASYMMETRIC MEMBRANES.

(iii) COATED FILM COMPOSITE (Fig 9c). In this case the nonporous thin film is coated on the surface of a suitable microporous backing. This coated film differs from a thin film laminate in that the coating penetrates into the pores of its backing material. An important advantage of coated film composites is that these can be prepared, so that the separation of gas mixtures permeating through the membrane occurs in the coated film or in its backing (i.e. a skinned membrane). An example of this is the Monsanto Prism Membrane for separation of hydrogen. These are skinned asymmetric hollow fibres of polysulfone coated with a thin film of silicone rubber (1μm thick). The separative membrane is the polysulfone skin (o.1μm thick) and the silicone rubber coating is to plug any micropores or defects in the skin. Other similar composite hollow fibres of this type are Polysulphone/Polyethylene, Poly(acrylonitrile)/Silicone Rubber or Poly(acrylnitrile)/Polyethylene.

The nonporous skin can be located on either the outer or inner diameter of the hollow fibre.

(c) MEMBRANE MODULE

In the design of membranes, three types have been used these are Plates, Spiral-Wound and Hollow Fibre. The selection of the type of membrane module depends on the application.

The factors to be considered are:-

Required maximum size and weight
Capital and Operating Cost
Required Pressure
Suitability for membrane material selected

Packing density (i.e. amount of membrane surface area packed into a given volume) is of great importance as illustrated by following approximate figures:-

Plate	-	300 cubic metres/square metre
Spiral-Wound	-	700 cubic metres/square metre
Hollow Fibre	-	20,000 cubic metres/square metre

However the thickness of the membrane film and the permeability of the material selected can have an important bearing on the amount of membrane area required and hence the compactness of any type of module.

For higher pressure operation and for low weight and compactness, hollow fibre membranes justify consideration if the hollow fibre membranes are manufactured with thin films of high permeability. This makes them the preferred design for marine, offshore platforms and large scale onshore applications.

(i) **PLATE MEMBRANE.** This module configuration has been used in the USSR and Japan for generation of oxygen enriched air due to the requirement for low pressure operation to minimise air compression cost.

Matsushita and Osaka Gas have developed a plate type air separation unit (11). The principle of operation is illustrated in Fig 10. The thin film and high permeability is achieved with the use of a copolymer, mainly composed of silicon polymer having a 0.1μm thickness coated on the surface layer of a polypropylene porous sheet. The air is drawn through the membrane cells by means of a suction blower on the oxygen enriched air permeate line, which supplies the low pressure combustion process. The nitrogen is vented from the chamber by an exhaust fan.
Fig 11 shows a photograph of the membrane air separation installation composed of five membrane chambers. The membrane process has been standardised into these multiple membrane chambers, each containing 100 membrane cells.

(ii) **SPIRAL- WOUND MEMBRANE** This type uses sheet material arranged in alternative layers of a feed gas channel, permeable membrane and permeate channel all rolled round a central permeate tube. This is shown in Fig 12, together with a typical assembly of several elements packaged in series in a pressure vessel (2). Fig 13 shows several Soviet Spiral-Wound elements being packaged in parallel into a pressure vessel at NPO Soyuzgaztekhnologiya, VNIIGAS, Moscow and Fig 14 shows a typical installation.

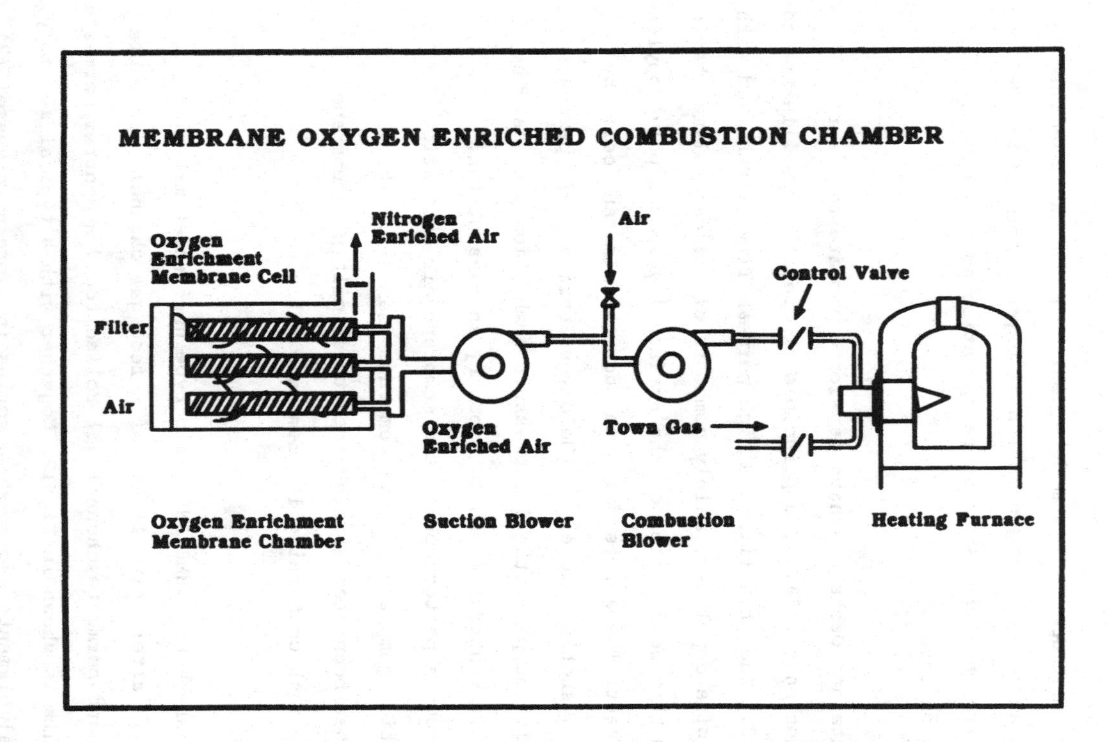

FIG. 10. PLATE TYPE MEMBRANE SCHEMATIC.

FIG. 11. PLATE TYPE MEMBRANES.

SCHEMATIC OF A SPIRAL-WOUND MEMBRANE ELEMENT AND ASSEMBLY.

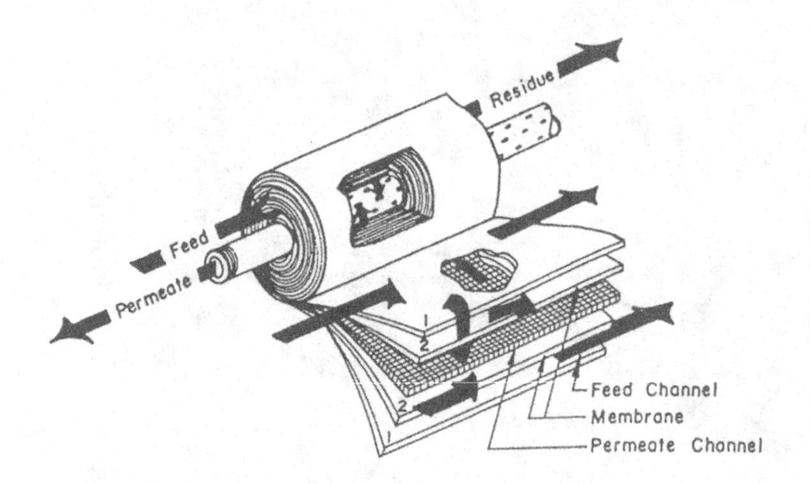

SPIRAL-WOUND MEMBRANE ELEMENT

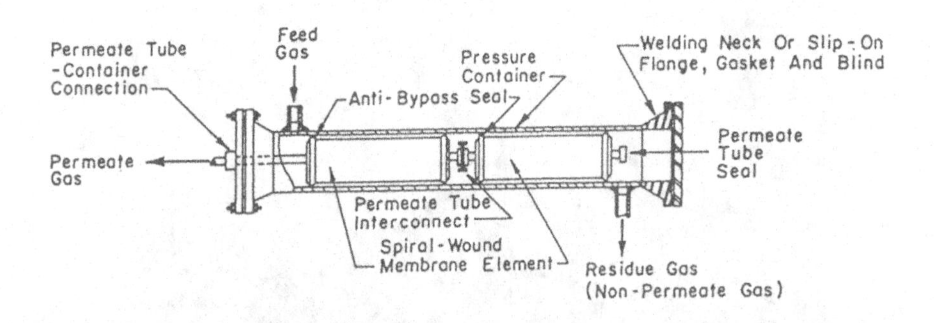

SPIRAL-WOUND MEMBRANE ELEMENT ASSEMBLY

FIG 12.

FIG. 13. SOVIET SPIRAL-WOUND MEMBRANE ELEMENTS.

FIG. 14. SOVIET SPIRAL-WOUND MEMBRANE INSTALLATION.

(iii) HOLLOW FIBRE MEMBRANE. Typical VNIIVPROEKT hollow fibres are illustrated in Fig 15. Their small diameter enables many millions of these hollow fibres to be wound into one module.

In air separation, Nitrogen is normally generated by a supply of compressed air to the membrane module and Oxygen Enriched Air is normally generated by use of a suction blower in the permeate line. Fig 16 shows an A/G Technology hollow fibre membrane (12). Air is supplied to the inside of each fibre. Oxygen permeates through the tube wall and nitrogen is discharged at the other end of the hollow fibre. The hollow fibres are potted into the tube sheets at each end. Fig 17 shows three of these modules being assembled into a pressure vessel, before final packaging of several of these separation pressure vessels into the skid-mounted membrane nitrogen unit shown in Figs 3 & 4.

Several typical VNIIVPROEKT Nitrogen Hollow Fibre Membrane modules are shown in Fig 18. The hollow fibres are wound at an angle of 45 degrees. The air enters inside the hollow fibres and nitrogen is produced by the oxygen permeating through the membrane wall enabling nitrogen to be discharged from the other end of each hollow fibre. A packaged multiple membrane module nitrogen unit is shown in Fig 19. These membrane modules are designed for nitrogen generation with pressure operation without the need for metal pressure vessels. If the oxygen enriched air can also be utilised, low pressure ductwork is all that is required.

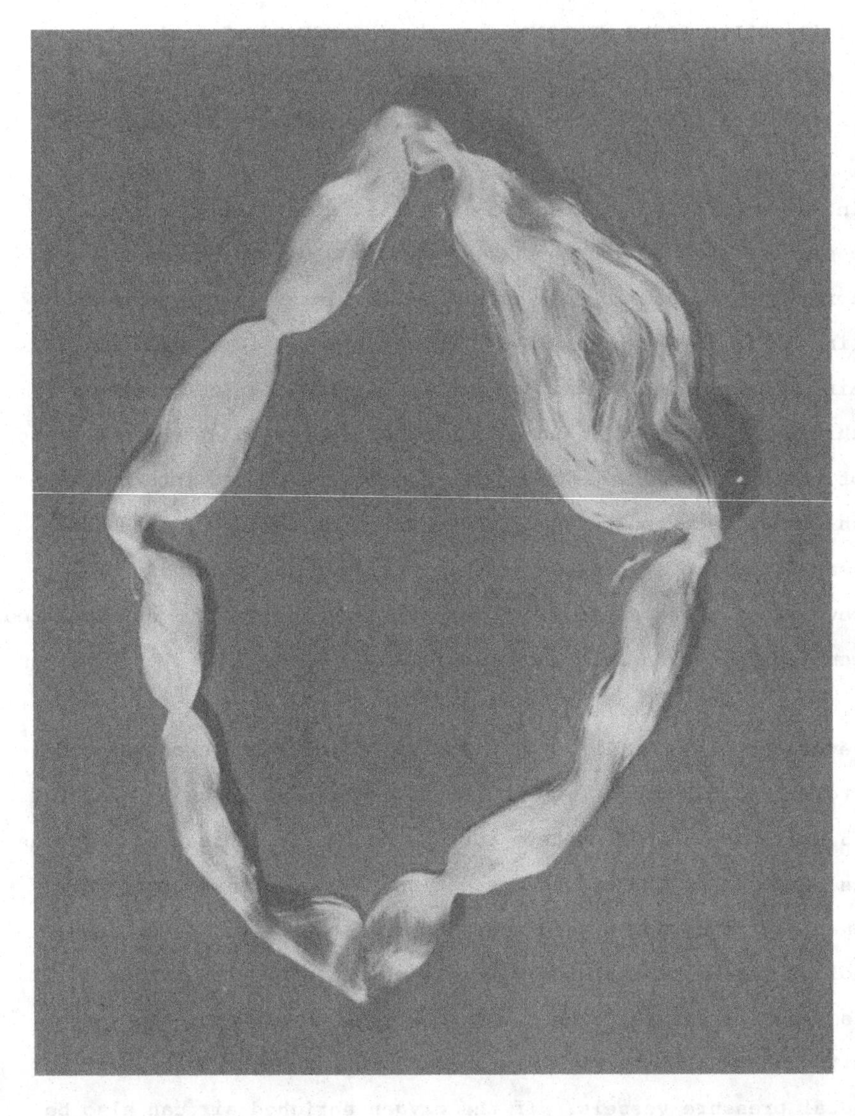

FIG. 15. VNIIVPROEKT MEMBRANE HOLLOW FIBRES.

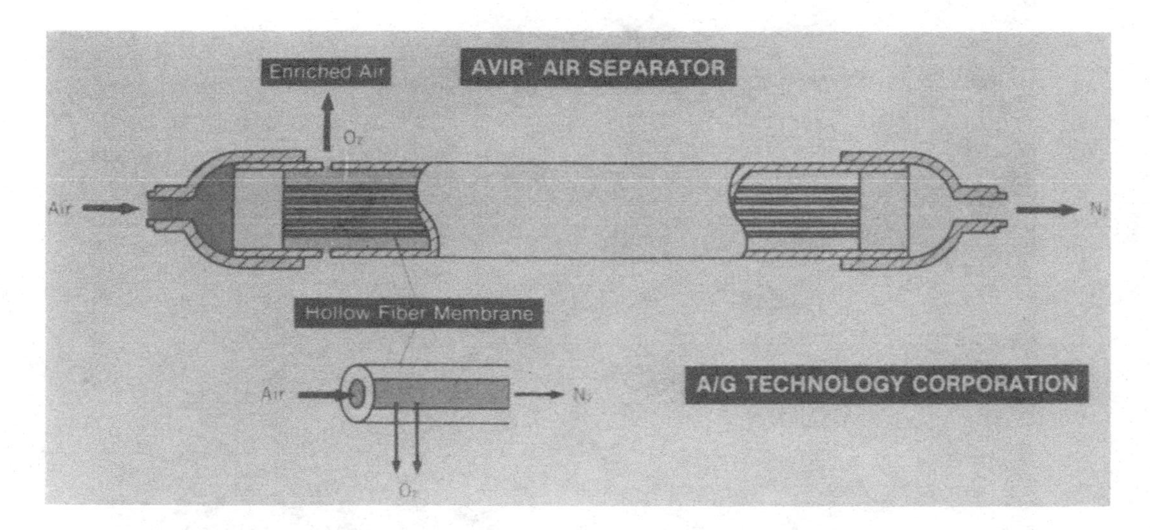

FIG. 16. PRIINCIPLE OF HOLLOW FIBRE MEMBRANE OPERATION.

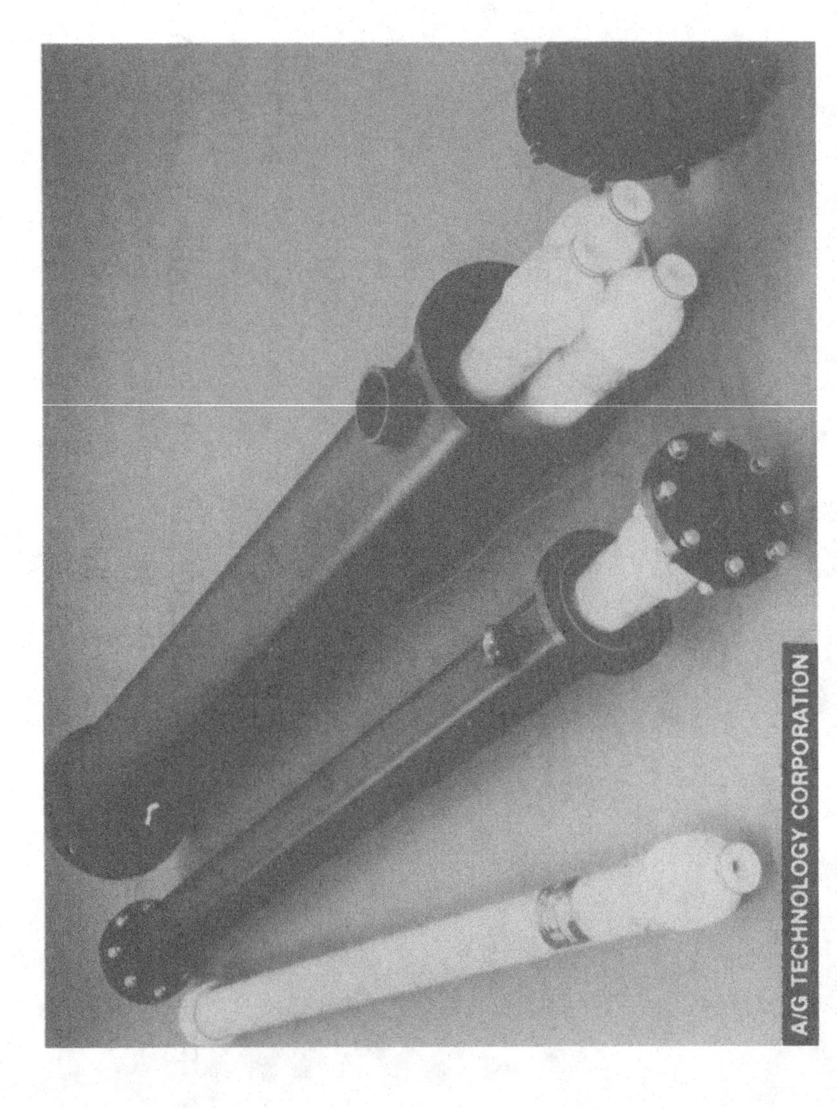

FIG. 17. THREE HOLLOW FIBRE MODULES ASSEMBLED IN PRESSURE VESSEL.

FIG. 18. VNIIVPROEKT HOLLOW FIBRE MEMBRANE MODULES.

FIG. 19. VNIIVPROEKT MULTIPLE MODULE MEMBRANE UNIT.

(d) MEMBRANE ECONOMICS FOR NITROGEN GENERATION

The feed gas (air) is essentially free and available on all sites.

The capital cost is appreciably reduced by high pressure operation but greatly increases the operating cost as the air has to be compressed.

Hollow fibre modules offer maximum membrane surface area per unit volume, therefore offer the best opportunity for minimum capital cost, plus savings in installation cost for offshore applications.

Specification of lower required nitrogen purities reduces capital cost by reduction in membrane required surface area.

Selection of membrane materials and adequate wall thickness to give long membrane life in order to reduce operating cost.

Prepackage and modulise as much as possible to minimise expensive on-site installation work.

Cost of on-site membrane Nitrogen is only 10 - 50% of delivered Nitrogen depending on location and volume of Nitrogen required.

CONCLUSIONS

The main requirements for gas separation membranes are:-

1. High selectivity, ensuring minimum energy expenditure

2. High permeability for low membrane capital cost.

3. Sufficient strength and life under operating conditions to give reliable service and low operating cost.

4. Resistance of membrane to chemical and physical attack.

5. Stability of separation characteristics of the membrane with varying flowrates, pressures.

The use of membranes for on-site air separation, particularly in the generation of nitrogen will progress rapidly as capital costs and operating costs decrease with new developments and all of the above requirements are met.

The important features of simple design, full turndown capability, continuous operation, low operator labour requirement, low maintenance, low weight and compactness makes the equipment of interest to the end user.

The High Energy Efficiency of Membrane Air Separation Systems is of increasing interest with the present Middle East Fuel Crisis.

REFERENCES

1. Economics of Gas Separation Membranes, Robert W. Spillmen, Chemical Engineering Progress, January 1989.

2. Developments in Membrane Technology for Gas Separation Processes, Fred Marlett and Graham Lock, 38th Canadian Chemical Engineering Conference, 2-5 October 1988.

3. Membrane Gas Separation, Current States and Future Developments Terence Foley, London and Southern Gas Association 11 June 1987.

4. Membrane C02 Separation, Sacroc Tertiary Recovery Project, David Parro, American Institute of Chemical Engineers, 1984 Spring National Meeting.

5. Membrane Gas Separation in Industry, V.P. Belyakov, L.N. Chekalov - USSR.

6. D. Mendeleyev, Journal of All-Union Chemical Society, 1987, Vol. XXXII, No. 6, USSR.

REFERENCES (CONTINUED)

7. Achievements of Chemistry, Science, 1988, Vol. LVII, No. 6.

8. Scopes of application of Membrane Type Gas Separation, L.N. Chekalov, and O.G. Talakin, International Symposium on Membrane for Gas and Vapour Separation Suzdal, USSR - 27 February - 5th March 1989.

9. New Developments in Membrane Processes for Gas Separations S. Alexander Stern, Department of Syracuse University NY USA.

10. Some New Applications of Membrane Gas Separating Systems J.M.S.Hennis, International Symposium on Membranes for Gas and Vapour Separation, Suzdal, USSR, 27 February - 5th March 1989.

11. The Economics of On-site Oxygen Production by Pressure Swing Adsorption (PSA) and Membrane Systems, J. Thurley, Osaka Gas, Japan, R&D Forum 88, 11 November 1988.

12. State of the Art Permeable Gas Separation, Dr Arye Gollen and Myles H. Kleper, The 1987 Fifth Annual Technology Planning Conference, 21 October 1987, USA.

SESSION K:

GAS SEPARATION IN THE PETROCHEMICAL INUSTRY II

NOVEL APPLICATIONS FOR GAS SEPARATION MEMBRANES

Ian W. Backhouse, Monsanto Permea, Brussels, Belgium

ABSTRACT

Monsanto successfully commercialised Prism® separators for industrial scale membrane gas separation system in 1980. Following a 6 year period of growth and development, Monsanto formed Permea Inc. in 1986 to manage the global gas separation membrane business. During the same year Monsanto introduced Prism® Alpha separators for air separation. Today 10 years after commercialisation over 10,000 Prism® separators have been used in more than 2,500 installations. These installations cover some 100 applications and are located in most parts of the world. The scale of system is from less than 1 Nm3/hr to over 80,000 Nm3/hr.

Membranes can be utilised in many industrial applications where they replace conventional gas separation technologies with cheaper and simpler solutions. Membranes are increasingly used in novel application areas where conventional technology has not utilised either because of the scale of production or the membrane performs a function that no other technology can duplicate. This paper reviews some of these novel applications for which gas separation membranes are suited.

INTRODUCTION

After a long period of research and laboratory development Monsanto's multi component gas separation membrane was ready for testing in 1975. By the end of year, the first pilot scale testing of the membranes on an industrial gas stream was completed.

Following those pilot trials, Monsanto was confident enough to authorise the installation of three industrial scale units during the next three years. The first of these units was installed in the Texas City plant, Texas in 1977. This membrane unit is used to adjust the ratio of hydrogen to carbon monoxide in synthesis gas for oxo alcohol production. The second unit was installed at the Pensacola plant, Florida in 1978 recovering hydrogen from a hydrogenator purge. The third unit was installed in 1979 at the Luling plant, Louisiana recovering hydrogen from the high pressure purge of an ammonia plant.

By late 1979 after a successful in house programme, Monsanto was ready to commercialise its gas separation membranes as Prism® separators. During the early part of the commercialisation programme Monsanto focused on process gas separations. Process gas separations encompass hydrogen and carbon dioxide separation from a whole host of contaminants such as nitrogen, argon, methane and other hydrocarbons. This

Effective Industrial Membrane Processes — Benefits and Opportunities, pp.383–389

covered a wide range of industries including the chemical, oil refinery and petrochemical industries.

By 1986 80 systems were operational for process gas separations and more were under construction. The majority of these systems were located in the ammonia production industry. The reason for the immediate popularity of ammonia plant purge gas recovery is that existing pressure differentials within the plant are high and the application required high recovery of hydrogen at relatively low purities which had not been addressed by conventional technologies.

Concurrent with the commercialisation of Prism® separators for process gas separation, Monsanto also embarked upon a development programme to produce a membrane tailored for air separation. As a result of this development program, Prism Alpha® separators were launched in 1986. Prism Alpha® separators not only gave a four fold increase in permeability but also a twenty per cent increase in selectivity over Prism® separators for air separation. Combined with a change in separator configuration, this development gave a product that could operate economically in the same pressure range as competing technologies.

Today more than 10,000 separators have been sold which have been incorporated into over 2,500 systems. These systems are installed throughout the world and have a cumulative system operating experience of over 500 years.

THE PRISM® SEPARATOR

Transport of gas molecules across a membrane occurs by solution of the molecule at the high partial pressure surface, diffusion of the molecule through the membrane under the partial pressure gradient and desorption at the low partial pressure surface.

Assuming Fickian diffusion of a Henrian dissolved gas in an isotropic dense membrane the following equation can be derived for each component i:

$$Q_i = (P/l)_i * A * (Pr_i - Pp_i)$$

where
- Q_i rate of flux
- $(P/l)_i$ permeability coefficient per unit thickness
- A membrane area
- Pr_i retentate partial pressure
- Pp_i permeate partial pressure

The permeability coefficient for a particular component can be described as the product of the solubility and diffusion coefficients. The membrane utilised in the Prism® separator is based on polysulphone which is formed as hollow fibres. Forming the membrane as hollow fibres has two distinct advantages one of which is its inherent capability to withstand high differential pressure and to provide an excellent self-supporting structure. The other is the high surface area to volume ratio obtained with small outside diameter fibres. This is an important factor in determining the ultimate size of a gas separation system.

Permea's Prism® separator design consists of a compact bundle of thousands of hollow fibres encapsulated in a high pressure shell similar in arrangement to a shell and tube heat exchanger. The fibre bundle is sealed at one end and encased in a tube sheet at the other and when mounted vertically the bundle fills the shell ensuring uniform gas distribution. Feed gas enters at the base of the shell side of the fibre bundle and exits at the top. As the gas passes over the membrane faster permeating gases concentrate into the fibre bore, exiting via the base of the fibre bundle, and the remaining slower permeating gases exit the top of the separator. Several separators may be connected in series or parallel, depending on velocity and pressure drop constraints, in order to achieve the requirements of a particular application.

The development of Prism Alpha® separators necessitated a change in separator configuration. These separators are fabricated with tube sheets at either end and encapsulated in a low pressure shell with higher pressure end caps. The method of operation is such that feed gas enters the inside of the hollow fibres at one of tube sheets. As the gas passes down the inside of the hollow fibre, faster permeating components are transported through the fibre wall and concentrate in the low pressure shell side. The slower permeating components concentrate at the exit of the tube sheet.

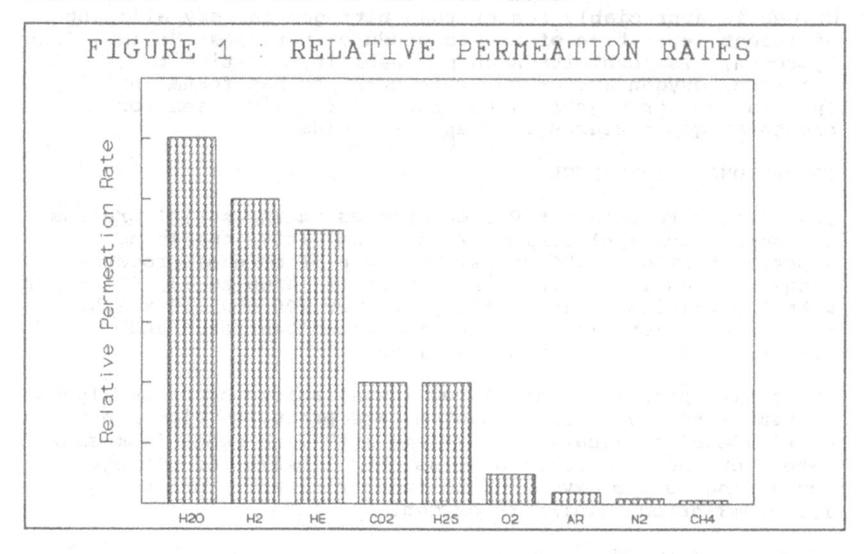

The permeability coefficient is a function of molecular size, molecular shape, polarity and chemical affinity. Table 1 above shows the spectrum of polymeric membrane gas permeabilities for some common gas molecules. As might be expected under diffusion control small molecules such as hydrogen and helium are "fast" whereas larger molecules such as methane and argon are "slow". Anomalies such as carbon dioxide and water, which are "faster" than would be expected for their molecular size, are polar molecules which have high solubilities in polymers. Solubility, as will be recalled, is one of the factors contributing to their permeability coefficient.

TRADITIONAL APPLICATIONS

Based on the separating characteristics of a gas separation membrane, hydrogen can be separated from nitrogen, methane and argon such as occurs in a purge stream from an ammonia plant. Hydrogen can also be separated from a mixture of hydrocarbons as occurs in many oil refinery purge streams and also separated from carbon monoxide which is frequently required on a petrochemical plant.

Due to the higher permeability of carbon dioxide relative to methane and higher hydrocarbons purification of the hydrocarbons or recovery of the carbon dioxide can be easily achieved. Purification of hydrocarbons from carbon dioxide is required for such applications as upgrading biogas or conditioning of natural gas streams. Recovery of carbon dioxide from hydrocarbons is encountered particularly in enhanced oil recovery where carbon dioxide is used to reduce oil viscosity in the oil well.

Oxygen is appreciably faster than nitrogen thereby allowing efficient separation of air to produce inert gas. Helium, like hydrogen, also exhibits high permeability relative to nitrogen, oxygen and argon. This property has found an application for upgrading the purity of helium used for breathing gas mixtures in deep sea diving.

OPERATIONAL EXPERIENCE

There are more than 125 Prism® systems in industrial process gas separation applications with a cumulative operating experience of over 350 years. The size of streams processed range from as little as 10 Nm3/hr to as large as 83,700 Nm3/hr with a total installed capacity of over 200,000,000 Nm3/hr. Systems are installed in every inhabited continent and virtually in every imaginable climate.

There have been more than 2,500 installations of Prism Alpha® systems since 1986. These systems are producing nitrogen for a multitude of applications. Systems installed range from small laboratory models producing less than 1 Nm3/hr of nitrogen containing 10 ppm oxygen up to industrial units producing 1,800 Nm3/hr containing 5% oxygen.

NOVEL APPLICATIONS

Many applications for gas separation membranes have been identified over the past 10 years. Initially there was focus on applications that could inherently provide the required partial pressure differentials. Typically this revolved around high pressure purges from petrochemical and oil refinery unit operations. These high pressure purges containing hydrogen could be efficiently passed over a membrane to provide a purified hydrogen stream for recycle at lower pressure to another part of the process.

Recent installations of membranes in these industries has included compression equipment to provide the required partial pressure differentials even when alternative gas separation technologies could have performed the separation without the additional compression equipment.

At the same time interest has been shown in the use of membranes for adjusting the ratio of hydrogen to carbon monoxide in synthesis gas. Unlike any other gas separation technique, membrane technology can achieve precise adjustment in ratio with negligible pressure drop on the synthesis gas.

Recently the state of the art for this separation has been the combination of membranes with another technology to provide a performance that neither technology in isolation can provide. A typical example of this hybrid technology has been the combination of membranes with pressure swing adsorption. The membranes perform the bulk separation and provide the required hydrogen to carbon monoxide ratio. The pressure swing adsorption system provides purification of hydrogen from the permeate stream from the membranes. The tail gas from the pressure swing adsorption system is recompressed and combined with the synthesis gas. Overall the hybrid system provides essentially 100% recovery of carbon monoxide in the synthesis gas and produces a second stream containing essentially 100% hydrogen.

Pressure swing adsorption/membrane hybrids have been used to enhance the recovery of a pressure swing adsorption system. The pressure swing adsorption system is ideal for producing high purity product but has the disadvantage of lower recovery. By recompressing the tail gas from the pressure swing adsorption system and feeding the gas to a membrane the valuable component can be purified and recycled to the feed of the pressure swing adsorption system.

Another example of hybrid technology is the combination of membranes and cryogenic technology. In these hybrids the membrane performs upstream bulk separation of a more permeable component prior to entering the cryogenic system. The component typically removed is hydrogen and, beneficially for the cryogenic system, water is also removed with the hydrogen due to its high permeation rate. This type of hybrid has been used to debottleneck existing cryogenic plants and has been included in the design of greenfield plants.

In the near future, membranes will become available for efficient dehydration of well head gas. This application will be especially beneficial for operators of North Sea platforms or remote location oil wells. Gas separation membranes have relatively high flux rates for water but as yet do not have sufficiently high fluxes for methane and other hydrocarbons to give efficient dehydration capabilities. However when that development becomes available, membranes will give operators an automatic, reliable, solvent and maintenance free dehydration unit.

Generally single stage membranes are only efficient for bulk gas separations although single stage membranes with recycle can achieve higher efficiencies with added power input. To this end, utilisation of single stage membranes for natural gas dehydration will be limited to applications requiring a minimum dew point of -10 to -20 °C. This range of application is not a disadvantage, as this range of dew point will allow safe transportation of natural gas to a centralised terminal without hydrate formation.

The capacity that can be obtained from the smallest Prism® separator has brought about several novel applications that can not covered economically by conventional gas separation technology purely because of the size. To give an idea of the scale, the surface area of the largest Prism® separator is more than 600 times more than the smallest Prism® separator. These small size separators have been incorporated into cabinet systems so that the separation system can be wall mounted. This cabinet mounted system has given rise to several applications that have been developed by Calor, one of Permea's partners, one of which is beer dispensing.

Traditionally keg beers have been dispensed with pressurised carbon dioxide. The use of 100% carbon dioxide for beer dispense has several disadvantages which will be familiar to most beer drinkers. These disadvantages include fobbing wastage, poor foam retention and slow dispense times. In more recent times there has been a move by the major brewers to dispense beer with so called mixed gas. Mixed gas is a precise mixture of nitrogen and carbon dioxide in a ratio determined by each brewer for each particular type of beer. The nitrogen content can vary from as little as 30 to as much as 70% depending on the type of beer. The major reason for the move to mixed gas is that it reduces or even eliminates the major disadvantages of beer dispense with pure carbon dioxide.

However there are two major disadvantages to the use of mixed gas. Primarily the mixed gas is delivered in cylinders. The capacity of the cylinders is limited to the fact that the mixture can only be stored as a gas under pressure unlike carbon dioxide which can be stored as a liquid. Thus a typical 14 lb mixed gas cylinder containing 70% nitrogen and 30% carbon dioxide will only dispense on average 4.25 kegs of beer. The other major disadvantage is that in a typical cellar there may well be a requirement for two or three ratios of mixed gas for several types of keg beer. This means that there is a either a need for different cylinders of mixed gas or a poor compromise of using one mixture for all the keg beers.

The advent of small capacity membrane nitrogen generators has allowed a revolutionary change in cellar management. Using membrane nitrogen generators in conjunction with cylinders of liquid carbon dioxide and several gas mixers, mixed gas of any specification can be generated in situ. The major benefit of this system is that the same 14lb cylinder of gas described above, now filled with liquid carbon dioxide, will now dispense 65 kegs of beer based on a 30% carbon dioxide and 70% nitrogen mixture. This clearly reduces the number of deliveries and handling requirements. The other benefit is that there is no limit to the number of gas mixtures that can be synthesised in the cellar.

Another application that will be addressed in the future by the small Prism Alpha® separators is the production of medical oxygen for asthmatic and hypoxic patients. Prism Alpha® separators have several advantages over the traditional methods of providing medical oxygen. The primary advantage is fail safe operation. At worst the patient can either receive air or 50% oxygen, there is no possibility of delivering a possibly dangerous 95 to 100% oxygen to the patient. The other advantages over other methods of in situ generation are that as the product has diffused through the membrane it is essentially sterile, humid and carbon dioxide enriched due to respective high water and carbon dioxide permeability. Also due to the nature of the process maintenance requirements are minimal with size and weight being much reduced.

IN CONCLUSION

Prism® separators typically tolerate most components at levels up to saturation in the process stream. Thus the necessity for pretreatment is reduced, if not eliminated, in most instances. This contrasts with other technologies that require much more stringent pretreatment.

Unlike most other gas separation technologies variations in both feed flowrate and composition to a Prism® system result in only minor changes in product flowrate and purity thus avoiding the need for operator attention.

As the technology is based on a truly continuous process utilising no moving parts, maintenance is low and on stream time is high. In the event that maintenance or isolation is required, shut down and start up can be achieved in minutes which compares favourably with other technologies.

The units are mechanically simple and compact ensuring that Prism® separators can be retrofitted in many existing facilities. Systems are skid mounted so that site installation is simple and rapid. As the units are of a modular nature then plant expansion can be easily achieved at a later date by simply adding more separators.

Prism® separators offer a simple solution to gas separation problems. One that is easy to operate and maintain which led one operator to comment that using a Prism® separator was "like operating a piece of pipe". Gas separation membranes have been commercially proven for over ten years in many applications, the next ten will see even more applications developed for this versatile technology.

MEMBRANE PERMEATION AND PRESSURE SWING ADSORPTION (PSA) FOR THE PRODUCTION OF HIGH PURITY HYDROGEN

G.W. Meindersma

DSM Research bv/PT-RU
Geleen, The Netherlands

SUMMARY

For a new hydrogenation process at DSM a hydrogen feed gas with a purity of at least 98% was required. There were four options in this case: separation of hydrogen by membranes or by Pressure Swing Adsorption (PSA), using hydrogen containing feed gas from either a naphtha cracker or a cryogenic recovery unit.

The naphtha cracker gas contains 90% hydrogen and 10% methane and has a pressure of 20 to 22 bar. By means of PSA it is possible to produce a hydrogen gas with a purity of 99.9% and a recovery of 84 to 89%. For a membrane unit the feed gas can be used at the available pressure or it can be compressed. By means of membranes it is possible to produce a hydrogen gas with a purity of 98 to 99%, with recoveries varying from 64 to 95%.

The product gas from the cryogenic unit contains 92.3% hydrogen and 6.2% nitrogen, the remainder being helium, argon and methane. The pressure is 70 bar. When membranes are used, the hydrogen product gas has a purity of 98.2% and the recovery is 93%. If the recovery of hydrogen has to be carried out by PSA, the feed pressure has to be lowered to 20 - 30 bar. The product gas has a purity of 99.6%, while the recovery is 80%.

Comparison of all options, including compression of product gas to 100 bar, shows that membrane permeation with cryogenic product gas as feed is the best choice.

NOMENCLATURE

A	membrane area	m^2
C_i	concentration of component i	Nm^3/m^3
D_i	diffusivity of component i in membrane	m^2/s
J_i	permeate flux of component i through membrane	$Nm^3/(m^2.s)$
P_i	permeability of component i through membrane	$Nm^3.m/(m^2.bar.s)$
P_f	feed pressure	bar
P_p	permeate pressure	bar
p_i	partial pressure of component i	bar
Q_i	flow rate of component i through membrane	Nm^3/s
S_i	solubility coefficient of component i in membrane	$Nm^3/(m^3.bar)$
x	distance of permeation in membrane	m
ℓ	thickness of effective separation layer	m
α	ideal separation factor	-

Effective Industrial Membrane Processes — Benefits and Opportunities, pp.391–400

INTRODUCTION

For a new hydrogenation process at DSM, which had to be in operation in November 1989, a hydrogen gas with a high purity was required in order to minimize purge losses - the hydrogenation takes place at 80 to 100 bar - and to achieve a higher conversion and selectivity. There are several methods of separating hydrogen gas from different feedstocks: cryogenic recovery, pressure swing adsorption, thermal swing adsorption, membrane permeation, absorption by metal hydrides, etc. Since the purity of the hydrogen feed gas to the hydrogenation process must be at least 98%, only two commercially available hydrogen separation processes were considered feasible: pressure swing adsorption (PSA) and membrane permeation.

An extension of the existing cryogenic recovery to include distillation was not considered because the product capacity of the cryogenic recovery is 12,000 Nm^3/hr, while the new hydrogenation plant required 1700 to 2000 Nm^3/hr only. Hybrid processes, e.g. membrane permeation and cryogenic recovery, were not considered because of the small scale of the process. The total investment costs for a hybrid installation would probably exceed the investment costs of the stand alone systems by a factor of 1.5 to 2.

At DSM's production complex in Geleen, The Netherlands, several hydrogen containing feedstocks are available: naphtha cracker gas (90% hydrogen), synthesis gas for ammonia production (74% hydrogen) and product gas from a cryogenic recovery unit fed with ammonia purge gas (89.6 to 92.3% hydrogen). Of these feed gases for the purification unit the naphtha cracker gas and the product gas from the cryogenic recovery unit were considered suitable. The hydrogen content of synthesis gas for ammonia production is too low. This feed gas would only be used if the other two feed gases were not available for this hydrogenation process due to priority use in other processes.

PRESSURE SWING ADSORPTION (PSA)

Separation through adsorption is based on selective adsorption of one or more components of a gas or liquid stream at the surface of a microporous adsorbent material. Since the attractive forces causing the adsorption are generally weaker

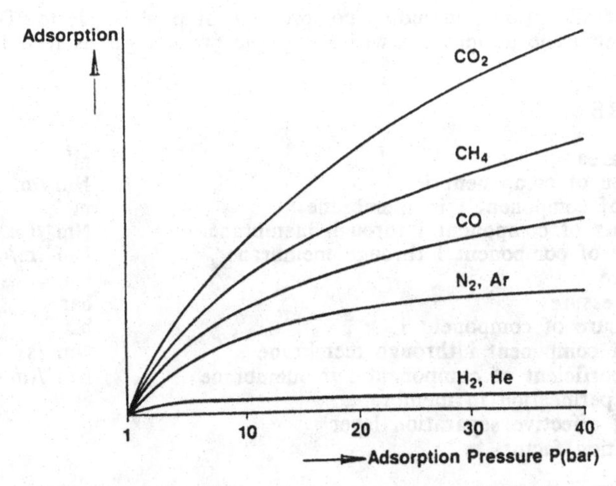

Figure 1 Adsorption of several gases on molecular sieves. Reprinted with permission from Knoblauch et al. (1979)

than the chemical bonds, the adsorbate can be desorbed by increasing the temperature or by reducing the partial pressure of the adsorbate. The PSA process is described in more detail by Wiessner (1988).

For hydrogen separations the commonly used adsorbent materials are zeolites and activated carbon. Typical adsorption curves are shown in Figure 1 (Knoblauch et al. 1979). It can be seen that hydrogen and helium are very weakly adsorbed, while methane and carbon dioxide are strongly adsorbed. It can also be seen that argon and nitrogen are weakly adsorbed.

In the adsorption step feed gas flows through a vessel filled with adsorbent material at a pressure of 5-40 bar. The most efficient operating pressure is in the range of 20 to 30 bar (Figure 2, Knoblauch et al. 1979). The adsorbable components are adsorbed and a purified gas leaves the vessel (Figure 3). Before the adsorbent material is completely saturated the feed gas is diverted to another adsorber vessel. Typical adsorber times are 6 to 7 minutes and the total cycle times vary from 20 to 30 minutes. Regeneration of the adsorber takes place at the lowest pressure of the system, determined by the required fuel gas system pressure, as the tail gases are normally used as a fuel. From Figure 2 it is clear that a low purge pressure favours a high hydrogen recovery.

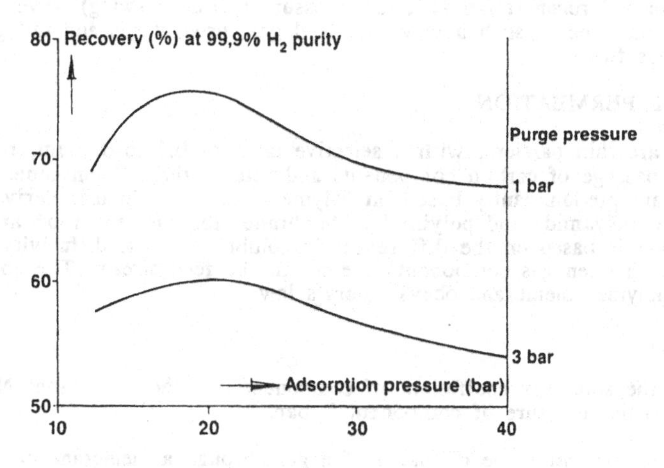

Figure 2 Hydrogen recovery at different adsorption and purge pressures
Feed: 60% H_2
Reprinted with permission from Knoblauch et al. (1979)

The depressurization of the adsorber is done in one or more concurrent flow steps, with each depressurization step being used for a corresponding pressurization step of another adsorber. Generally speaking, the greater the number of pressure equalization steps carried out, the better the efficiency of the unit. However, increasing the number of pressure equalization steps means that more adsorbers units will have to be installed.

The highest possible hydrogen recovery is also dependent on the composition of the feed gas. When nitrogen is present the recovery is low, since nitrogen is relatively weakly adsorbed; in the presence of about 25% nitrogen in the feed the maximum hydrogen recovery is 68%. In the presence of methane the recovery can be much higher, 85 to 89%, depending on the number of adsorbers.

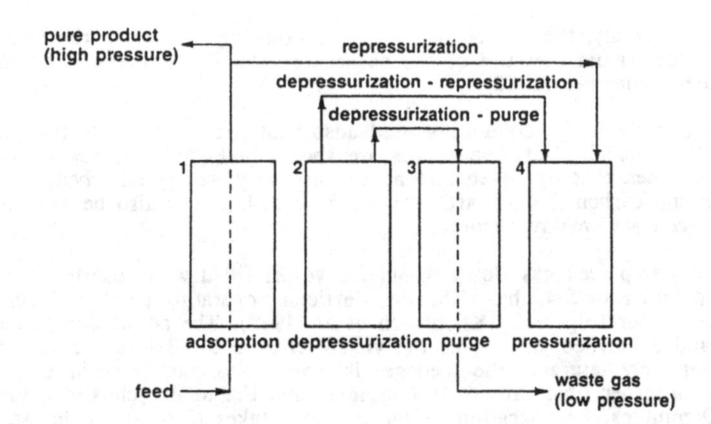

Figure 3 PSA system using four beds.

Since each adsorber runs through the same cycle, characterized by adsorption at a high pressure and regeneration at a low pressure (pressure swing), several adsorber vessels are combined in such a way that feed gas, product gas and tail gas streams are continuous flows.

MEMBRANE PERMEATION

Membranes are thin barriers, with a selective layer of 0.1 to 0.6 μm, that allow preferential passage of certain components and retain others. Membranes used for gas separations are predominantly based on polymers such as cellulosic derivatives, polysulphone, polyamide and polyimide. Membranes for gas separation are non-porous and separation is based on the differences in solubility in and diffusivity through a membrane between gas components present in the feed stream. The solubility of a gas in the polymer membrane obeys Henry's law:

$$C_i = S_i * p_i, \qquad (1)$$

where S_i is the solubility coefficient of component i in the membrane, $Nm^3/(m^3.bar)$, and p_i the partial pressure of component i, bar.

As a first approximation the diffusion of a gas through a membrane can be described by Fick's law:

$$J_i = -D_i * dC_i/dx, \qquad (2)$$

where J_i is the flux of component i through the membrane, $Nm^3/(m^2.s)$, D_i the diffusivity or diffusion coefficient of component i, m^2/s, C_i the concentration of component i in the membrane, Nm^3/m^3 and x the distance of permeation in the membrane, m. However, the diffusion coefficient is in most cases not a constant and therefore usually an average value of the diffusion coefficient is used.

The solubility of a component increases with increasing molecular weight, while the diffusivity decreases with increasing molecular weight. The product of the solubility and the diffusivity is called the permeability of a gas component.

$$\bar{P}_i = S_i * \bar{D}_i \qquad (3)$$

The ratio of $\bar{P}_i$ and $\bar{P}_j$ for two different gases i and j, defined as

$$\alpha_{ij} = \bar{P}_i/\bar{P}_j \tag{4}$$

is the ideal separation factor of the membrane. The actual flow rate of component i is given by:

$$Q_i = \bar{P}_i/\ell \,.A.(P_f - P_p), \tag{5}$$

where Q_i is the flow rate of component i through the membrane, Nm^3/s, ℓ the thickness of the effective separation layer, m, A the membrane surface area, m^2, P_f and P_p the feed and permeate pressures respectively, bar.

The maximum obtainable purity of the product gas is dependent on the separation factor α, the pressure ratio P_f/P_p, the composition of the feed gas and the recovery required. The pressure difference over the membrane determines the necessary membrane area.

As the permeabilities of hydrogen and helium are about equal for most membranes, these two gases cannot be separated from each other by membrane permeation. The separation factors for H_2/N_2 and H_2/CH_4 depend on the membrane material and range from 40 to 200.

ECONOMIC EVALUATION

In November 1989 a new plant was installed at DSM for a hydrogenation process. For this particular process hydrogen gas with a high purity is necessary to minimize purge losses - the hydrogenation takes place at a high pressure (80 - 100 bar) - and to achieve a higher conversion and selectivity.

There were four options: recovery by PSA or recovery by membranes, using hydrogen gas from either the naphtha cracker or the cryogenic unit as feed.

Basis Of Calculation

All options were compared on the same basis. The fixed and variable costs shown in Table 1 and 2 are relative costs with the economically most feasible option as base case. The investments include the costs of equipment and installation for PSA, membrane units, heaters and compressors. The figures for the equipment costs were obtained from the respective vendors. For calculation purposes the fixed costs were set at 40% of the total investment costs, including installation, to account for depreciation, interest, maintenance, etc. Also the membrane costs are included in this figure, as the lifetime of the membranes is expected to be at least five years. For the purification of ammonia purge gas a membrane unit was installed at DSM in 1983 and no membrane replacement has taken place to date (August 1990) (Meindersma 1990). The pressure differences over the membrane are larger in the case of the ammonia purge gas than in the present application. Therefore, the assumption of a membrane lifetime of five years seems to be reasonable.

The following variable costs were taken into account:
-value of the feed gas (hydrogen content only);
-return value of the tail gases of the purifier;
-return value of the waste gases from the process;
-cost of precompression to 50 bar for membrane units;
-cost of compression of the product gas to 100 bar;
-cost of cooling water;
-cost of heating the feed gas, if applicable;
-cost of instrument air.

Case A: Naphtha Cracker Gas As Feed Gas
The naphtha cracker gas contains 90% hydrogen and 10% methane and has a pressure of 20 to 22 bar and a temperature of 30°C. By means of PSA it is possible to produce from this gas a hydrogen gas with a purity of 99.9%, with a recovery of 84 or 89%, depending on the number of adsorbers. The purified gas is available at 19 bar and 35°C. The tail gas has a pressure of 1.3 bar and can only be used as a fuel.

From Figure 4 and Table 1 it can be seen that increasing the recovery from 85% to 89% for PSA does not affect the purity of the product gas, but the investment costs are higher, as more adsorbers will have to be installed. If the recovery is 89% a smaller quantity of feed gas is needed, resulting in reduced operating costs. As a result of this the sum of the fixed and variable costs are about equal for both options.

For a membrane unit the feed gas can be used at the available pressure or it can be compressed. By means of membranes it is possible to produce a hydrogen gas with a purity of 98 or 99%, with recoveries varying from 64 to 95%. If the feed gas is compressed to 50 bar a smaller quantity of feed gas is needed and a higher recovery can be achieved. In all cases tail gas will be used as a fuel. With membrane permeation a higher recovery leads to a lower purity, unless the permeate pressure is much lower. If no precompression is used, the permeate pressure of the product with the higher purity (99%) must be lower than that of the lower purity product (98%), in order to still have a reasonable recovery.

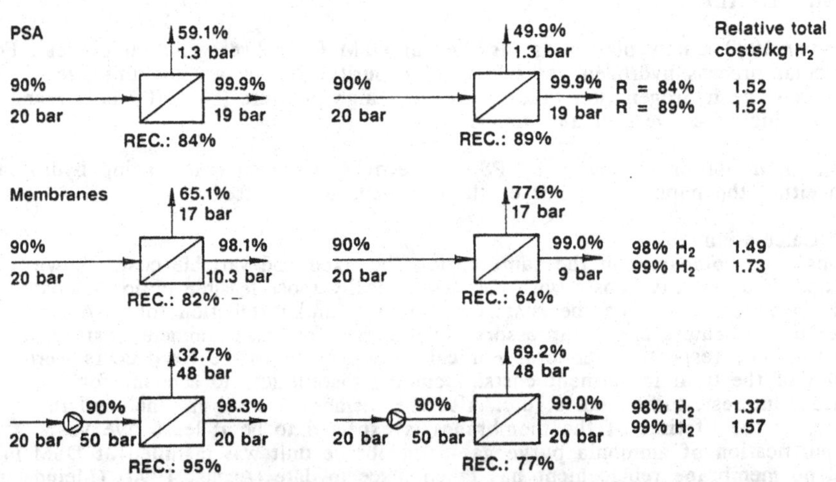

Figure 4 Hydrogen purification using naphtha cracker gas as feed
Product gas is compressed to 100 bar.

As the permeate pressure for the unit yielding higher purity product is lower, the unit can be smaller, but of course the compressor must be larger in order to compensate for the lower pressure. In the end the investment costs for both options are comparable. Since the recovery of hydrogen is lower when a high purity gas is produced, the variable costs are also higher. From formula 5 it is apparent that, if precompression is used and a larger pressure difference over the membranes is generated, the membrane units can be smaller than in the first two cases. Of course, the costs of the precompressors have to be taken into account. It appears that the sum of the equipment costs is comparable to those of the first two cases. With precompression of the feed gas to the membrane unit the recovery of hydrogen is higher than in the cases without precompression of the feed gas.

The results of the calculations in Table 1 show that when the recovery of the hydrogen is increased, the variable costs will be lower. With a high hydrogen recovery less feed gas is needed to produce a certain amount of product gas and less feed gas is degraded to tail gas, which is used as a fuel.

Table 1 Comparison of PSA and membrane permeation for hydrogen purification with naphtha cracker gas as feed (90% H_2)
Product gas is compressed to 100 bar

	Feed		Product			Tail			Rec	Relative costs/Nm^3		
	Q_{H_2}	P	Q_{H_2}	H_2	P	Q_{H_2}	H_2	P		Fix.	Var.	Tot.
	Nm^3/hr	bar	Nm^3/hr	mole%	bar	Nm^3/hr	mole%	bar	%	-	-	-
PSA	2490	20	2090	99.9	19	400	59.1	1.3	84	0.49	1.03	1.52
	2350	20	2090	99.9	19	260	49.9	1.3	89	0.54	0.98	1.52
Membr	2560	20	2100	98.1	10.3	460	65.1	17	82	0.43	1.06	1.49
	3280	20	2100	99.0	9.0	1180	77.6	17	64	0.42	1.31	1.73
	2190	50	2090	98.3	20	100	32.7	48	95	0.43	0.94	1.37
	2710	50	2090	99.0	20	620	69.2	48	77	0.44	1.13	1.57

Comparison of several options with naphtha cracker gas as feed gas (Table 1) shows that the best choice is membrane permeation with precompression to 50 bar and a hydrogen purity of 98%. The sum of the fixed and variable costs for this option was lower than for any of the others.

Case B: Cryogenic Product Gas As Feed Gas
The total flow scheme of the ammonia reactor, cryogenic unit and PSA is shown in Figure 5. The feed of the ammonia reactor contains about 74% hydrogen and 25% nitrogen. The purge gas of the ammonia reactor contains approximately 60% hydrogen and 20% nitrogen and has a pressure of about 130 to 140 bar. As the cryogenic unit is not suitable for these high pressures, the pressure has to be lowered to 76 bar.

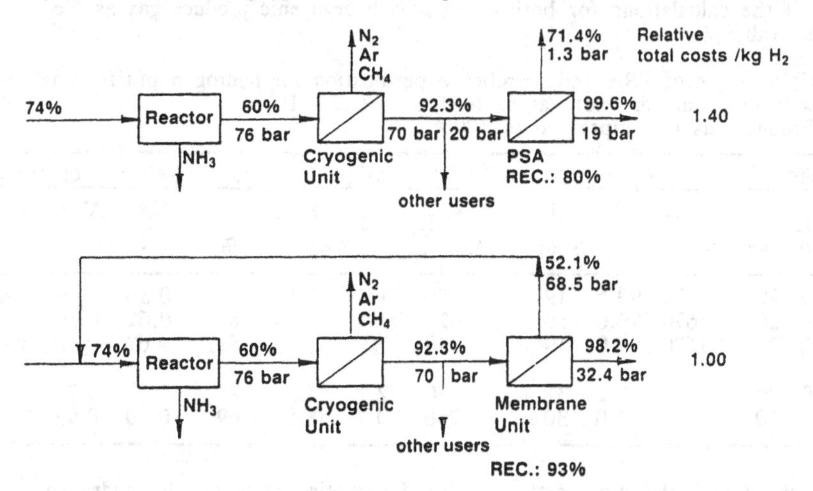

Figure 5 Hydrogen purification using cryogenic product gas as feed
Product gas is compressed to 100 bar.

The product gas of the cryogenic unit has a hydrogen content of 89.6 to 92.3%, depending on the amount of cryogenic product gas being recycled to the ammonia reactor. As helium cannot be removed from the purge gas by cryogenic recovery the helium concentration will increase if all the cryogenic product gas is recycled to the ammonia reactor. The maximum helium concentration allowed is 3.2%. However, the product gas is usually applied for other hydrogenation reactions and is not recycled to the ammonia reactor. Therefore, the helium concentration is lower (0.3%) and the hydrogen content of the cryogenic product gas is 92.3%. The other components are nitrogen (6.2%), argon and methane. The pressure is 70 bar and the temperature is 33°C.

If the recovery of hydrogen has to be carried out by PSA, the feed pressure has to be lowered to 20 - 30 bar, as this is the most efficient pressure for PSA (see Figure 2), but then the product gas has to be recompressed to 100 bar. The maximum purity of the product gas is 99.6% hydrogen, because also helium is present (0.3%). In addition the hydrogen recovery is lower, 80%, due to the presence of 6.2% nitrogen. This means a larger volume of tail gas, which can only be used as a fuel gas, as its pressure is 1.3 bar. It is possible to increase the recovery by lowering the product purity. However, the recovery is raised to only 81%, when the product purity is decreased to 98% hydrogen. Therefore, this option hardly leads to lower cost, as is shown in Table 2.

In Figure 5 also the flow scheme of the ammonia reactor, the cryogenic unit and a membrane unit is shown. If membranes are used, no precompression is needed as the feed gas is already available at a high pressure. As a result of this the permeate pressure can also be higher: 32.4 bar and a relatively small membrane area is needed. Therefore, the fixed cost for this option are lower than for any other option. The hydrogen product gas has a purity of 98.2% and the recovery is 93%. The retentate gas remains at a high pressure and is recycled to the first stage of the syngas compressor. In this way no synthesis gas is degraded to fuel gas. The overall recovery of hydrogen is 100% in this case. Thence also the variable cost are lower than for any of the other options. By means of membranes it is also possible to produce a hydrogen gas with a purity of 99%, but the recovery will be lower compared to the membrane process yielding 98% hydrogen gas. The total costs are about 29% higher than those of the unit delivering the low purity hydrogen gas.

The results of the calculations for both options with cryogenic product gas as feed are shown in Table 2.

Table 2 Comparison of PSA and membrane permeation for hydrogen purification with cryogenic product gas as feed gas (92.3% H_2)
Product gas is compressed to 100 bar

	Feed		Product			Tail			Rec	Relative costs/Nm^3		
	Q_{H_2}	P	Q_{H_2}	H_2	P	Q_{H_2}	H_2	P		Fix.	Var.	Tot.
	Nm^3/hr	bar	Nm^3/hr	mole%	bar	Nm^3/hr	mole%	bar	%	-	-	-
PSA	2500	20	2000	99.6	19	500	71.4	1.3	80	0.56	0.84	1.40
	2090	20	1670	99.6	19	420	71.4	1.3	80	0.62	0.84	1.46
	2060	20	1670	98.0	19	390	67.9	1.3	81	0.62	0.83	1.45
Membr	1790	70	1660	98.2	32.4	130	52.1	68.5	93	0.27	0.73	1
	2440	70	1680	99.0	30	760	80.2	68.5	69	0.30	0.99	1.29

As a result of all this, the sum of the fixed and operating costs for the hydrogen recovery are at least 40% higher for PSA than for the membrane process yielding 98.2% hydrogen.

The ultimate choice for the production of pure hydrogen gas was membrane permeation delivering a product gas with 98.2% hydrogen with cryogenic product gas as feed. The reasons for this choice were the better availability of the cryogenic product gas (two other hydrogenation processes have a priority use of the naphtha cracker hydrogen gas) and the site of the hydrogenation plant (shorter supply lines). And, of course, the total costs were lower than that of any other option considered.

However, if DSM had opted for a different hydrogenation process for the same product, PSA with naphtha cracker gas as feed would have been the best choice. This alternative hydrogenation process required a higher hydrogen pressure, about 200 bar. Because of the high purity of the hydrogen produced by PSA (99.9%) an exit gas stream could have been recycled to the reactor, saving feed gas. If membrane permeation with precompression to 50 bar producing a hydrogen gas of 98% purity had been used, the fixed costs would have been 30% lower but the variable costs would have been 38% higher compared to the PSA process. The total process costs using PSA as the purification process and with an internal process recycle would have been 8% lower compared to the process using membrane permeation.

At this relatively small scale of 1700 Nm^3/hr product gas membrane permeation is economically more feasible than PSA. At DSM also a comparative study was carried out for the production of 15,000 Nm^3 H_2/hr (1350 kg H_2/hr) at a pressure of 20 bar by means of PSA, membranes and metal hydrides.

With naphtha cracker gas (90% H_2) as feed the investment costs for PSA would be 40% lower compared to those for membrane permeation with feed gas compression to 50 bar. Due to the higher hydrogen recovery for the membrane permeation (95%) compared to that for PSA (84%) the variable costs of membrane permeation are lower. However, the sum of the fixed and variable costs of both processes are about equal. PSA has then the advantage of a higher purity, 99.9%, vs 98.1% for membrane permeation.

Since the capacity of the cryogenic recovery unit is too small to produce 15,000 Nm^3 H_2 per hour, an alternative feedstock is ammonia synthesis gas (74% H_2). With this gas as feed the investment costs for PSA were 1.5 to 2.5 times higher than for membrane permeation, depending on the purity of the product gas (99 or 98% H_2, respectively). Because the retentate gas of the membrane unit remains at a high pressure, it can be recycled to the ammonia reactor. In this way the retentate gas is not degraded to fuel gas. If PSA is used as the purification process the hydrogen recovery is a mere 68%, due to the presence of 25% nitrogen, and the tail gas can only be used as a fuel gas, as it has a pressure of 1.3 bar. Therefore, the variable costs of PSA are also higher than those of membrane permeation. The total costs of the hydrogen gas produced by PSA will be 35 to 40% higher than those of the gas produced by membrane permeation, depending on the product purity (99 and 98%, respectively).

If all the cryogenic product gas (90% H_2, 12,000 Nm^3/hr) is used as feed for either a PSA unit or a membrane unit the investment costs for PSA would be twice as high as those for the membrane unit. The variable costs for PSA would also be higher than those for membrane permeation because the tail gas has only fuel gas value, while the retentate gas of the membrane unit can be recycled to the ammonia reactor. The total costs of the hydrogen gas produced by PSA would be 12.5% higher than those for the gas produced by membrane permeation.

For the production of a large quantity of pure hydrogen gas, 10,000 to 15,000 Nm^3/hr hydrogen, the economically most feasible option would still be membrane permeation with cryogenic product gas as feed gas. The sum of the fixed and variable costs of hydrogen gas produced by PSA with naphtha cracker gas as feed would be 8% higher than the costs of hydrogen gas produced by membrane permeation.

The purification of hydrogen gas for this hydrogenation process is in fact a hybrid process of cryogenic recovery and membrane permeation, although the capacities of both systems do not match. Hybrid systems are a new development, where the advantages of the individual units are combined to produce a purer gas than is possible with any of the stand-alone systems, and at lower costs (Agrawal 1988). Most cases reported are only model studies, as no real systems have been installed yet. In each case the exact combination, with or without recycle streams, will have to be determined. The choice will depend on the scale of operation and the possible use of the non-product gas. Several joint ventures have been formed between manufacturers of cryogenic units and membrane units.

CONCLUSIONS

The choice of a separation process for a given application is not only dependent on the desired purity and recovery of the product gas, but also on other process conditions like integration with other processes, the desired capacity and the composition of the available feed gas, as is shown here.

In the case where a minimum purity of 98% hydrogen is required, membrane permeation with cryogenic product gas as feed gas is the most favourable option.

However, if a hydrogenation process is chosen, where very high hydrogen pressures are required, a hydrogen purity of 99.9% would be the most favourable option. In that case PSA with naphtha cracker gas as feed gas would be the optimum choice.

In the future hybrid systems will probably play an increasingly important role in hydrogen separation processes, as they promise to combine a higher hydrogen purity with lower overall costs.

REFERENCES

- Agrawal, R., Auvil, S.R., DiMartino, S.P., Choe, J.S. and Hopkins, J.A. (1988) 'Cryogenic hybrid processes for hydrogen purification', *Gas Sep. Purif.*, vol 2, no 1, pp. 9-15.
- Knoblauch, K., Jüntgen, H. and Peters, W. (1979) 'Gastrennung mit Kohlenstoffmolekularsieben', *Erdol Kohle, Erdgas, Petrochem.*, vol 32, no 12, pp. 551-556.
- Meindersma, G.W. (1990) 'Comparison of several hydrogen separation processes', Process Technology Proceedings, 8, *Gas Separation Technology*, Ed. by Vansant, E.F. and Dewolfs, R., Elsevier Science Publishers B.V., Amsterdam, pp. 623-630.
- Wiessner, F.G. (1988) 'Basics and industrial applications of pressure swing adsorption (PSA), the modern way to separate gas', *Gas Sep. Purif.*, vol. 2, no 3, pp. 115-119.

MEMBRANE GAS SEPARATION IN PETROCHEMISTRY: PROBLEMS OF THE POLYMERIC MEMBRANE SELECTION

V.Teplyakov (A.V.Topchiev Institute of Petrochemical Synthesis, Academy of Sciences, USSR)

ABSTRACT
Although the solubility and diffusivity of each gas and vapour in each polymer are temperature and in some systems pressure (concentration) dependent, regular trends are noted when either many gases are studied in a single polymer or a single gas is studied in many polymers. Several empirical though scientifically based correlations of such data have been proposed which are at best semi-quantitative. In this paper improved correlations are described to correlate data on diffusivities *D* and solubilities *S* of rare gases, multiatomic gasesand lower hydrocarbons in polymers. The use of these, demonstrated by modelling of the variation of the membrane permselectivity to hydrocarbon containing gas mixtures with variation ofconcentration of the most condensable component this is of importance for petrochemistry applications.

The procedures rely on determining an effective Lennard-Jones {6-12} potentialforce constant and molecular diameter for each gas which hold in all polymers above T_g. Slightly smaller diameters are needed for multiatomic gases below T_g. Each polymer is characterized by four temperature and in some systems concentration dependent parameters, two for diffusivity and two for solubility, which hold with all gases.

The procedure described may be used to predict *D* and *S*, and hence permeabilities, in cases where data do not already exist. The values and ratios of the predicted permeabilities are a valuable guide when seeking polymers for separating gas mixtures by membrane processing and also for modelling of the permselectivity profile under concentration-dependent conditions.

INTRODUCTION

In the last few years membrane separation of different liquid and gaseous mixtures took on great significance as highly effective technological processes realized in different branches of industry. Prospects of use membrane technology development in petrochemistry include the following (Durgaryan 1983):

1. The extraction of valuable components from petrochemical streams either for returning into the process or for application in other processes.
2. The opportunities of process operation by membrane control of compositions of the circulating streams that increases the productivity of existing technology.
3. The solution of the problems of the gas mixture separation in cases where criogenic and other traditional techniques are not effective enough.
4. Removing of deleterious components from gaseous and liquid streams for solution of the ecology problems.

Considering the advantageous economic, ergonomic and

Effective Industrial Membrane Processes — Benefits and Opportunities, pp.401–415

ecological features possessed by membrane technology the very slow extention of the membrane gas separation methods into different branches of petrochemistry is surprising. An exception is the recovery of hydrogen from itsmixtures with carbon monoxide, methane, nitrogen, argon (Henis 1980) as is also, in some cases, the carbon dioxide/methane separation (Fabiani 1986). The limitations of membrane technology applications in petrochemistry are seen to be connected with the selection of polymeric membranes possessing desirable permeability, selectivity and stability to petrochemical mixtures which, as a rule, consist of organic and inorganic gases, organic vapours and liquids. Mixtures of low molecular weight gases and vapours need especially the application of a fine membrane separation by use of non-porous polymeric membranes. Such membrane materials should have high permeability coefficients, high selectivity for certain gas mixtures and be suitable for the production of very thin and stable films i.e. membranes (Crank 1968, Reytlinger 1974, Hwang 1975, Meares 1976, Nikolaev 1980, Lloyd 1985, Bungay 1986).

Despite considerable theoretical and experimental research in the field of gas transport through various classes of polymers, existing theoretical approaches cannot satisfactorily explain all of the observed facts. This is especially true of gas and vapour separations when the permeability parameters are concentration dependent. Nor does the current state of knowledge allow the formulation of detailed criteria of chemical structure, physical and phase states and other physico-chemical properties to guide the screening or synthesis of homo- and copolymers for use in membranes with the required properties for petrochemistry applications. The problem is especially acute in the case of the understanding of the selectivity of membranes towards organic gases and vapours which modify the properties of the membrane materials. This is not adequately explained by present theories.

An attempt is made in this paper to analyze known correlations of permeability parameters for membranes that are alredy in use in order to select the chemical properties most important in controlling gas transport selectivity. Concentration dependent diffusivity and solubility coefficients as well as predictions of permeabilities to inorganic gases and C_1-C_4 hydrocarbons are included.

METHODS OF THE CHOICE OF GAS PERMEABILITY PARAMETERS.

At the present time experimental techniques and mathematical methods are available to allow the analysis of simple and moderately complex diffusion phenomena in polymers. The methods of analyzing experimental data on diffusion are scattered throughout the scientific literature (e.g. Crank 1968, Crank 1975, Nikolaev 1980, Shvyryaev 1981, Beckman 1983, Palmai 1984).

Modern techniques permit the main gas transfer parameters i.e. the coefficients of diffusivity D, solubility S and permeability P to be determined reliable for single gases and vapours in polymers. In the simplest case

$$P = DS = Qh/A\Delta p \qquad (1)$$

where Q is the steady state flux of gas under the pressure difference Δp through an area A of membrane thickness h.

The experimental difficulties connected with the determination of diffusion parameters are not covered here since they have been widely discussed in the abovementioned literature. It is necessary however to bear in mind the possibility that gas permeability parameters depend on concentration, space and time. In particular, they may depend on the extent of crosslinking (Crank 1968, Reytlinger 1974), extent of crystallinity in partly crystalline polymers (Michaels 1961), annealing temperature and plasticization of glassy polymers (Crank 1968, Reytlinger 1974), morphology and organization in copolymers and in blends of polymers (Odani 1975, Ievlev 1982, Barrie 1983, Petropoulos 1985). With mixtures of permeants one penetrant may influence the permeability of another and, especially in glassy polymers under high gas pressures, a fraction of the permeant may be immobilazed (e.g.Whyte 1983).

In this paper the permeability parameters, derived mainly from permeability experiments, for rare gases, a number of multiatomic gases and hydrocarbons from C_1-C_4 in number of polymeric materials are considered. The influence of the permeant concentration on these characteristics are also taken into account.

Polymers on which data have been taken into consideration include polydienes, polysiloxanes, polyolefins, halogen and silicon-containing carbochain polymers, polymeric esters and ketones, polyamides and polyaromatic compounds. In addition, random and block copolymers of polydienes, polydimethyl siloxane (PDMS), polyvinyltrimethylsilane (PVTMS) and polyolefins have been included. The sources of data are (Brandrup 1975, Allen 1977, Volkov 1979, Durgaryan 1979, Nakagawa 1986, Teplyakov 1987). On the base of known dependencies of diffusion, solubility (and as result permeability) coefficients on concentration the estimation of the membrane permselectivity to gas mixtures consisting of abovementioned gases is evaluated in dependence on increasing amounts of C_4-hydrocarbons in the feed.

In the ideal case the selectivity F_P of membrane for a pair of gases i and j is given by

$$F_P = P_i/P_j = (D_i/D_j)(S_i/S_j) = F_D/F_S \tag{2}$$

where F_D and F_S are the selectivities of the diffusion and solution processes respectively. These ratios of the permeabilities of a particular pair of gases vary widely over a range of polymers. For example, F_P for **He/Kr** varies from less than unity in PDMS, NR & PE (Lundstrom 1974) to greater than 10^5 in PAN (Allen 1977). (Abbreviations of polymer names are listed in Table 2).

Such large permeability ratios may be compared with the behaviour of porous membranes. Flow in microporous membrane is characterized by the Knudsen diffusion mechanism and selectivity is determined by the square root of the molecular weights of the permeants. In the absence of surface flow it is independent of the chemical nature of the polymer. In macroporous membrane viscous flow occurs and there is no separation (Hwang 1975).

Such great variability in the permeability and selectivity towards gases of organic polymers which, as a class, have rather similar densities and elemental compositions, has led to considerable research into the mechanism of gas diffusion as well as on the influence of the physical and chemical properties of the polymer on its gas separation parameters.

When a membrane has similar physical properties to the relaxed bulk polymer from which it was cast, its selectivity is close to the ratio of the permeabilities in the starting polymeric material. Imperfections in the membrane resulting for example from poor processing or poor film-forming properties lead to lower selectivity. Furthermore the effectiveness of gas separation by real membranes depends on the equipment, process design and operating strategies, for example the organization of gas fluxes, cascades etc (Hwang 1975).

Since the intrinsic permeation parameters of different gases in polymers form the basis for membrane gas separation, the selection of suitable polymeric materials has become of paramount importance. Consequently a knowledge of the general trends in polymer/gas transport properties, and their dependency on the concentration of active permeant is of value particularly if such trends can be quantified by correlations.

THE CORRELATION ASPECTS OF PERMEABILITY PARAMETERS.

Several correlations of the coefficients of diffusivity D, solubility S and permeability P of gases and vapours related to their properties and those of polymers have been advanced already.The correlation of permeabilities has been proposed with polymer properties such as density and extent of crystallinity in polyolefins, glass transition temperatures and melting points of glassy polymers, cohesive energy density of polyaromatic compounds. Correlations of P with gas properties such as molecular size and critical temperature have been suggested (e.g. Peterlin 1975). Such correlations are valid only for particular groups of homopolymers, often suffer from errors even in the order of magnitude of the predicted values and do point the way to the need for a more general approach to the understanding of the gas permeabilities of polymeric materials and membranes.

There are two approaches to relating permeability data with polymer chemical structure. They are the Permachor and the solubility space factor concepts. The Permachor correlation makes use of a scheme of additive (positive or negative) contributions from the chemical structural elements of the polymer to its permeability (e.g.Reytlinger 1974, Hwang 1975). These structural contributions however, unlike classical additive properties, are not the same for all gases nor for all polymers. The Permachor concept can be applied in terms of known experimental data only to homopolymers. Thus it is not a reliable predictive tool for copolymers, for new polymers or for polymeric blends, composites and so on.

The solubility space factor method has been suggested in an attempt to establish a more general basis for permeability prediction. The final result was an equation which contained only three additive terms (Hwang 1975)

$$\ln P = a + b(T_c/T) + c(d/T)^n \qquad (3)$$

Here a, b and c are constants depending on the polymer, T is the absolute temperature and T_c is the critical temperature of the permeant. d is the molecular diameter of the permeant which is a function of temperature, and n is the space factor. A feature of this method is that equation (3) contains separately thermodynamic $[b(T_c/T)]$ and kinetic $[c(d/T)^n]$ components.

The space factor n allows for the effect of the thermal expansion of the polymer membrane. The thermal expansion coefficient can be estimated by a method of additive contributions (Van Krevelen 1976). While this method has some promise it has the same disadvantages as the Permachor.

From a theoretical standpoint the correlations of diffusivity and solubility of gases in polymers would be better considered separately.

The diffusion coefficient D of penetrants in polymer usually decreases with increasing dimensions of the penetrant molecule. The critical volume, Van der Waals volume, collision diameter, cross-section or any combination of these may be selected as a measure of the molecular dimension (e.g.Reytlinger 1974, Koros 1988). In a number of papers difficulties have been noted in analyzing diffusion data using such properties and several scales of molecular diameters obtained by different calculation methods have been explored (Stuart 1967, p.84).

An analysis of the correlation between D and the molecular diameter has been carried out for more than fifty polymeric materials (Teplyakov 1984, 1985). It was found that the dependence of D on molecular diameter for the rare gases and a number of multiatimic gases was optimally expressed by equation (4) below

$$\log D_i = K_1 - K_2 d_i^2 \quad (4)$$

In this equation D_i is in [m^2/s]. d_i in [nm] is the effective molecular diameter listed in Table 1. These effective diameters were obtained by comparing the diffusion coefficients of the various gases in different polymers with those of the rare gases . It should be noted that the more experimental permeability data are considered the more reliable value of d_{ef} are obtained. From this point of view the reliability of d_{ef} values decreases in sequence of rare gases, permanent gases, C_1-C_4 hydrocarbons and acid gases.

Equation (4) is valid for homopolymers and copolymers under different phase and physical condition. In contrast to equation (3), although both coefficients of equation (4) depend on temperature they are isothermally constant for any one polymer and, for example, can depend on the plasticizer concentration (Evseenko 1979).

Equation (4) is based on the hypothesis that bi- and multi-atomic molecules are oriented during diffusion jumps in the direction of the maximum molecular dimension of the molecule (Michaels 1961, Berens 1982, Teplyakov 1984). In support of this hypothesis it may be noted that the effective cross-sections of multi-atomic gases obtained by comparison with the rare gases are lower than the gas phase equilibrium values shown in the second column of Table 1. In Figure 1 the correlation of D with these effective molecular cross-sections is illustrated for CO_2 and C_3H_8.

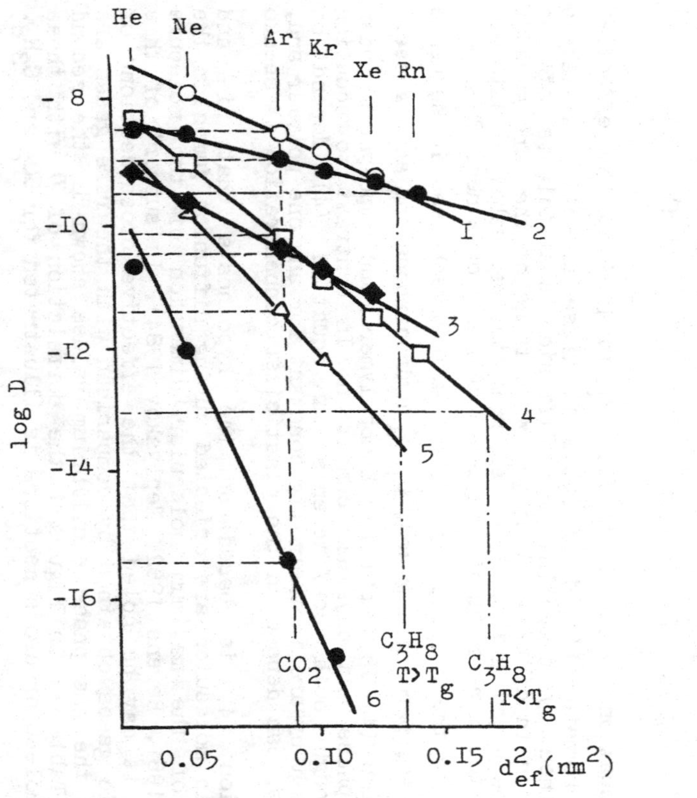

Fig.1. The dependence of diffusivity (as $log\ D$) on d^2_{ef} of rare and multiatomic gases at 298 K: 1 PTMSP; 2 - PDMS; 3 - PE; 4 - PVTMS; 5 - PEMA; 6 - PAN. The estimation of the effective cross-section of CO_2 and C_3H_8 is shown. [D, (m^2/s); d_{ef}, (nm)].

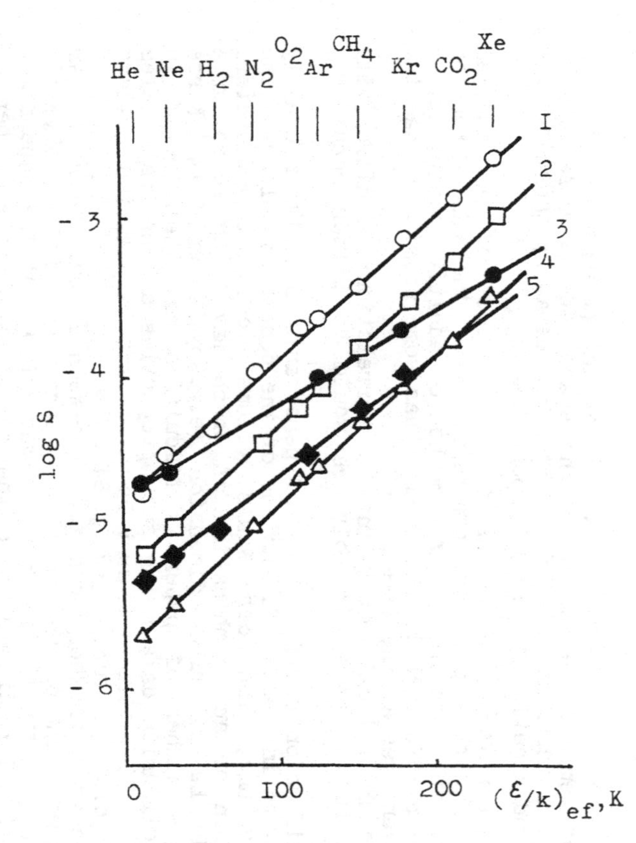

Fig.2. The dependence of the coefficients of gas solubility (as $log S$) at 298 K on the {6-12}-potential force constants ε/k: 1 - PVTMS; 2 - NR; 3 - PA-11; 4 - PE; 5 - PVA. [S, (mol/m^3Pa)].

TABLE 1

Effective molecular diameters (d_{ef}) and force constants $(\varepsilon/k)_{ef}$ of inorganic gases and C_1-C_4 hydrocarbons obtained from published diffusion data (taken from Teplyakov 1984, 1987)

Permeant	d, nm from (Stuart 1967, de Ligni 1972, Reid 1977)	d_{ef}, nm		(ε/k),K from (Hirshfelder 1954, de Ligni 1972, Reid 1977)	$(\varepsilon/k)_{ef}$,K
		$T<T_g$	$T>T_g$		
He	0.182-0.376	0.178	0.178	1.5- 10.2	9.5
Ne	0.225-0.305	0.230	0.230	27.5- 32.8	27.1
Ar	0.300-0.377	0.297	0.297	93.3- 116.0	122.3
Kr	0.320-0.404	0.322	0.322	179 - 190	176.6
Xe	0.355-0.439	0.352	0.352	229 - 231	232.9
Rn	0.370-0.470	0.377	0.377	290	-
H_2	0.218-0.362	0.214	0.214	39 - 60	62.2
O_2	0.254-0.375	0.289	0.289	88 - 107	112.7
N_2	0.258-0.402	0.304	0.304	71.4- 80	83.0
CO_2	0.319-0.414	0.302	0.302	189 - 213	213.4
CO	0.323-0.369	0.304	0.304	92 - 100	102.3
SO_2	0.429	0.344	0.344	252	291
H_2S	-	0.324	0.324	-	190
CH_4	0.330-0.420	0.318	0.318	143 - 149	154.7
C_2H_6	0.370-0.520	0.369	0.346	216 - 229	250
C_3H_8	0.410-0.580	0.409	0.367	237 - 277	305
C_4H_{10}	0.460-0.690	0.440	0.369	263 - 531	364
C_2H_4	0.360-0.550	0.357	0.338	166 - 225	225
C_3H_6	0.410-0.540	0.386	0.352	225 - 299	294
C_4H_8-1	0.450-0.520	0.418	0.369	270 - 345	356
C_2H_2	0.350-0.570	0.338	0.329	190 - 232	223
C_3H_4(m)	0.400-0.480	0.362	0.339	236 - 302	321
C_4H_6(e)	0.440-0.500	0.381	0.350	257 - 371	360
C_3H_4(a)	0.460-0.490	0.364	0.345	266 - 332	335
C_4H_6(b)	0.500	0.385	0.345	253 - 350	327

(m) - methylacetylene, (e) - ethylacetylene, (a) - allen, (b)-butadien.

The need to correlate with these effective molecular diameters is especially apparent for the C_2–C_4 hydrocarbons. For these two values of d_{ef} are proposed; one calculated from diffusion data on glassy polymers and the other for rubber-like polymers (Teplyakov 1984).

It is interesting to note that, while two values of d_{ef} are required for the C_2–C_4 hydrocarbons, a single value is sufficient for the permanent gas molecules although the molecular masses of these can be appreciably larger than those of the hydrocarbons [e.g.**Kr** = 84 Dalton, C_3H_8 = 44 Dalton]. In all cases increasing the effective cross-section of the permeant leads to decreasing of D as seen in Figure 1.

The form of the diffusion selectivity for any chosen pair of gases in a polymer is determined by the exponential index in equation (4). Thus one obtains

$$F_D = D_i/D_j = 10^{[K_2(d_j^2 - d_i^2)]} \qquad (5)$$

The diffusion selectivity is seen to be higher when there is a greater difference between the effective cross-sections of the permeants i.e. a gas property, and a higher K_2 value i.e. a polymer property.

When K_1 and K_2 are known for a polymer, coefficients of diffusivity of more than ten gases and a number of hydrocarbons up to C_4 can be estimated in that polymer. A large number of K_1 and K_2 values have been published already (Teplyakov 1987). Some more, for different polymeric materials, are listed in Table 2. The values of K_1 and K_2 for new polymer can be obtained by collecting diffusion data on two or three gases. Although the correlation equation (4) can be used satisfactorily to predict the D values of many gases and vapours in different polymers, including polymers of unknown composition, on the basis of only minimum experimental data, a simple empirical approach is not sufficient to estimate the values of K_1 and K_2 from polymer structure. These coefficients are determined not only by the chemical structure of the polymer but also by physical properties such as density, extent of crystallinity, glass temperature, interchain interactions, macrochain packing and supra-molecular organization. Interchain interactions and macrochain flexibility can have a major effect on gas diffusivity and these features can predominate under concentration dependent permeability.

Although it has to be expected that K_1 and K_2 are parameters depending on many polymer properties, it may noted that there are predominating trends. Thus K_1 varies relatively little from one polymer to another while K_2 increases continuously with increasing cohesive energy density (CED) of polymer. For example, in PDMS (CED = 238 MJ m^{-3}) K_2 is +0.1, in PAN (CED = 1033 MJ m^{-3}) K_2 is + 0.947. It is seen also from Table 2 that K_2 decreases continuously with

TABLE 2
The correlation coefficients of equations (4), (8) for estimation of gas-permeability parameters of polymers (Teplyakov 1984, 1987)

No	Polymer	K_1	K_2	K_3	10^2K_4
	Homopolymers				
1	Polydimethylsiloxane (PDMS)	-7.85	0.100	-4.70	0.77
2	Polytrimethylsilylpropyne (PMSP)*	-6.94	0.185	-4.07	0.85
3	Polybutadien (PB)	-8.15	0.211	-5.47	0.85
4	PE (ρ=0.914 g/ccm)	-8,42	0.225	-5.74	0.92
5	PE (ρ=0.968 g/ccm)	-8.71	0.250	-5.95	0.79
6	Polymethylvinylether (PMVE)	-7.83	0.324	-5.56	0.80
7	Polyvinyltrimethylsilane (PVTMS)	-7.22	0.347	-4.84	0.93
8	Polyphenylenoxide (PPO)	-7.96	0.373	-4.42	0.94
9	Polycarbonate (PC)	-8.24	0.423	-5.14	1.31
10	Polyamide-11 (PA-11)	-8.16	0.427	-5.87	1.02
11	Polyvinylacetate (>T_g) (PVA)	-7.40	0.451	-5.41	0.77
12	Polyvinylchloride (PVC)	-8.09	0.470	-5.64	0.94
13	Polyethylmetacrylate (PEMA)	-7.29	0.471	-5.38	0.98
14	Polyethylenterephtalate (PETPH)	-7.73	0.556	-5.52	0.98
15	Polyvinylacetate (<T_g) (PVA)	-7.15	0.578	-5.59	1.01
16	Polymethylvinylketone (PMVK)	-7.49	0.603	-5.60	0.74
17	Polyacrylonitrile (PAN)	-7.09	0.947	-5.65	1.47
	Random copolymers**				
18	Butadien:acrylonitrile =80:20	-8.04	0.248	-5.60	1.06
19	=73:27	-8.05	0.282	-5.66	1.09
20	=68:32	-7.96	0.316	-5.74	1.14
21	=61:39	-7.97	0.357	-5.78	1.17
	Block-copolymers				
22	Polybutadien-polystyren (ABA) =73:27	-8.47	0.142	-5.50	1.10
23	=59:41	-8.43	0.171	-5.64	1.16
24	Polyarylate-polydimethylsiloxane $(AB)_x$ =46:54	-8.19	0.121	-4.82	0.75
25	Polyvinyltrimethylsilane-polydimethylsiloxane (AB) =80:20	-7.62	0.282	-4.53	0.76
26	=64:36	-7.84	0.219	-4.51	0.88
27	=60:40	-7.83	0.188	-4.86	0.80
	Plasticized polymers***				
31	Polyvinyltrimethylsilane (PVTMS) + oligovinyltrimethylsilane = 90:10	-7.44	0.320	-4.88	0.90
32	= 80:20	-7.50	0.280	-4.85	0.88
33	= 60:40	-7.02	0.220	-4.77	0.90
	+ oligovinyloctyldimethylsilane				
34	= 90:10	-7.27	0.324	-4.89	0.89
35	= 85:15	-7.36	0.310	-4.86	0.88
36	= 75:25	-7.26	0.280	-4.70	0.85
37	+ dioctylsebacinate = 95:5	-7.24	0.323	-4.88	0.90
38	= 85:15	-7.40	0.305	-4.89	0.87
39	= 80:20	-7.11	0.283	-4.76	0.86

* Nakagawa 1986, ** wt:wt parts, ***Evseenko 1979.

concentration of plasticizer (see Tabl.2).

It should be noted that K_2 and hence F_D are largest for glassy polymers with polar side groups attached to the main chain. K_2 has medium values in glassy polymers with non-polar side groups and the smallest values are found with rubber-like polymers.

It is very important for petrochemistry application aims to know variation of K_1 and K_2 with concentration of diffusant. Unfortunately the published data describe as a rule individual dependences of diffusivity for small group of polymers without taking into consideration the effect of other gases in the diffusing medium. The diffusion coefficients of different gases in polymers can vary with concentration of permeant in accordance with the empirical equation (Nikolaev 1980):

$$D = D_0 exp(\alpha(p/p^*)) \qquad (6)$$

where α is a factor dependent on temperature, p/p^* is the activity of gas (vapour). Additionally it may be noted that plasticizer action on glassy polymers leads to decreasing of T_g (Reytlinger 1974, Evseenko 1979) with appreciable changing D values. It follows that the concentration of diffusant in polymeric membrane can control K_1 and especially K_2 values to a similar extent as a variation from glassy to rubber-like state (see Table 2). Furthermore it can be supposed that more soluble (condensable) permeant (for example, C_4 hydrocarbon) can most easily plasticize the polymeric material of the membrane. In this case with consideration of above correlations the diffusivity of gases and vapours can be estimated under the concentration dependent ones. The factor α should be known for concrete gas-polymer system.

While the kinetic factor D in the permeability coefficient decreases with increasing cross-section of the migrating molecule, the thermodynamic factor S normally increases. More correctly, the solubilities S of gases and vapours in different polymers (e.g.Volkov 1979), by analogy with solubilities in liquids, increase in accordance with T_c, T_b or the force constant of the {6-12}intermolecular potential. Such correlations can be derived theoretically by equating the chemical potentials of the permeant in the gas and in the solid polymeric phase (Michaels 1961) and the most popular of that is

$$\log S = 0.011(\varepsilon/k) + const \qquad (7)$$

Equation (7) suggests that the index of the exponent,0.011, will be independent of the nature of the polymer and that solubilities will depend only on the Lennard-Jones potential ε/k. It means that the solubility selectivity should only weakly depend on the permeant concentration in polymeric membrane.

There is no established view on which property to choose as characteristic of the sorbent power of a polymer. It is known from experimental data that the greater the free volume fraction of the polymer the greater the solubility of gases in it (e.g. Peterlin 1975).

On generalizing this approach the correlation equation

$$\log S_i = K_3 + K_4(\varepsilon/k) \qquad (8)$$

is proposed. Its success with the rare and some simple gases is illustrated in Figure 2. (S in [mol(gas)/m^3(polymer)*Pa].

The coefficients K_3 and K_4 obtained from such correlation lines are listed in Table 2. The values for any polymer can be estimated by determining the solubilities of two or three simple gases. The effective values of ε_i/k which determine the solubilities of other gases in polymers can be obtained from these correlation plots if the solubility in a single polymer is known. By comparing the last two columns in Table 1 it can be seen that the effective values of ε/k lie within or very close to the range from entirely different data that can be found in the literature.

Regarding the values of K_3 and K_4 in Table 2, K_4 is seen to vary relatively little among different polymers and to be quite to the theoretical value 0.011. It can be seen however that K_4 shows a slow upward trend with increasing polarity of the polymer. Although the values of K_3 are not constant the variations are not large and show no clear trend with the chemical composition of the polymer including blends with plasticizers (see Table 2).

The selectivity due to the solubility difference for any pair of gases can be represented by the relation

$$F_S = S_i/S_j = 10^{[K_4(\varepsilon_i - \varepsilon_j)/k]} \qquad (9)$$

It is determined therefore by the difference between the ε/k values and the correlation coefficient K_4. As expected, the solubility selectivity depends only weakly on the chemical nature of the polymer.

By combining the correlation equations (4) and (8) one obtains general relations for the permeability coefficients and the selective permeation of gases. They are equations (10) and (11) below

$$\log P_i = \log D_i + \log S_i = K_1 + K_3 - K_2 d_i^2 + K_4 \varepsilon_i/k \qquad (10)$$

$$F_P = F_D F_S = 10^{[K_2(d_j^2 - d_i^2) + K_4(\varepsilon_i - \varepsilon_j)/k]} \qquad (11)$$

Equation (11) is similar in form to that given by Michaels et.al (1961) for polyolefins except that all the values in it have been substantiated statistically and it can be applied to homopolymers and copolymers under different phase and physical conditions.

The selective permeation of many gases and some hydrocarbons in polymers depends on the difference between the cross-sections of the permeant molecules and the force constants of the Lennard-Jones {6-12} potential as well as on the properties of the polymer expressed through K_2 and K_4. These coefficients appear to be controlled by the cohesive energy density and the free volume fraction of the polymer respectively as predominant properties. The balance of these properties regulates the balance of the kinetic and thermodynamic components of the permeability. As a result the selectivity of gas permeation for a given pair of gases can be either greater than unity or less than unity depending on the polymer. For this reason the choice of polymeric material for use as a membrane may depend on the aim of the separation. For example, if it is wished to separate the more soluble component the search should be made among polymers possessing a high free volume fraction and low cohesive energy density. When it is required to separate the lighter gases with low solubilities the search should be among polymers possessing the contrary balance of properties i.e. low free volume fraction and high CED values. The acid gases can be in intermediate position. In practice, of course, the choice of membrane materials depends on many requirements including the composition of the feed mixture, the aim of the separation, the boundary conditions of the operation, the required purity of the permeant product and economic factors (Lloyd 1985, Bungay et al 1986).

It may be noted that in rubber-like polymers, including partly crystalline ones and one glassy polymer PTMSP, the solubility of the gases and vapours is the predominant factor and in these cases the permeability of the more soluble gas is greater than that of the less soluble one. On the other hand the contrary holds with regard to the selectivity of the glassy polymers.

All these considerations should also hold for concentration dependent membrane systems. A good example of this can be seen in Figure 3, where the variation of the permselectivity reliefs are shown for mixture of 16 gases and vapours for PVTMS membrane with predominant plasticization by butane as active permeant. The surfaces of Fig.3 were modelled by using K_1-K_4 factors for PVTMS + C_4H_{10} system with T_g above, at, or below experimental temperature and also for the viscosus condition caused by increasing the concentration of butane in feed.

It is seen that the related selectivity of gas separation under influence of the hydrocarbon components can transform a membrane system from one for separating the light and acid gases from the hydrocarbons into one which separates the heavier hydrocarbon molecules from the othe gases. The control of the stability level of the highly plasticized polymeric membrane can be accomplished by the cross-linking techniques, regulation of hydrofillic-hydrofobic balance etc.

CONCLUSION

The proposed improved correlations of permeability parameters can be used for empirical analysis of the membrane permselectivity of the hydrocarbons containing gas mixtures under the concentration-dependent conditions for petrochemistry aims.

It should be noted that once coefficients K_1-K_4 in a polymer have been established an estimate can be made of the permeability of any gas for which the effective diameter and force constant are known. Thus the potential of any polymer for separating a particular pair of gases (or a number of those) can be assessed without carrying out additional experimental work. Since the values

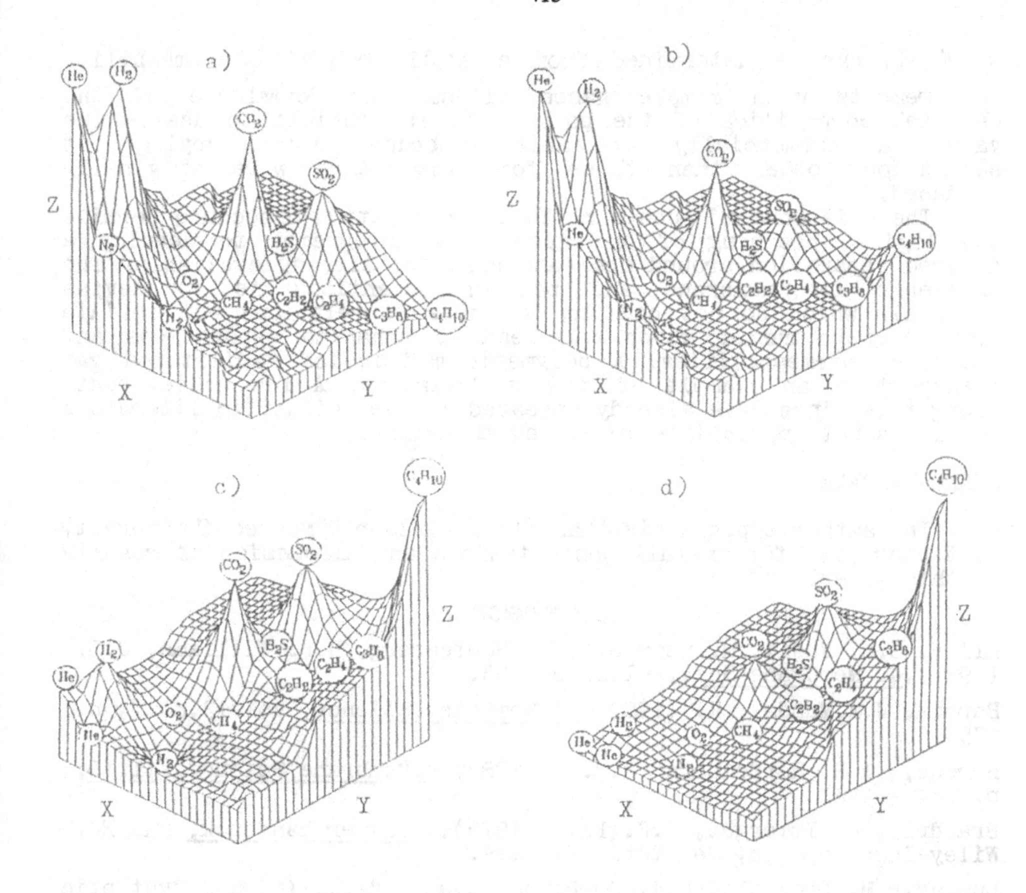

Fig.3. The variation of the permselectivity relief *vs* increasing of the C_4 hydrocarbon fraction (as decreasing of polymer-gas system T_g) in PVTMS membrane (the model calculation by using the correlation equations). The meanings of coordinates (arbitrary units): X is the effective cross-section of permeant molecule; Y is the {6-12}-potential force constant of permeant molecule; Z is the coefficient of gas permeability (or productivity) of PVTMS membrane.

a) $T_{exp} < T_g$; b) $T_{exp} \sim T_g$; c) $T_{exp} > T_g$; d) the viscous condition caused by butane plasticization.

of K_1-K_4 can be determined from a small nomber of permeability measurements with simple gases without any knowledge of the chemical composition of the polymer it is possible to assess the value of commercially avaliable membranes for carrying out separations other than those for which they were originally produced.

The existence of such a reliable correlation procedure permits the logical planning of an experimental programme to develop a desired separation around membrane polymers already avaliable. The challenge for membrane and polymer scientisis is to devise materials which do not confirm to these correlations but deviate from them in manner which can lead to enhanced separations and increased permeabilities. A polymeric medium for facilitated gas transport is an example of such a deviation. A few achievements along this lines have already appeared in the scientific literature but it is not appropriate to review them here.

ACKNOWLEDGEMENT

The author express his thank to Professor P.Meares (University of Exeter, UK) for usefull consultations and discussion of results of this paper.

REFERENCES

Allen, S.M. Fujii, M. Stannett, V. Hopfenberg, H.P. Williams, J.L. (1977), J. Membrane Sci., vol.2, p. 153.

Barrie, J.A.Munday, K. (1983), J.Membrane Science, vol.13, no.1, p. 175.

Berens, A.R. Hopfenberg H.B. (1982) J.Membrane Science, vol.10, p.283.

Brandrup, J. Immergut, E.H.(Ed.) (1975), Polymer Handbook, 2nd Ed. Wiley-Interscience, New-York, III-229.

Bungay, P.M. Lonsdale, H.K. de Pinho, M.N. (eds.), (1986), Synthetic Membranes: Science, Engineering and Applications (NATO ASI Series C, vol.181), D.Reidel Publishing Company: Dordrecht/Boston/lancaster/Tokyo, p.733.

Crank, J. Park, G.S.(1968) Diffusion in polymers, NY.: Acad.Press, p.445.

Crank, J., (1975), The Mathematics of Diffusion, 2nd Edition, Clarendon Press, Oxford.

Dourgaryan, S.G., Yampolsky, Y.P. (1983), Petrochemistry, vol.XXIII no.5, pp. 579-95.

Evseenko, A.L.Teplyakov, V.V.Dourgaryan, S.G. Nametkin, N.S. (1979) Vysokomolek.soed., vol.21B, no.2, p.153.

Fabiani, C. Pizzichini, M. Bimbi. L. Visentin, L. (1986), Separation Sci.Techn., vol. 21, p.1111.

Henis, J.M. Tripodi, M.K. (1980) Separation Sci.Techn., vol. 15, no. 4, pp.1059-68.

Hirshfelder, J.O. Curtiss, Ch.F. Bird, R.B. (1954) Molecular Theory of Gases and Liquids, Wiley, New-York, p.1100.

Hwang, S.-T. Kammermeyer, K. (1975) Membranes in Separations, Wiley-Interscience: New-York, , Chapter 3.

Ievlev, A.L. Teplyakov, V.V. Durgaryan, S.G. Nametkin, N.S. (1982) Dokl. of the USSR Academy of Sciences, vol.264, no.6, p.1421.

Koros, W.J. Fleming, G.K. Jordan, S.M. Kim, T.H. Hoehn, H.H. (1988) Polymeric membrane maerials for solution-diffusion based permeation separations, Prog. Polym. Sci., vol. 13, p.339-401.

de Ligni, C.L. and van der Veen, N.G. (1972) Chem.Eng.Sci., vol.27, p.391.

Lloyd, D.R., Ed.(1985) Materials Science of Synthetic Membranes, ASC Sump Ser 269 American Chemical Society, p.492.

Lundstrom, J.E. and Bearman, R.J, (1974), J. Polymer Sci: Polym. Phys., vol.12, p.97.

Meares, P. Ed. (1976), Membrane Separation Processes, Elsevier : Amsterdam - Oxford - New-York, p.600.

Michaels, A.S. Bixler, H.J.(1961) J.Polymer Sci., vol.50,p.393,413.

Nakagawa, T. (1986) Membranes in gas separation and enrichment, Proc 4th BOC-Priestley Conf Specl Publ 62, Royal Society of Chemistry, London, UK, p.351.

Nikolaev, N.I.(1980), Diffusion in membranes. Moscow: Khimiya, p.232.

Odani, H. Taira, K. Nemoto, N. Kuraba, M. Bull. Inst.Chem.Res. (Japan) (1975) vol.53, p.216.

Palmai,G. Olah, K. (1984) J. Membrane Science, vol. 21,p. 161.

Peterlin, A. (1975) J.Macromolec.Sci.,vol. 11(B),no. 1,p.57.

Petropoulos, J.H.(1985) Journal of Polymer Science: Polymer Physics Edition, vol.23, p. 1309.

Reytlinger, S.A. (1974) Permeability of polymeric materials,Moscow: Khimiya, p.268.

Reid, R.C. Prausnitz, J.M. Sherwood, T.K. (1974) The properties of Gases and Liquids McGrow-Hill, New-York, 3rd Edn.

Shvyryaev, A.A. Beckman, I.N. (1981) Vestnik MGU, Chemistry, vol. 22, p. 517.

Stuart, H.A. (1967) Molekulstruktur, Berlin: Springer, p.84.

Teplyakov, V.V. Durgaryan, S.G. (1984), Vysokomolek. soed., vol. 26A, no.7, p.1498; (1986), vol.28A, no.3, p.504.

Teplyakov, V.V.(1987), D.I.Mendeleev J. Vses.Khim.Ob., vol.32, no.6, pp.693.

Teplyakov, V.V. Ievlev, A.L. and Durgaryan, S.G. (1985) Vysokomolek. soed., vol.27A, p.818.

Van Krevelen, D.W.(1976) Properties of Polymers, Elsevier, Amsterdam.

Volkov, V.V. Nametkin, N.S. Novitsky, E.G. Durgaryan, S.G. (1979), Vysokomolek. soed., vol.21A, no.4, pp.920, 927.

Whyte, T.E. Ed. (1983) Industrial Gas Separations: ASC Symp. Ser. American Chemical Society, Washington, D.C. p.467.

INDEX OF CONTRIBUTORS

Printed in the United States
By Bookmasters